Graduate Texts in Mathematics 14

Graduate Texts in Mathematics

1 TAKEUTI/ZARING. Introduction to Axiomatic Set Theory. 2nd ed.
2 OXTOBY. Measure and Category. 2nd ed.
3 SCHAEFFER. Topological Vector Spaces.
4 HILTON/STAMMBACH. A Course in Homological Algebra.
5 MACLANE. Categories for the Working Mathematician.
6 HUGHES/PIPER. Projective Planes.
7 SERRE. A Course in Arithmetic.
8 TAKEUTI/ZARING. Axiomatic Set Theory.
9 HUMPHREYS. Introduction to Lie Algebras and Representation Theory.
10 COHEN. A Course in Simple Homotopy Theory.
11 CONWAY. Functions of One Complex Variable. 2nd ed.
12 BEALS. Advanced Mathematical Analysis.
13 ANDERSON/FULLER. Rings and Categories of Modules.
14 GOLUBITSKY/GUILLEMIN. Stable Mappings and Their Singularities.
15 BERBERIAN. Lectures in Functional Analysis and Operator Theory.
16 WINTER. The Structure of Fields.
17 ROSENBLATT. Random Processes. 2nd ed.
18 HALMOS. Measure Theory.
19 HALMOS. A Hilbert Space Problem Book. 2nd ed., revised.
20 HUSEMOLLER. Fibre Bundles. 2nd ed.
21 HUMPHREYS. Linear Algebraic Groups.
22 BARNES/MACK. An Algebraic Introduction to Mathematical Logic.
23 GREUB. Linear Algebra. 4th ed.
24 HOLMES. Geometric Functional Analysis and its Applications.
25 HEWITT/STROMBERG. Real and Abstract Analysis.
26 MANES. Algebraic Theories.
27 KELLEY. General Topology.
28 ZARISKI/SAMUEL. Commutative Algebra. Vol. I.
29 ZARISKI/SAMUEL. Commutative Algebra. Vol. II.
30 JACOBSON. Lectures in Abstract Algebra I: Basic Concepts.
31 JACOBSON. Lectures in Abstract Algebra II: Linear Algebra.
32 JACOBSON. Lectures in Abstract Algebra III: Theory of Fields and Galois Theory.
33 HIRSCH. Differential Topology.
34 SPITZER. Principles of Random Walk. 2nd ed.
35 WERMER. Banach Algebras and Several Complex Variables. 2nd ed.
36 KELLEY/NAMIOKA et al. Linear Topological Spaces.
37 MONK. Mathematical Logic.
38 GRAUERT/FRITZSCHE. Several Complex Variables.
39 ARVESON. An Invitation to C^*-Algebras.
40 KEMENY/SNELL/KNAPP. Denumerable Markov Chains. 2nd ed.
41 APOSTOL. Modular Functions and Dirichlet Series in Number Theory.
42 SERRE. Linear Representations of Finite Groups.
43 GILLMAN/JERISON. Rings of Continuous Functions.
44 KENDIG. Elementary Algebraic Geometry.
45 LOÈVE. Probability Theory I. 4th ed.
46 LOÈVE. Probability Theory II. 4th ed.
47 MOISE. Geometric Topology in Dimensions 2 and 3.

continued after Index

M. Golubitsky *V. Guillemin*

Stable Mappings and Their Singularities

Springer-Verlag
New York · Berlin · Heidelberg · Tokyo

Martin Golubitsky

Department of Mathematics
University of Houston
Houston, Texas 77004

Victor Guillemin

Department of Mathematics
Massachusetts Institute of Technology
Cambridge, Massachusetts 02139

AMS Subject Classification (1973)
Primary: 57D45, 58C25, 57D35, 57D40
Secondary: 57D50-57D70, 58A05, 58C15, 58D05, 58D15, 57E99

Library of Congress Cataloging in Publication Data
Golubitsky, M 1945–
Stable mappings and their singularities.
(Graduate texts in mathematics, 14)
1. Differentiable mappings. 2. Singularities (Mathematics) 3. Manifolds (Mathematics)
I. Guillemin, V., 1937– joint author. II. Title.
III. Series.
QA613.64.G64 516.36 73-18276

Printed in the United States of America.

9 8 7 6 5 4 3 (Corrected Third Printing, 1986)
ISBN 0-387-90073-X Springer-Verlag New York Berlin Heidelberg Tokyo (soft cover)
ISBN 0-387-90072-1 Springer-Verlag New York Berlin Heidelberg Tokyo (hard cover)
ISBN 3-540-90073-X Springer-Verlag Berlin Heidelberg New York Tokyo (soft cover)
ISBN 3-540-90072-1 Springer-Verlag Berlin Heidelberg New York Tokyo (hard cover)

PREFACE

This book aims to present to first and second year graduate students a beautiful and relatively accessible field of mathematics—the theory of singularities of stable differentiable mappings.

The study of stable singularities is based on the now classical theories of Hassler Whitney, who determined the generic singularities (or lack of them) for mappings of $R^n \to R^m$ ($m \geq 2n - 1$) and $R^2 \to R^2$, and Marston Morse, who studied these singularities for $R^n \to R$. It was René Thom who noticed (in the late '50's) that all of these results could be incorporated into one theory. The 1960 Bonn notes of Thom and Harold Levine (reprinted in [42]) gave the first general exposition of this theory. However, these notes preceded the work of Bernard Malgrange [23] on what is now known as the Malgrange Preparation Theorem—which allows the relatively easy computation of normal forms of stable singularities as well as the proof of the main theorem in the subject—and the definitive work of John Mather. More recently, two survey articles have appeared, by Arnold [4] and Wall [53], which have done much to codify the new material; still there is no totally accessible description of this subject for the beginning student. We hope that these notes will partially fill this gap. In writing this manuscript, we have repeatedly cribbed from the sources mentioned above—in particular, the Thom-Levine notes and the six basic papers by Mather. This is one of those cases where the hackneyed phrase "if it were not for the efforts of . . . , this work would not have been possible" applies without qualification.

A few words about our approach to this material: We have avoided (although our students may not always have believed us) doing proofs in the greatest generality possible. For example, we assume in many places that certain manifolds are compact and that, in general, manifolds have no boundaries, in order to reduce the technical details. Also, we have tried to give an abundance of low-dimensional examples, particularly in the later chapters. For those topics that we do cover, we have attempted to "fill in all the details," realizing, as our personal experiences have shown, that this phrase has a different interpretation from author to author, from chapter to chapter, and—as we strongly suspect—from authors to readers. Finally, we are aware that there are blocks of material which have been included for completeness' sake and which only a diehard perfectionist would slog through—especially on the first reading although probably on the last as well. Conversely, there are sections which we consider to be right at the "heart of the matter." These considerations have led us to include a Reader's Guide to the various sections.

Chapter I: This is elementary manifold theory. The more sophisticated reader will have seen most of this material already but is advised to glance through it in order to become familiar with the notational conventions used elsewhere in the book. For the reader who has had some manifold theory before,

Chapter I can be used as a source of standard facts which he may have forgotten.

Chapter II: The main results on stability proved in the later chapters depend on two deep theorems from analysis: Sard's theorem and the Malgrange preparation theorem. This chapter deals with Sard's theorem in its various forms. In §1 is proved the classical Sard's theorem. Sections 2–4 give a reformulation of it which is particularly convenient for applications to differentiable maps: the Thom transversality theorem. These sections are essential for what follows, but there are technical details that the reader is well-advised to skip on the first reading. We suggest that the reader absorb the notion of k-jets in §2, look over the first part of §3 (through Proposition 3.5) but assume, without going through the proofs, the material in the last half of this section. (The results in the second half of §3 would be easier to prove if the domain X were a compact manifold. Unfortunately, even if we were only to work with compact domains, the stability problem leads us to consider certain noncompact domains like $X \times X - \Delta X$.) In §4, the reader should probably skip the details of the proof of the multijet transversality theorem (Theorem 4.13). It is here that the difficulties with $X \times X - \Delta X$ make their first appearance.

Sections 5 and 6 include typical applications of the transversality theorem. The tubular neighborhood theorem, §7, is a technical result inserted here because it is easy to deduce from the Whitney embedding theorem in §5.

Chapter III: We recommend this chapter be read carefully, as it contains in embryo the main ideas of the stability theory. The first section gives an incorrect but heuristically useful "proof" of the Mather stability theorem: the equivalence of stability and infinitesimal stability. (The theorem is actually proved in Chapter V.) For motivational reasons we discuss some facts about infinite dimensional manifolds. These facts are used nowhere in the subsequent chapters, so the reader should not be disturbed that they are only sketchily developed. In the remaining three sections, we give all the elementary examples of stable mappings. The proofs depend on the material in Chapter II and the yet to be proved Mather criterion for stability.

Chapter IV gives the second main result from analysis needed for the stability theory: the Malgrange preparation theorem. Like Chapter II, this chapter is a little technical. We have provided a way for the reader to get through it without getting bogged down in details: in the first section, we discuss the classical Weierstrass preparation theorem—the holomorphic version of the Malgrange theorem. The proof given is fairly easy to understand, and has the virtue that the adaptation of it to a proof of the Malgrange preparation theorem requires only one additional fact, namely, the Nirenberg extension lemma (Proposition 2.4). The proof of this lemma can probably be skipped by the reader on a first reading as it is hard and technical.

In the third section, the form of the preparation theorem we will be using in subsequent chapters is given. The reader should take some pains to understand it (particularly if his background in algebra is a little shaky, as it is couched in the language of rings and modules).

Chapter V contains the proof of Mather's fundamental theorem on stability. The chapter is divided into two halves; §§1–4 contain the proof that infinitesimal stability implies stability and §§5 and 6 give the converse. In the process of proving the equivalence between these two forms of stability we prove their equivalence with other types of stability as well. For the reader who is confused by the maze of implications we provide in §7 a short summary of our line of argument.

It should be noted that in these arguments we assume the domain X is compact and without boundary. These assumptions could be weakened but at the expense of making the proof more complicated. One pleasant feature of the proof given here is that it avoids Banach manifolds and the global Mather division theorem.

Chapters VI and VII provide two classification schemes for stable singularities. The one discussed in Chapter VI is due to Thom [46] and Boardman [6]. The second scheme, due to Mather and presented in the last chapter, is based on the "local ring" of a map. One of the main results of these two chapters is a complete classification of all equidimensional stable maps and their singularities in dimensions ≤ 4. (See VII, §6.) The reader should be warned that the derivation of the "normal forms" for some stable singularities (VII, §§4 and 5) tend to be tedious and repetitive.

Finally, the *Appendix* contains, for completeness, a proof of all the facts about Lie groups needed for the proofs of Theorems in Chapters V and VI.

This book is intended for first and second year graduate students who have limited—or no—experience dealing with manifolds. We have assumed throughout that the reader has a reasonable background in undergraduate linear algebra, advanced calculus, point set topology, and algebra, and some knowledge of the theory of functions of one complex variable and ordinary differential equations. Our implementation of this assumption—i.e., the decisions on which details to include in the text and which to omit—varied according to which undergraduate courses we happened to be teaching, the time of day, the tides, and possibly the economy. On the other hand, we are reasonably confident that this type of background will be sufficient for someone to read through the volume. Of course, we realize that a healthy dose of that cure-all called "mathematical sophistication" and a previous exposure to the general theory of manifolds would do wonders in helping the reader through the preliminaries and into the more interesting material of the later chapters.

Finally, we note that we have made no attempt to create an encyclopedia of known facts about stable mappings and their singularities, but rather to present what we consider to be basic to understanding the volumes of material that have been produced on the subject by many authors in the past few years. For the reader who is interested in more advanced material, we recommend perusing the volumes of the "Proceedings of Liverpool Singularities" [42, 43], Thom's basic philosophical work, "Stabilité Structurelle et Morphogenèse" [47], Tougeron's work, "Ideaux de Fonctions Differentiables" [50], Mather's forthcoming book, and the articles referred to above.

There were many people who were involved in one way or another with the writing of this book. The person to whom we are most indebted is John Mather, whose papers [26–31] contain almost all the fundamental results of stability theory, and with whom we were fortunately able to consult frequently. We are also indebted to Harold Levine for having introduced us to Mather's work, and, for support and inspiration, to Shlomo Sternberg, Dave Schaeffer, Rob Kirby, and John Guckenheimer. For help with the editing of the manuscript we are grateful to Fred Kochman and Jim Damon. For help with some of the figures we thank Molly Scheffe. Finally, our thanks to Marni Elci, Phyllis Ruby, and Kathy Ramos for typing the manuscript and, in particular, to Marni for helping to correct our execrable prose.

Cambridge, Mass.
August, 1973

Martin Golubitsky
Victor W. Guillemin

TABLE OF CONTENTS

Chapter VI: *Classification of Singularities, Part I: The Thom-Boardman Invariants*

Chapter VII: *Classification of Singularities, Part II: The Local Ring of a Singularity*

Appendix

Stable Mappings and Their Singularities

Chapter I

Preliminaries on Manifolds

§1. Manifolds

Let $\mathbf{R}$ denote the real numbers and $\mathbf{R}^n$ denote n-dimensional Euclidean space. Points of $\mathbf{R}^n$ will be denoted by n-tuples of real numbers $(x_1, \ldots, x_n)$ and $\mathbf{R}^n$ will always be topologized in the standard way.

Let U be subset of $\mathbf{R}^n$. Then denote by $\overline{U}$ the closure of U, and by Int (U) the interior of U.

Let U be an open set, $f\colon U \to \mathbf{R}$, and $x \in U$. Denote by $(\partial f/\partial x_i)(x)$ the partial derivative of f with respect to the ith variable x_i at x. To denote a higher order mixed partial derivative, we will use multi-indices, i.e., let $\alpha = (\alpha_1, \ldots, \alpha_n)$ be an n-tuple of non-negative integers. Then

$$\frac{\partial^{|\alpha|}}{\partial x^\alpha} f = \frac{\partial^{|\alpha|}}{\partial x_1{}^{\alpha_1}\, \partial x_2{}^{\alpha_2} \cdots \partial x_n{}^{\alpha_n}} f \quad \text{where} \quad |\alpha| = \alpha_1 + \cdots + \alpha_n$$

and $f\colon U \to \mathbf{r}$ is *k-times differentiable* (or of *class* C^k, or C^k) if $(\partial^{|\alpha|} f/\partial x^\alpha)(x)$ exists and is continuous for every n-tuple of non-negative integers α with $|\alpha| \le k$. (Note that when $\alpha = (0, \ldots, 0)$, $\partial^{|\alpha|} f/\partial x^\alpha$ is defined to be f.) f is *real analytic* on U if the Taylor series of f about each point in U converges to f in a neighbourhood (nbhd) of that point.

Suppose $\phi\colon U \to \mathbf{R}^m$ where U is an open subset of $\mathbf{R}^n$ and f is some real-valued function defined in the range of ϕ; then $\phi^* f \equiv f \cdot \phi$ (where $\cdot$ denotes composition of mappings) is called the *pull-back function* of f by ϕ.

Definition 1.1. *Let $\phi\colon U \to \mathbf{R}^m$, U an open subset of $\mathbf{R}^n$.*

(a) ϕ *is* differentiable of class C^k *if the pull-back by ϕ of any k-times differentiable real-valued function defined on the range of ϕ is k-times differentiable.*

(b) ϕ *is* smooth (*or* differentiable of class C^∞) *if for every non-negative integer k, ϕ is differentiable of class C^k.*

(c) ϕ *is* real analytic *if the pull-back by ϕ of any real analytic real-valued function defined on the range of ϕ is real analytic.*

Let $\phi\colon U \to \mathbf{R}^m$ be C^1 differentiable in U and x_0 a point in U. Then by Taylor's theorem there exists a unique linear map $(d\phi)_{x_0}\colon \mathbf{R}^n \to \mathbf{R}^m$ and a function $\rho\colon U \to \mathbf{R}^m$ such that

$$\phi(x) = \phi(x_0) + (d\phi)_{x_0}(x - x_0) + \rho(x)$$

for every x in a nbhd V of x_0, where

$$\lim_{x \to x_0} \frac{|\rho(x)|}{|x - x_0|} = 0.$$

Note that we will use $|x|$ to denote the Euclidean norm $(\sum x_i^2)^{1/2}$. Let $(d\phi)_{x_0}: \mathbf{R}^n \to \mathbf{R}^m$ be the *Jacobian of* ϕ *at* x_0; it is given with respect to the coordinates $x_1, \ldots, x_n$ on $\mathbf{R}^n$ and $y_1, \ldots, y_m$ on $\mathbf{R}^m$ by the $m \times n$ matrix

$$\left(\frac{\partial \phi^i}{\partial x^j}(x_0)\right)_{1 \le j \le n}^{1 \le i \le m}$$

where $\phi^i : \mathbf{R}^n \to \mathbf{R}(1 \le i \le m)$ are the m coordinate functions defining ϕ.

The chain rule holds, of course. That is, if $\phi : U \to \mathbf{R}^m$ and $\psi : V \to \mathbf{R}^p$ are both C^1 differentiable where $U \subset \mathbf{R}^n$ and $V \subset \mathbf{R}^m$ are open and $V \supset \phi(U)$, then $d(\psi \cdot \phi)_{x_0} = (d\psi)_{\phi(x_0)} \cdot (d\phi)_{x_0}$ for every x_0 in U.

Theorem 1.2. (*Inverse Function Theorem*). *Let $U \subset \mathbf{R}^n$ be open and p be a point in U. Let $\phi: U \to \mathbf{R}^n$ be a C^k differentiable mapping. Assume that $(d\phi)_p : \mathbf{R}^n \to \mathbf{R}^n$ is invertible. Then there exists an open set V in R^n contained in the range of ϕ and a mapping $\psi: V \to U$, differentiable of class C^k, such that $\phi \cdot \psi(x) = x$ for every x in V, and $\psi \cdot \phi(x) = x$ for every x in $\psi(V)$.*

Proof. See appendix of Sternberg; or Lang. □

Definition 1.3. *A* local homeomorphism *of $\mathbf{R}^n$ is a homeomorphism of some open subset of $\mathbf{R}^n$ onto another.* (*So the domain of a local homeomorphism need not be all of $\mathbf{R}^n$.*)

Let ϕ be a mapping. Denote by *dom* ϕ the domain of ϕ. Also, if $U \subset$ dom ϕ denote by $\phi|U$ the restriction of ϕ to U. If X is a set, then $\mathrm{id}_X : X \to X$ denotes the identity mapping on X.

Definition 1.4. *A* pseudogroup on $\mathbf{R}^n$ *is a collection Γ of local homeomorphisms on $\mathbf{R}^n$ with the following properties:*

(a) *$id_{\mathbf{R}^n}$ is in Γ,*

(b) *if ϕ and ψ are in Γ with* dom ψ *= range of ϕ then $\psi \cdot \phi$ is in Γ, i.e., Γ is closed under composition for all pairs of elements for which this operation makes sense.*

(c) *if ϕ is in Γ, then ϕ^{-1} is in Γ (where ϕ^{-1} denotes the inverse function of ϕ)*

(d) *if ϕ is in Γ and U is an open subset of* dom ϕ, *then $\phi|U$ is in Γ, and*

(e) *if $\{U_\alpha\}_{\alpha \in I}$ (I some index set) is a collection of open subsets of $\mathbf{R}^n$, ϕ is a local homeomorphism of $\mathbf{R}^n$ defined on $U = \bigcup_{\alpha \in I} U_\alpha$, and $\phi|U_\alpha$ is in Γ for every α in I, then ϕ is in Γ.*

Some examples of pseudogroups are:

(a) $(\mathrm{diff})^k$ = the set of all local diffeomorphisms on $\mathbf{R}^n$ (n fixed) which are differentiable of class C^k.

(b) $(\mathrm{diff})^\infty$ = the set of local diffeomorphisms of $\mathbf{R}^n$ (n fixed) which are smooth.

(c) $(\mathrm{diff})^\omega$ = the set of all local diffeomorphisms of $\mathbf{R}^n$ (n fixed) which are real analytic.

To show that (a) and (b) satisfy the conditions of the definition you need to use only the chain rule, the inverse function theorem, and the local character of differentiability. For (c) you need the strengthened versions of the above theorems for analytic functions.

A more general class of pseudogroups can be given as follows:

(d) Let G be a group of linear mappings of $\mathbf{R}^n \to \mathbf{R}^n$. Then the pseudogroup $\Gamma_G{}^k$ is the set

$$\{\phi \in (\text{diff})^k \mid \forall x \in \text{dom}\,\phi,\ (d\phi)_x \in G\}$$

(i) $G =$ all linear maps on $\mathbf{R}^n$ with positive determinant. Then $\Gamma_G{}^k = (\text{diff})_0^k$ consists of orientation preserving C^k mappings.

(ii) $G =$ all linear maps on $\mathbf{R}^n$ with determinant equal to 1. Then $\Gamma_G{}^k$ consists of all volume preserving C^k mappings.

(iii) Let $(\ ,\ )$ be an inner product on $\mathbf{R}^n$. Let G be the group of orthogonal matrices relative to $(\ ,\ )$; namely, $A \in G$ iff $(x, y) = (Ax, Ay)$ for every x, y in $\mathbf{R}^n$. Then $\Gamma_G{}^k$ consists of all C^k isometries in $\mathbf{R}^n$.

Definition 1.5. *Let Γ be a pseudogroup on $\mathbf{R}^n$ and X a Hausdorff topological space which satisfies the second axiom of countability. Let A be a subset of all local homeomorphisms of X into $\mathbf{R}^n$, i.e., homeomorphisms which are defined on an open subset of X and whose range is an open subset of $\mathbf{R}^n$. Then*

(i) *A is a* Γ-atlas *on X if*

(a) $X = \bigcup_{\phi \in A} \text{dom}\,\phi$

(b) *if ϕ, ψ are in A, then $\psi \cdot \phi^{-1} | \phi(\text{dom}\,\phi \cap \text{dom}\,\psi)$ is in Γ.*

(ii) *The elements of A are called* charts *on X.*

(iii) *Two Γ-atlases A_1 and A_2 on X are* compatible *if $\psi \cdot \phi^{-1} | \phi(\text{dom}\,\phi \cap \text{dom}\,\psi)$ is in Γ whenever ϕ is in A_1 and ψ is in A_2, and vice-versa.*

(iv) *A Hausdorff space X together with an equivalence class of compatible Γ-atlases is called a* Γ-structure *on X.*

Notes. (1) Recall that X satisfies the second axiom of countability if the topology on X has a countable base.

(2) If X has a Γ-structure, then X is locally compact, since it is locally Euclidean.

Definition 1.6. *Let X have a Γ-structure.*

(a) *If $\Gamma = (\text{diff})^k$ and $k > 0$, then X is a* differentiable manifold of class C^k.

(b) *If $\Gamma = (\text{diff})^0$, then X is a* topological manifold.

(c) *If $\Gamma = (\text{diff})^\infty$, then X is a* smooth manifold *or a* manifold of class C^∞.

(d) *If $\Gamma = (\text{diff})^\omega$, then X is a* real analytic manifold.

(e) *If $\Gamma = (\text{diff})_0^k$ and $k > 0$ then X is an* oriented *C^k differentiable manifold. Any differentiable manifold which has a $(\text{diff})_0^1$ structure in which the charts are elements of the original $(\text{diff})^1$ structure is* orientable.

Examples

(1) $S^{n-1} = \left\{x = (x_1, \ldots, x_n) \in \mathbf{R}^n \;\middle|\; \sum_{i=1}^{n} x_i^2 = 1\right\}$.

Let $N = (1, 0, \ldots, 0)$ and $S = (-1, 0, \ldots, 0)$.

Let $\phi_N : \{S^{n-1} - \{N\}\} \to \mathbf{R}^{n-1}$ be stereographic projection via N, i.e., $\phi_N(x_1, \ldots, x_n) = (1/(1 - x_1))(x_2, \ldots, x_n)$ and $\phi_S : \{S^{n-1} - \{S\}\} \to \mathbf{R}^{n-1}$ be stereographic projection via S, i.e., $\phi_S(x_1, \ldots, x_n) = (1/(1 + x_1))\,(x_2, \ldots, x_n)$. Then $\phi_S \cdot \phi_N{}^{-1} : \mathbf{R}^{n-1} - \{0\} \to \mathbf{R}^{n-1} - \{0\}$ is given by $y \to y/|y|^2$ for all y in $\mathbf{R}^{n-1} - \{0\}$. Since $(\phi_S \cdot \phi_N{}^{-1}) \cdot (\phi_S \cdot \phi_N{}^{-1}) = id$ we see that $\det(d\phi_S \cdot \phi_N{}^{-1})_y = \pm 1$. Evaluate at $y = (1, 0, \ldots, 0)$ to see that, in fact, $\det(d\phi_S \cdot \phi_N{}^{-1}) = -1$. To show that S^{n-1} is an oriented analytic manifold we can change the last coordinate of ϕ_N to $-x_n/(1 - x_1)$ thus changing the determinant to $+1$.

(2) $\mathbf{P}^n$ = real projective n-space.

To define $\mathbf{P}^n$ we introduce the equivalence relation $\sim$ on $\mathbf{R}^{n+1} - \{0\}$: $(x_0, \ldots, x_n) \sim (x'_0, \ldots, x'_n)$ iff there is a real constant c such that $x_i = cx'_i$ for all i.

$\mathbf{P}^n = \mathbf{R}^{n+1} - \{0\}/\sim$ is the set of these equivalence classes.

Let $\pi : \mathbf{R}^{n+1} - \{0\} \to \mathbf{P}^n$ be the canonical projection. $\mathbf{P}^n$ is given the standard quotient space topology and note that with this topology π is an open mapping. To show that $\mathbf{P}^n$ has a manifold structure it is necessary to produce local homeomorphisms of $\mathbf{P}^n$ into $\mathbf{R}^n$ which overlap properly.

Let $V_i = \mathbf{R}^{n+1} - \{\text{hyperplane } x_i = 0\}$ for $0 \le i \le n$. V_i is open in $\mathbf{R}^{n+1} - \{0\}$, hence $\pi(V_i) = U_i$ is open in $\mathbf{P}^n$. Clearly $\mathbf{P}^n = U_0 \cup \cdots \cup U_n$. Define $\phi_i : U_i \to \mathbf{R}^n$ by

$$\phi_i(p) = \frac{(-1)^i}{x_i}(x_0, \ldots, \hat{x}_i, \ldots, x_n) \quad \text{where} \quad p = \pi(x_0, \ldots, x_n)$$

and ^ indicates that coordinate is to be omitted. Using the equivalence relation defining $\mathbf{P}^n$ and the fact that p is in U_i, one sees that ϕ_i is a well-defined homeomorphism onto $\mathbf{R}^n$.

$$\phi_i(U_i \cap U_j) = \mathbf{R}^n - \{\text{hyperplane } y_j = 0\} \qquad (i > j)$$

$$\phi_i(U_i \cap U_j) = \mathbf{R}^n - \{\text{hyperplane } y_{j-1} = 0\} \qquad (i < j)$$

where we assume $y_1, \ldots, y_n$ are the coordinates on $\mathbf{R}^n$. So for $i < j$ $\phi_i \cdot \phi_j^{-1} : \mathbf{R}^n - \{\text{hyperplane } y_i = 0\} \to \mathbf{R}^n - \{\text{hyperplane } y_{j-1} = 0\}$. A computation yields for $i < j$

$$\phi_i \cdot \phi_j^{-1}(y_1, \ldots, y_n) = \frac{(-1)^{i+j}}{y_{i+1}}(y_1, \ldots, y_i, y_{i+2}, \ldots, y_j, 1, y_{j+1}, \ldots, y_n)$$

which is a real analytic mapping so $\mathbf{P}^n$ becomes a real analytic manifold. When $i < j$ another computation yields

$$\det\left(d\phi_i \cdot \phi_j^{-1}\right)_{(y_1, \ldots, y_n)} = \left(\frac{1}{y_{i+1}}\right)^{n+1} (-1)^{(n+1)(i+j)}$$

from which we see that real projective space in any odd dimension ($\mathbf{P}^{2n+1}$, $n \ge 0$) is orientable. It can be proved that $\mathbf{P}^{2n}$ is not orientable.

(3) $G_{k,n}$ = Grassmannian space of k-planes through the origin in $\mathbf{R}^n$.
= set of all k-dimensional subspaces of Euclidean n-space.

Note that $G_{1,n+1} = \mathbf{P}^n$.

We will give $G_{k,n}$ a decomposition space topology. Let W = all ordered k-tuples $P = (P_1, \ldots, P_k)$ of k linearly independent vectors in R^n. W is an open subset of

$$\underbrace{\mathbf{R}^n \oplus \cdots \oplus \mathbf{R}^n}_{k-\text{times}}.$$

Define an equivalence relation $\sim$ on W as follows:

$$P \sim Q \quad \text{if} \quad \{P_1, \ldots, P_k\} \quad \text{and} \quad \{Q_1, \ldots, Q_k\}$$

span the same k-dimensional subspace of $\mathbf{R}^n$.

Clearly $G_{k,n}$ can be identified with $W/\sim$ as sets so we may give $G_{k,n}$ the topology induced by this identification. We now give $G_{k,n}$ an analytic structure. Equip $\mathbf{R}^n$ with an inner product (,). Then given a subspace V of $\mathbf{R}^n$, there is an orthogonal projection π_V of $\mathbf{R}^n$ onto V. Suppose V is a k-dimensional subspace of R^n. Let $\pi_{U,V}$ = restriction of π_V to U. Let $W_V = \{U \in G_{k,n} \mid \pi_{U,V} \text{ is a bijection onto } V\}$.

Let $V^\perp$ = the orthogonal complement of V in $\mathbf{R}^n$. Define

$$\rho_V : W_V \to \operatorname{Hom}(V, V^\perp)$$

as follows: Let $U \in W_V$. Then $\rho_V(U) = \pi_{U,V^\perp} \cdot \pi_{U,V}^{-1} \in \operatorname{Hom}(V, V^\perp)$. We leave it to the reader to check that ρ_V is a homeomorphism. Now make the identification $\operatorname{Hom}(V, V^\perp) \cong \mathbf{R}^{k(n-k)}$, to get a chart $\phi_V : W_V \to R^{k(n-k)}$. Again it is left to the reader to check that $\rho_V \cdot \rho_{V'}^{-1} : \mathbf{R}^{k(n-k)} \to \mathbf{R}^{k(n-k)}$ is real analytic. Hence $G_{k,n}$ is a real analytic manifold of dimension $k(n-k)$. Note that for $k = 1$ this is the same atlas that we constructed for $\mathbf{P}^{n-1}$. Thus $G_{1,n} = \mathbf{P}^{n-1}$.

Definition 1.7. *Let X and Y be C^k differentiable manifolds of dimension n and m, respectively. Then $X \times Y$ can be made into a C^k differentiable manifold of dimension $n + m$ in the following natural way. Let A_X and A_Y be atlases on X and Y. Let $\phi \in A_X$, $\psi \in A_Y$. Then $\phi \times \psi : \operatorname{dom} \phi \times \operatorname{dom} \psi \to \mathbf{R}^n \times \mathbf{R}^m = \mathbf{R}^{m+n}$ is given by $\phi \times \psi(x, y) = (\phi(x), \psi(y))$ $x \in X, y \in Y$. $\phi \times \psi$ is clearly a local homeomorphism of $X \times Y \to \mathbf{R}^{n+m}$. Then $A_{X\times Y} = \{\phi \times \psi \mid \phi \in A_X, \psi \in A_Y\}$ is an atlas for $X \times Y$.*

Applications

(1) The r-Torus,

$$\underbrace{S^1 \times \cdots \times S^1}_{r-\text{times}}$$

is a smooth manifold of dimension r.

(2) If X and Y are oriented manifolds, then so is $X \times Y$.

Definition 1.8. *Let X be a topological n-manifold, and p a point in X. A set of* local coordinates on X based at p *is a collection of n real-valued functions $\{\phi_1, \ldots, \phi_n\}$ defined on an open nbhd U of p, (i.e., $\phi_i : U \to \mathbf{R}$) so that $\phi_i(p) = 0$ $(1 \le i \le n)$ and $\phi : U \to \mathbf{R}^n$ defined by $\phi(q) = (\phi_1(q), \ldots, \phi_n(q))$ is a chart in the manifold structure on X.*

Clearly if ϕ is a chart of X based at p (i.e., ϕ is defined on a nbhd of p and $\phi(p) = 0$) then the coordinate functions of ϕ define a system of local coordinates on X based at p.

The common domain of a set of local coordinates based at p is a *coordinate nbhd of p*.

§2. Differentiable Mappings and Submanifolds

Definition 2.1. *Let Y be a C^k-differentiable manifold of dimension m.*

(a) *Let $f\colon Y \to \mathbf{R}$ be a function. f is* C^k-differentiable *if for every chart $\phi\colon \operatorname{dom}\phi \to \mathbf{R}^m$, $f\cdot\phi^{-1}\colon \operatorname{range}\phi \to \mathbf{R}$ is a C^k-differentiable mapping. f is* smooth *if f is C^k-differentiable for every k.*

(b) *Let X be a C^k-differentiable manifold. Then $\phi\colon X \to Y$ is* C^k-differentiable *if for every C^k-differentiable function $f\colon Y \to \mathbf{R}$, the pullback $f\cdot\phi$ is C^k-differentiable. ϕ is* smooth *if ϕ is C^k-differentiable for every k.*

(c) *We will use* differentiable *to mean C^k-differentiable for k at least* 1.

Remark. Suppose that $\phi\colon X \to Y$ is a mapping with p in X and $q = \phi(p)$ in Y. Let U and V be coordinate nbhds of X and Y based at p and q respectively, and assume that $\phi(U) \subset V$. Suppose $\rho\colon V \to \mathbf{R}^m$ and $\tau\colon U \to \mathbf{R}^n$ are charts. Then ϕ is C^k-differentiable iff $\rho\cdot\phi\cdot\tau^{-1}\colon \operatorname{range}\tau \subset \mathbf{R}^n \to \mathbf{R}^m$ is C^k-differentiable. This shows that differentiability of a function between manifolds is a local question and is independent of the particular local representation used.

Definition 2.2. *Let X and Y be differentiable manifolds of dimension n and m, respectively. Let $\phi\colon X \to Y$ be differentiable. Let p be in X, ρ a chart on X with p in* dom ρ, *and τ a chart on Y with $\phi(\operatorname{dom}\rho) \subset \operatorname{dom}\tau$.*

Then $(d\tau\cdot\phi\cdot\rho^{-1})_{\rho(p)}\colon \mathbf{R}^n \to \mathbf{R}^m$ is a linear mapping. Define rank of ϕ at p *to be rank $(d\tau\cdot\phi\cdot\rho^{-1})_{\rho(p)}$.*

Note. The definition of rank does not depend on which charts are selected. Let ρ', τ' be charts with the above properties. Then on a nbhd of p and $f(p)$,

$$\begin{aligned}\operatorname{rank}(d\tau'\cdot\phi\cdot(\rho')^{-1})_{\rho'(p)} &= \operatorname{rank}(d\tau'\cdot\tau^{-1}\cdot\tau\cdot\phi\cdot\rho^{-1}\cdot\rho\cdot(\rho')^{-1})_{\rho'(\rho)}\\ &= \operatorname{rank}(d\tau\cdot\phi\cdot\rho^{-1})_{\rho(p)}\end{aligned}$$

by the chain rule and the fact that $\tau'\cdot\tau^{-1}$ and $\rho\cdot(\rho')^{-1}$ are in $(\text{diff})^1$.

Definition 2.3. *Let X and Y be differentiable manifolds. Let $\phi\colon X \to Y$ be a differentiable mapping. Suppose that at the point p in X, ϕ has the maximum possible rank. Then*

(a) *if* $\dim X \leq \dim Y$, *ϕ is an* immersion at p,

(b) *if* $\dim X \geq \dim Y$, *ϕ is a* submersion at p,

(c) *if for every p in X, ϕ is an immersion* (*submersion*) *at p, then ϕ is an* immersion (submersion).

(d) *if* $\dim X = \dim Y = n$, ϕ *is bijective, and the rank of* ϕ *is* n *at every point of* X, *then* ϕ *is a* diffeomorphism.

(e) *if* $\phi : X \to Y$ *is an immersion and a homeomorphism (into), then it is an* embedding.

(f) *if there exists a diffeomorphism of* $X \to Y$, *then* X *and* Y *are* diffeomorphic.

Note. If $\phi : X \to Y$ is a diffeomorphism, then $\phi^{-1} : Y \to X$ is well-defined and is as differentiable as ϕ is by the inverse function theorem (Theorem 1.2.)

We will show that locally immersions "look like" linear injections, submersions "look like" projections, and diffeomorphisms "look like" the identity mapping. (The notion of "looks like" will be made precise in 2.5 and 2.6.) To do this we use the implicit function theorem.

Let $U_1 \subset \mathbf{R}^k$ and $U_2 \subset \mathbf{R}^l$ be open sets. Let $\phi : U_1 \times U_2 \to \mathbf{R}^l$ be differentiable. Define $(d_y\phi)_{(x_0,y_0)} \equiv (d\phi_{x_0})_{y_0}$ where x_0 in U_1, y_0 in U_2 and $\phi_{x_0} : U_2 \to \mathbf{R}^l$ is given by $\phi_{x_0}(y) = \phi(x_0, y)$ for all y in U_2.

Theorem 2.4. (*Implicit Function Theorem*). *Suppose* $\phi : U_1 \times U_2 \to \mathbf{R}^l$ *is* C^s*-differentiable and* $\phi(x_0, y_0) = y_0$. *If* $(d_y\phi)_{(x_0,y_0)}$ *is of rank* l, *then there exist open sets* $U_1' \subset U_1$ *and* $U_2' \subset U_2$, *with* x_0 *in* U_1' *and* y_0 *in* U_2', *and a* C^s*-differentiable function* $\psi : U_1' \times U_2' \to U_2$ *such that* $\phi(x, \psi(x, y)) = y$ *for every* x *in* U_1' *and* y *in* U_2'. *Moreover* ψ *can be chosen so that* $\psi(x_0, y_0) = y_0$.

Proof. Define $\hat{\phi} : U_1 \times U_2 \to \mathbf{R}^k \times \mathbf{R}^l$ to be the graph of ϕ, i.e., $\hat{\phi}(x, y) = (x, \phi(x, y))$ for all $x \in U_1$, $y \in U_2$. In the standard coordinates $x_1, \ldots, x_k$ on $\mathbf{R}^k$ and $y_1, \ldots, y_l$ on $\mathbf{R}^l$

$$(d\hat{\phi})_{(x,y)} = \left(\begin{array}{c|c} I_k & 0 \\ \hline * & (d_y\phi)_{(x,y)} \end{array}\right)$$

where I_k is the $k \times k$ identity matrix. The assumption on $(d_y\phi)_{(x_0,y_0)}$ implies that the rank of $(d\hat{\phi})_{(x_0,y_0)}$ is $k + l$, i.e., $(d\hat{\phi})_{(x_0,y_0)}$ is invertible. Apply the inverse function theorem to find U_1', U_2' so that $\hat{\psi} = \hat{\phi}^{-1} \mid U_1' \times U_2'$ is C^s-differentiable. Let $\hat{\psi}(x, y) = (\psi_1(x, y), \psi_2(x, y))$ be in $\mathbf{R}^k \times \mathbf{R}^l$. Since $\hat{\phi} \cdot \hat{\psi} = id_{U_1' \times U_2'}$, we have that

$$(x, y) = \hat{\phi}(\hat{\psi}(x, y)) = (\psi_1(x, y), \phi(\psi_1(x, y), \psi_2(x, y)))$$

Hence $\psi_1(x, y) = x$ and $y = \phi(x, \psi_2(x, y))$. Take $\psi = \psi_2$. ☐

Corollary 2.5. *Let* $U \subset \mathbf{R}^n$ *be open,* x_0 *in* U, *and* $\phi : U \to \mathbf{R}^m$ *an immersion at* x_0. *Then there exists an open set* U' *in* U *with* x_0 *in* U', *an open set* V *in* $\mathbf{R}^m$ *with* $\phi(U') \subset V$, *and a map* $\tau : V \to \mathbf{R}^m$ *which is a diffeomorphism onto its image so that* $\lambda = \tau \cdot \phi$ *is the standard injection of* $\mathbf{R}^n \to \mathbf{R}^n \times \mathbf{R}^{m-n}$ *restricted to* U'. *(Thus by a change of coordinates in the range,* ϕ *can be linearized locally.)*

Proof. Since $(d\phi)_{x_0}$ has rank n, there is an $n \times n$ minor which is nonsingular. Let $\phi_1, \ldots, \phi_m$ be the coordinate functions defined by ϕ. Then

$$(d\phi)_{x_0} = \begin{pmatrix} \frac{\partial\phi_1}{\partial x_1} & \frac{\partial\phi_1}{\partial x_2} & \cdots & \frac{\partial\phi_1}{\partial x_n} \\ \vdots & \vdots & & \vdots \\ \frac{\partial\phi_m}{\partial x_1} & \frac{\partial\phi_m}{\partial x_2} & \cdots & \frac{\partial\phi_m}{\partial x_n} \end{pmatrix}$$

The appropriate minor is determined by n columns $i_1, \ldots, i_n$.

Let τ_1 be a linear isomorphism of $\mathbf{R}^m$ which maps $\varepsilon_{i_j} \mapsto \varepsilon_j$ $(1 \le j \le n)$ where ε_j is the unit vector along the jth coordinate. Then $\tau_1 \cdot \phi$ has the property that $(d\tau_1 \cdot \phi)_{x_0}$ has rank n and the appropriate $n \times n$ minor which is nonsingular is given by the first n-columns. By including τ_1 in the definition of τ we assume that ϕ has this property.

Write $\mathbf{R}^m = \mathbf{R}^n \times \mathbf{R}^l$ where $l = m - n$ and $\mathbf{R}^n$ is given by the first n-coordinates $x_1, \ldots, x_n$ and $\mathbf{R}^l$ by the last l-coordinates $y_1, \ldots, y_l$. $\phi : U \to \mathbf{R}^n \times \mathbf{R}^l$ is given by $\phi = (\phi_1, \phi_2)$ where $\phi_1 : U \to \mathbf{R}^n$, $\phi_2 : U \to \mathbf{R}^l$, and $(d\phi_1)_{x_0}$ has rank n.

Since U is in $\mathbf{R}^n$, we may construct $\tilde{\phi} : U \times \mathbf{R}^l \to \mathbf{R}^n \times \mathbf{R}^l$ given by $(x, y) \mapsto \phi(x) + (0, y)$ where x is in U and y is in $\mathbf{R}^l$.

Then

$$(d\tilde{\phi})_{(x_0, y)} = \left(\begin{array}{c|c} (d\phi_1)_{x_0} & 0 \\ \hline * & I_l \end{array}\right)$$

which has rank m. By the inverse function theorem, there exists a differentiable inverse τ to $\tilde{\phi}$ defined on a nbhd of $\tilde{\phi}(x_0, 0) = \phi(x_0)$. Let $\lambda(x) = \tau \cdot \tilde{\phi}(x, 0) = (x, 0)$. Then $\lambda : \mathbf{R}^n \to \mathbf{R}^n \times \mathbf{R}^l = \mathbf{R}^m$ is given by $\lambda(x) = (x, 0)$ which is a linear map of rank n. □

Corollary 2.6. *Let $U \to \mathbf{R}^n$ be open, z_0 a point in U, and $\phi : U \to \mathbf{R}^m$ a submersion at z_0. Then there exists a nbhd U' of z_0 in U, a diffeomorphism $\sigma : U' \to \mathbf{R}^n$ (onto its image), and λ a linear mapping of rank m so that $\phi = \lambda \cdot \sigma$ on U'. (In fact, λ can be taken to be the standard projection of $\mathbf{R}^m \times \mathbf{R}^{n-m} \to \mathbf{R}^m$. Thus by a change of coordinates in the domain, ϕ can be linearized.)*

Proof. Let $\mathbf{R}^n = \mathbf{R}^m \times \mathbf{R}^l$ with coordinates $x_1, \ldots, x_m$ on $\mathbf{R}^m$ and $y_1, \ldots, y_1$ on $\mathbf{R}^l$. By an appropriate choice of bases on $\mathbf{R}^n$, this decomposition can be done so that $(d_x\phi)_{z_0}$ has rank m.

Define $\tilde{\phi} : U \to \mathbf{R}^m \times \mathbf{R}^l$ by $\tilde{\phi}(x, y) = (\phi(x, y), y)$. Then

$$(d\tilde{\phi})_{z_0} = \left(\begin{array}{c|c} (d_x\phi)_{z_0} & * \\ \hline 0 & I_l \end{array}\right)$$

which has rank n. By the inverse function theorem $\tilde{\phi}$ is locally a diffeomorphism. Let $\sigma = \tilde{\phi}$ and $\lambda: \mathbf{R}^m \times \mathbf{R}^l \to \mathbf{R}^m$ be given by $\lambda(x, y) = x$. Then $\lambda \cdot \tilde{\phi}(x, y) = \lambda(\phi(x, y), y) = \phi(x, y)$. □

Definition 2.7. *Let X be a C^k-manifold of dimension n. Let Y be a subset of X. Then Y is a* submanifold *of X of dimension m if for every point p in Y, there exists a chart ϕ : dom $\phi \to \mathbf{R}^n$ of the differentiable structure on X so that $\phi^{-1}(V) = Y \cap \text{dom } \phi$ where*

$$V = \{(x_1, \ldots, x_n) \in \mathbf{R}^n \mid x_{m+1} = \cdots = x_n = 0\}$$

and $x_1, \ldots, x_n$ are the canonical coordinates on $\mathbf{R}^n$.

Note. If Y is a submanifold of a C^k-differentiable manifold, then it itself is a C^k-differentiable manifold. Give Y the induced topology from X. (Warning: There are weaker definitions of submanifold in which Y does *not* bear the subspace topology. See Definition 2.9.) For each p in Y, let ϕ_p be the chart on X, given in the definition of submanifold. $Y \cap \text{dom } \phi_p$ is an open set of Y and $\phi_p|Y : Y \cap \text{dom } \phi_p \to \mathbf{R}^m$ is a local homeomorphism. The set of mappings $\{\phi_p|Y\}_{p \in Y}$ give Y a C^k-differentiable structure of dimension m.

Theorem 2.8. *Let X and Y be C^k-differentiable manifolds of dimensions n and m respectively with $n \geq m$. Let $\phi: X \to Y$ be a C^k-mapping. Then*

(1) *If ϕ is a submersion, then $\phi(X)$ is an open subset of Y. In fact, ϕ is an open mapping.*

(2) *Let Z be a submanifold of Y. If ϕ is a submersion at each point in $\phi^{-1}(Z)$, then* $\phi^{-1}(Z)$ is a C^k submanifold of X with codim $\phi^{-1}(Z)$ = codim Z *where* codim Z = dim Y − dim Z.

Proof.

(1) Let U be an open set in X and V an open set in Y with $\phi(U) \subset V$ and y_0 in V. Let $\psi : U \to \mathbf{R}^n$ and $\rho : V \to \mathbf{R}^m$ be charts. Choose x_0 in $U \cap \phi^{-1}(y_0)$. All of this is possible since ϕ is continuous.

Now $\rho \cdot \phi \cdot \psi^{-1} : U' \to \mathbf{R}^m$ is a submersion where $U' = \psi(U)$ is open in $\mathbf{R}^n$. By Corollary 2.6 there exists a nbhd U'' of $\psi(x_0)$ in U' and a diffeomorphism $\sigma : U'' \to \sigma(U'') \subset \mathbf{R}^n$ and a linear mapping λ of rank m so that $\rho \cdot \phi \cdot \psi^{-1} = \lambda \cdot \sigma$ on U''. Let $\psi' = \sigma \cdot \psi$. Then ψ' is a chart on X with x_0 in dom ψ' and $\rho \cdot \phi \cdot (\psi')^{-1} = \lambda$. Since $\lambda : \mathbf{R}^n \to \mathbf{R}^m$ has rank m, it maps open sets to open sets. Choose W an open nbhd of x_0 in X so that $\psi'(W) \subset \sigma(U'')$. Then $\lambda(\psi'(W))$ is open in $\mathbf{R}^m$ and $\rho^{-1}(\lambda(\psi'(W))) = \phi(W)$ is open in Y. So $\phi(X)$ is open in Y.

(2) Note that $\lambda: \mathbf{R}^m \times \mathbf{R}^{n-m} \to \mathbf{R}^m$ can be given by $\lambda(x, y) = x$. Let ρ be a chart which makes Z into a submanifold, i.e., one for which $\rho(Z \cap \text{dom } \rho)$ is a hyperplane in R^m. Now $\lambda \cdot \psi'(\text{dom } \psi' \cap \phi^{-1}(Z)) \subset \rho \cdot \phi \cdot \phi^{-1}(Z) \subset \rho(Z) =$ hyperplane by the choice of ρ. Thus $\psi'(\phi^{-1}(Z) \cap \text{dom } \psi') \subset \lambda^{-1}(\text{hyperplane})$ = hyperplane, since λ is linear. Thus ψ' is a chart near x_0 making $\phi^{-1}(Z)$ into a C^k-submanifold of codimension = codim Z. □

Example. Let $\phi : \mathbf{R}^n \to \mathbf{R}$ be given by $\phi(x_1, \ldots, x_n) = x_1^2 + \cdots + x_n^2$. This is a submersion on $S^{n-1} = \phi^{-1}(1)$. Thus S^{n-1} is an $n - 1$ dimensional submanifold of $\mathbf{R}^n$.

Note. Let X be a differentiable manifold, Y a set, and $f: X \to Y$ a bijection. Then there is a natural way to make Y into a differentiable manifold. First declare that the topology on Y is the one which makes f a homeomorphism. Then define the charts on Y, to be the pull-backs via f^{-1} of the charts on X.

Definition 2.9. *The image of a* 1-1 *immersion, made into a manifold in the manner just described, is an* immersed submanifold. (*Warning: this definition of immersed submanifold is not the same, in general, as that of a submanifold. In particular, the topology of the immersed submanifold need not be the same as the induced topology from the range.*)

Proposition 2.10. *Let $\phi: X \to Y$ be an immersion. Then for every p in X, there exists a nbhd U of p in X such that*

(1) *$\phi|U: U \to \phi(U)$ is a homeomorphism where $\phi(U)$ is given the induced topology from Y and*

(2) *$\phi(U)$ is a submanifold of Y.*

Proof. Given p in X, there exist open nbhds U of p in X and V of $\phi(p)$ in Y with $\phi(U) \subset V$, charts $\rho: U \to \mathbf{R}^n$ and $\tau: V \to \mathbf{R}^m$, and a linear map $\lambda: \mathbf{R}^n \to \mathbf{R}^m$ of rank n so that the diagram

$$\begin{array}{ccc} U & \xrightarrow{\phi} & V \\ {\scriptstyle\rho}\downarrow & & \downarrow{\scriptstyle\tau} \\ \mathbf{R}^n & \xrightarrow{\lambda} & \mathbf{R}^m \end{array}$$

commutes. This is possible by Corollary 2.5.

Now (1) follows since $\lambda: \mathbf{R}^n \to \mathbf{R}^m$ is a homeomorphism onto its image. For $\tau(\phi(U))$ is homeomorphic to $\phi(U)$ with the induced topology since τ is a local homeomorphism defined on V. $\tau(\phi(U)) \subset \operatorname{Im} \lambda$ since the diagram commutes, thus $\lambda^{-1}(\tau(\phi(U)))$ is homeomorphic to $\phi(U)$ with the induced topology from Y. Finally $\rho^{-1}(\lambda^{-1}(\tau(\phi(U)))) = \phi^{-1}\phi(U) = U$ is homeomorphic to $\phi(U)$ with the induced topology from Y.

To see that $\phi(U)$ is a submanifold, use the chart τ. Choose coordinates on $\mathbf{R}^m$ so that $\lambda(x) = (x, 0)$. This decomposes $\mathbf{R}^m$ into $\mathbf{R}^n \times \mathbf{R}^{m-n}$. Then $\tau|\phi(U): \phi(U) \to \mathbf{R}^n \times \{0\}$. ∎

Notes. (1) Proposition 2.10 is only a local result since not every immersion is 1:1. For instance, the mapping of $\mathbf{R} \to \mathbf{R}^2$ given pictorially by

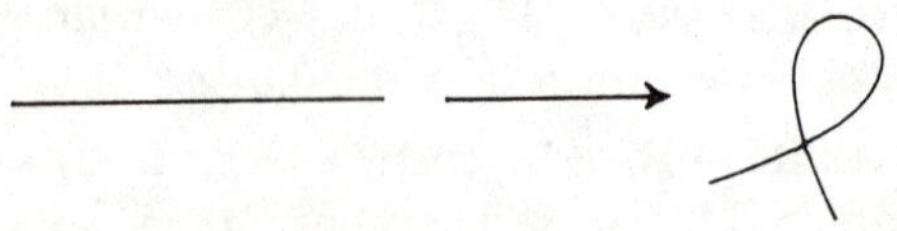

is an immersion (when drawn smoothly enough!).

(2) The image of an immersion need not be a submanifold even if the immersion is 1:1. For example, consider

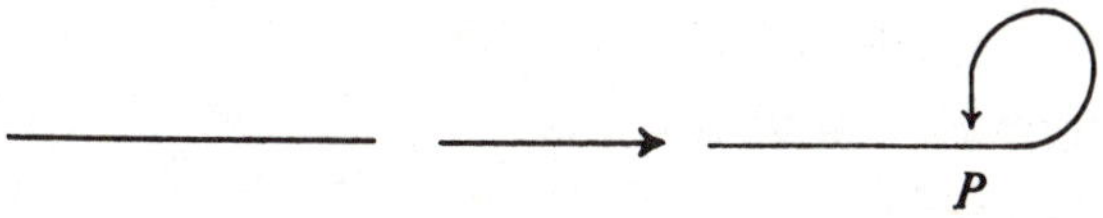

where $P = \text{Lim}_{t\to\infty} \phi(t)$. The induced topology on $\phi(\mathbf{R})$ from $\mathbf{R}^2$ is not the same (near P) as the induced manifold topology on $\phi(\mathbf{R})$. The following corollary is left as an exercise.

Corollary 2.11. *Let $\phi: X \to Y$ be an immersion. Then*

(1) *For every y in Y, $\phi^{-1}(y)$ is a discrete subset of X.*
(2) *$\phi(X)$ is a submanifold of Y iff the topology induced on $\phi(X)$ from its inclusion in Y is the same as its topology as an immersed submanifold.*

Clearly, in the second example above, open nbhds of P in the two relevant topologies on $\phi(\mathbf{R})$ are different.

Definition 2.12. *Let X and Y be topological spaces with $\phi: X \to Y$ continuous. Then ϕ is* proper *if for every compact subset K in Y, $\phi^{-1}(K)$ is a compact subset of X.*

Theorem 2.13. *Let X and Y be C^k manifolds and let $\phi: X \to Y$ be a C^k 1:1 proper immersion. Then $\phi(X)$ is a C^k submanifold of Y.*

Proof. Using Corollary 2.11 (2) we see that $\phi(X)$ is a submanifold iff $\phi: X \to \phi(X)$ is a homeomorphism where $\phi(X)$ is given the topology induced from Y. Clearly $\phi: X \to \phi(X)$ is continuous and bijective, so we need only show that ϕ^{-1} is continuous. Let $y_1, y_2, \ldots$ be a sequence in $\phi(X)$ converging to y in $\phi(X)$. Let $x_i = \phi^{-1}(y_i)$ and $x = \phi^{-1}(y)$. It is enough to show that $\text{Lim}_{i\to\infty} x_i = x$. Let K be a compact nbhd of y in Y. Since $\phi(X)$ has the topology induced from Y, $K \cap \phi(X)$ is a nbhd of y in $\phi(X)$ and we may assume, without loss of generality, that each y_i is in K. Since ϕ is proper, $\phi^{-1}(K)$ is compact and $\phi|\phi^{-1}(K): \phi^{-1}(K) \to \phi(X) \cap K$ is a homeomorphism. Thus $\text{Lim}_{i\to\infty} x_i = x$ by the continuity of $\phi^{-1}|\phi(X) \cap K$. □

Note. A 1:1 immersion can be a submanifold even if the immersion is not proper. Consider the spiral of $\mathbf{R}^+ \to \mathbf{R}^2$ given pictorially by

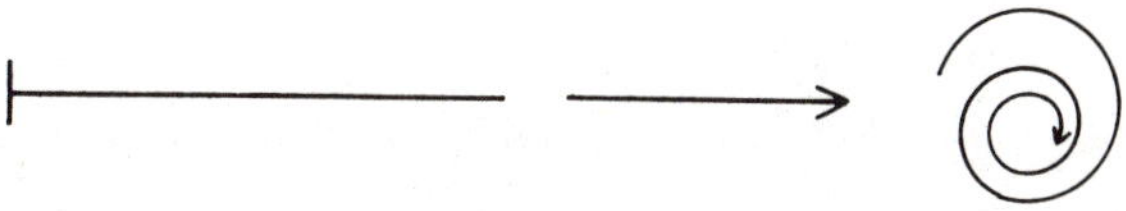

and analytically by $f(r) = (\cos(r)/r, \sin(r)/r)$. Clearly, f is a 1-1 immersion and f is not proper since $f^{-1}(B_1) = [1, \infty)$ where B_1 is the closed disk of radius 1 centered at the origin. But the two possible topologies on $f(\mathbf{R}^+)$ are the same so $f(\mathbf{R}^+)$ is a submanifold of $\mathbf{R}^2$.

Exercises:

(1) Let $f: \mathbf{R}^n \to \mathbf{R}^2$ be defined by

$$(x_1, \ldots, x_n) \mapsto (x_1^2 + \cdots + x_n^2, x_1^2 - (x_2^2 + \cdots + x_n^2))$$

(a) For which x in $\mathbf{R}^n$ is f a submersion at x?

(b) Let f_1 and f_2 be the coordinate functions of f. For which r, s in $\mathbf{R}$ is $f_1^{-1}(r) \cap f_2^{-1}(s)$ a smooth submanifold of $\mathbf{R}^n$.

(2) Let M_n be the set of $n \times n$ real matrices. Let M_n^k be the set of matrices in M_n of rank k. Prove that M_n^k is a submanifold of M_n and compute its dimension. (Hint: Let $S = \begin{pmatrix} A & B \\ C & D \end{pmatrix}$ be in M_n where $A \in M_k^k$. Show that $S \in M_n^k$ iff $D - CA^{-1}B = 0$.)

§3. Tangent Spaces

Definition 3.1. *Let X be a differentiable n-manifold.*

(1) *Let $c: \mathbf{R} \to X$ be differentiable with $c(0) = p$. Then c is a* curve on X based at p.

(2) *Let c_1 and c_2 be curves on X based at p. Then c_1* is tangent to c_2 at p *if for every chart ϕ on X with p in* dom ϕ,

(*) $$(d\phi \cdot c_1)_0 = (d\phi \cdot c_2)_0.$$

This makes sense since $\phi \cdot c_1$ and $\phi \cdot c_2$ are mappings of open nbhds of 0 in $\mathbf{R}$ into $\mathbf{R}^n$.)

Lemma 3.2. *If (*) holds for one chart ϕ, then it holds for every chart.*

Proof. Let ψ be another chart defined near p. Then

$$\begin{aligned}(d\psi \cdot c_1)_0 &= (d\psi\phi^{-1}\phi \cdot c_1)_0 \\ &= (d\psi \cdot \phi^{-1})_{\phi(p)}(d\phi \cdot c_1)_0 \\ &= (d\psi \cdot \phi^{-1})_{\phi(p)}(d\phi \cdot c_2)_0 = (d\psi \cdot c_2)_0 \end{aligned}$$ □

Definition 3.3. *Let S_pX denote the set of all curves on X based at p, p a point in X. Let $c_1, c_2 \in S_pX$. $c_1 \simeq c_2$ if c_1 is tangent to c_2 at p. $\simeq$ is clearly an equivalence relation. The set $T_pX \equiv S_pX/\simeq$ is called the* tangent space to X at p. *If c_1 is in S_pX, let $\hat{c}_1$ denote the equivalence class of c_1 in T_pX.*

Let ϕ be a chart on X with p in dom ϕ. *Note that $c_v(t) = \phi^{-1}(\phi(p) + tv)$ is a curve on X based at p where v is some vector in $\mathbf{R}^n$. Define $\lambda_\phi^p: R^n \to T_pX$ by $\lambda_\phi^p(v) = \hat{c}_v$.*

Lemma 3.4. *Let X be a differentiable n-manifold, p a point in X. Let ϕ be a chart on X near p. Then $\lambda_\phi^p: \mathbf{R}^n \to T_pX$ is bijective.*

Proof.

(a) λ_ϕ^p is 1:1. Let $v_1, v_2 \in \mathbf{R}^n$ and $\lambda_\phi^p(v_1) = \lambda_\phi^p(v_2)$. Then c_{v_1} and c_{v_2} are tangent at p; i.e., $(d\phi \cdot c_{v_1})_0 = (d\phi \cdot c_{v_2})_0$. Now

$$(d\phi \cdot c_{v_1})_0 = (d\phi \cdot \phi^{-1}(\phi(p) + tv_1))_0 = (d(\phi(p) + tv_1))_0$$

but $t \mapsto \phi(p) + tv_1$ has derivative at $t = 0$ equal to v_1. Similarly for v_2, so $v_1 = v_2$.

(b) $\lambda_\phi{}^p$ is onto. Let α be in T_pX. Let c be a curve representing the equivalence class α. Let $v = (d\phi \cdot c)_0$ be a vector in $\mathbf{R}^n$. By the calculation in part (a), $(d\phi \cdot c_v)_0 = v$ so $(d\phi \cdot c_v)_0 = (d\phi \cdot c)_0$ which implies that c and c_v are tangent at p. Stated differently, $\lambda_\phi{}^p(v) = \hat{c}_v = \hat{c} = \alpha$. □

Proposition 3.5. *There exists a unique vector space structure on T_pX such that for every chart ϕ on X with p in* dom ϕ, *the mapping $\lambda_\phi{}^p : \mathbf{R}^n \to T_pX$ is a linear isomorphism.*

Proof. Let ϕ, ψ be charts with p in dom $\phi \cap$ dom ψ. Then

(*) $$(\lambda_\psi{}^p)^{-1}\lambda_\phi{}^p = (d\psi \cdot \phi^{-1})_{\phi(p)} : \mathbf{R}^n \to \mathbf{R}^n.$$

Assuming this formula, it is clear that if $\lambda_\phi{}^p$ is linear for some chart ϕ, then $\lambda_\psi{}^p$ is linear for any other chart ψ. Let the vector space structure on T_pX be the one induced by $\lambda_\phi{}^p$ from $\mathbf{R}^n$, i.e., if α and β are in T_pX, then

$$\alpha + \beta = \lambda_\phi{}^p[(\lambda_\phi{}^p)^{-1}(\alpha) + (\lambda_\phi{}^p)^{-1}(\beta)]$$

We now prove the formula (*). Let v be in $\mathbf{R}^n$ and let $A = (d\psi \cdot \phi^{-1})_{\phi(p)}$. Then

$$\begin{aligned}(d\phi \cdot c_v)_0 &= (d(\phi(p) + tv))_0 = v \\ &= A^{-1}Av = (d\phi)_p \cdot d(\psi^{-1}(\psi(p) + tAv))_0\end{aligned}$$

Therefore $\lambda_\phi{}^p(v) = \lambda_\psi{}^p(Av)$, which is what was to be shown. □

Definition 3.6. *Let $f: X \to Y$ be a differentiable mapping with p in X and $q = f(p)$. Then f induces a linear map $(df)_p: T_pX \to T_qY$ called the* Jacobian of f at p *as follows: Let c be in S_pX; then $f \cdot c$ is in S_qY. To induce a map from $T_pX \to T_qY$ we need to know that if $c_1 \simeq c_2$ in S_pX, then $f \cdot c_1 \simeq f \cdot c_2$ in S_qY. Let ϕ be a chart on X near p and ψ a chart on Y near q. Then $c_1 \simeq c_2$ implies that $(d\phi \cdot c_1)_0 = (d\phi \cdot c_2)_0$. Hence*

$$\begin{aligned}(d\psi \cdot f \cdot c_1)_0 &= (d\psi \cdot f \cdot \phi^{-1})_{\phi(p)}(d\phi \cdot c_1)_0 \\ &= (d\psi \cdot f \cdot \phi^{-1})_{\phi(p)}(d\phi \cdot c_2)_0 = (d\psi \cdot f \cdot c_2)_0\end{aligned}$$

using the chain rule. So by definition, $f \cdot c_1 \simeq f \cdot c_2$. This defines $(df)_p: T_pX \to T_qY$. To check that $(df)_p$ is linear, we have the following formula:

(**) $$(df)_p = \lambda_\psi{}^q(d\psi \cdot f \cdot \phi^{-1})_{\phi(p)}(\lambda_\phi{}^p)^{-1}$$

Let $\hat{c}$ be in T_pX. Then we may take $c(t) = \phi^{-1}(\phi(p) + tv)$ for some v in $\mathbf{R}^n$. Now

$$\lambda_\psi{}^q(d\psi \cdot f \cdot \phi^{-1})_{\phi(p)}(\lambda_\phi{}^p)^{-1}\hat{c} = \lambda_\psi{}^q(d\psi \cdot f \cdot \phi^{-1})_{\phi(p)}(v)$$

which is equal to the equivalence class of the curve

$$c_1(t) = \psi^{-1}(\psi(q) + t(d\psi \cdot f \cdot \phi^{-1})_{\phi(p)}(v)).$$

Thus $(df)_p(\hat{c})$ is the equivalence class of the curve

$$c_2(t) = f \cdot \phi^{-1}(\phi(p) + tv).$$

To see that c_1 and c_2 are tangent at q, we compute

$$(d\psi \cdot c_1)_0 = (d\psi \cdot f \cdot \phi^{-1})_{\phi(p)}(v)$$

and

$$(d\psi \cdot c_2)_0 = (d\psi \cdot f \cdot \phi^{-1})_{\phi(p)}(d(\phi(p) + tv))_0 = (d\psi \cdot f \cdot \phi^{-1})_{\phi(p)}(v).$$

Remark. Using (**) and the fact that $\lambda_\psi{}^q$ and $\lambda_\phi{}^p$ are isomorphisms we have that f is an immersion at p if rank $(df)_p = \dim X$ and that f is a submersion at p if rank $(df)_p = \dim Y$.

***Definition* 3.7.** *Let X be a differentiable manifold. Then*

$$TX = \bigcup_{p \in X} T_pX = \text{tangent bundle to } X$$

Let $\pi : TX \to X$ denote the natural projection.

***Proposition* 3.8.** *Let X be a C^k-differentiable n-manifold* ($k > 0$). *Then TX has, in a natural way, the structure of a C^{k-1} manifold of dimension $2n$.*

Proof. Let p be a point in X, U an open nbhd of p in X, and ϕ a chart with domain U. Let $T_UX = \pi^{-1}(U)$. Define $\tilde{\phi} : T_UX \to \phi(U) \times \mathbf{R}^n$ by $\tilde{\phi}(\alpha) = (\phi \cdot \pi(\alpha), (\lambda_\phi{}^{\pi(\alpha)})^{-1}(\alpha))$ for every α in T_UX. $\tilde{\phi}$ is bijective. We claim that if $\{\phi_\beta\}$ is an atlas on X, then TX can be topologized so that $\{\tilde{\phi}_\beta\}$ is an atlas on TX. Note that

$$\begin{aligned} \tilde{\phi} \cdot \tilde{\psi}^{-1}(a, v) &= (\phi \cdot \psi^{-1}(a), (\lambda_\phi{}^q)^{-1} \lambda_\psi{}^q(v)) \\ &= (\phi \cdot \psi^{-1}(a), (d\phi \cdot \psi^{-1})_a(v)) \end{aligned}$$

where $q = \psi^{-1}(a)$, by using the formula (*) in Proposition 3.5. Now $\phi \cdot \psi^{-1} : \mathbf{R}^n \to \mathbf{R}^n$ is C^k-differentiable and $(d\psi \cdot \phi^{-1}) : U \times \mathbf{R}^n \to U \times \mathbf{R}^n$ is C^{k-1}-differentiable since it is given by a matrix whose coefficients are first partial derivatives of $\psi \cdot \phi^{-1}$ on U. Define the topology on TX so that all the $\tilde{\phi}_\beta$ are homeomorphisms. Then TX has the structure of C^{k-1}-differentiable manifold. □

Notes. (1) Let V be a (finite dimensional) vector space with p in V. It is obvious that there is a canonical identification of V with T_pV given by $v \mapsto \hat{c}$ where $c(t) = p + tv$.

(2) Let V be a vector space and let $G(k, V)$ be the Grassmann manifold of k-dimensional subspaces of V. Let W be in $G(k, V)$. (We shall view W both as a point in $G(k, V)$ and a subspace of V.) We show that there is a canonical identification of $T_WG(k, V)$ with Hom $(W, V/W)$. Choose a complementary subspace S to W in V. Let $C(t)$ be a curve in $G(k, V)$ based at W. Define $A_t : W \to V/W$ by $A_t(w) = \pi(s_t)$ where $\pi : V \to V/W$ is the obvious projection and $w = s_t + c_t$ where $c_t \in C(t)$ and $s_t \in S$. (Note that for t small, writing $w = s_t + c_t$ is always possible.) First show that if $C(t)$ and $C'(t)$ are two curves on $G(k, V)$ tangent at W, then

$$\left.\frac{dA_t}{dt}(w)\right|_{t=0} = \left.\frac{dA'_t}{dt}(w)\right|_{t=0}$$

as mappings of $W \to V/W$. Thus we have a linear mapping $\phi: T_W G(k, V) \to \mathrm{Hom}(W, V/W)$ given by

$$\frac{dC}{dt}(0) \mapsto \frac{dA_t}{dt}(w)\bigg|_{t=0}.$$

Next show that ϕ is, in fact, an isomorphism. Finally show that ϕ is independent of the choice S. Hint: Let S' be another complementary subspace to W in V. Then $s_t - s_t' = c_t' - c_t$ is in $C(t)$. Thus there is an α_t in $C(t)$ such that $s_t - s_t' = t\alpha_t$. Now show that

$$\frac{d(A_t - A_t')}{dt}(w)\bigg|_{t=0} = 0.$$

§4. Partitions of Unity

Manifolds are geometric objects that locally "look like" Euclidean space. It would then be convenient to be able to do whatever analysis or calculus that we have to do locally; i.e., in Euclidean space. The use of partitions of unity is the technique to accomplish this goal.

Definition 4.1. *Let X be a topological space.*

(1) *$\{U_\alpha\}_{\alpha \in I}$ (I some index set) is a* covering of *X if each U_α is contained in X and $X = \bigcup_{\alpha \in I} U_\alpha$.*

(2) *Let $\{U_\alpha\}_{\alpha \in I}$ and $\{V_\beta\}_{\beta \in J}$ be coverings of X. Then $\{V_\beta\}_{\beta \in J}$ is a* refinement of *$\{U_\alpha\}_{\alpha \in I}$ if for every β in J, there is an α in I so that $V_\beta \subset U_\alpha$.*

(3) *Let $\{V_\beta\}_{\beta \in J}$ be a covering of X. Then $\{V_\beta\}_{\beta \in J}$ is* locally finite *if for every p in X, there is a nbhd U of p in X so that $U \cap V_\beta = \varnothing$ for all but a finite number of β's in J.*

(4) *X is* paracompact *if every open covering of X has a locally finite refinement.*

Proposition 4.2. *Let X be a Hausdorff topological space which is locally compact and satisfies the second axiom of countability. Then X is paracompact. In particular, all manifolds are paracompact.* (Recall that X satisfies the second axiom of countability if the topology on X has a countable base.)

Proof. We first construct a sequence of compact sets $K_1, K_2, \ldots$ such that

(1) $K_i \subset \mathrm{Int}(K_{i+1})$ for all i, and

(2) $X = \bigcup_{i=1}^{\infty} K_i$.

Since X is locally compact and second countable, we may choose a sequence of open sets $N_1, N_2, \ldots$ each of which has compact closure and such that the N_i's cover X. Let $M_k = \bigcup_{i=1}^{k} N_i$. Let $K_1 = \overline{M}_1$. Since K_1 is compact there exists $M_{i_1}, \ldots, M_{i_r}$ so that $K_1 \subset M_{i_1} \cup \cdots \cup M_{i_r}$. Let $K_2 = \overline{M}_{i_1} \cup \cdots \cup \overline{M}_{i_r}$. Thus K_2 is compact and $K_1 \subset \mathrm{Int}(K_2)$. Proceed inductively.

Now let $\{W_\alpha\}_{\alpha \in I}$ be an open covering of X. We construct a locally finite refinement. For each i, let $W_{\alpha_1}{}^i, \ldots, W^i_{\alpha_{k_i}}$ be a finite subcovering of the compact subset $K_i - \operatorname{Int}(K_{i-1})$. Let $V_j{}^i = W_{\alpha_j}{}^i \cap \operatorname{Int}(K_{i+1} - K_{i-2})$. Then the collection $\{V_j{}^i\}$ is a locally finite refinement of the covering $\{W_\alpha\}$. □

Corollary 4.3. *Let X be a differentiable manifold and let $\{U_\alpha\}_{\alpha \in I}$ be an open covering of X. Then there is a countable locally finite refinement $\{V_\beta\}_{\beta \in J}$ of $\{U_\alpha\}_{\alpha \in I}$ such that*

(a) *For every β in J, there is a chart $\phi_\beta : V_\beta \to B_3$ which is onto, and*

(b) *the sets $V_\beta{}^1 = \phi_\beta{}^{-1}(B_1)$ form a countable open covering of X, where $B_r = \{x \in \mathbf{R}^n \mid |x| < r\}$.*

Proof. Choose $K_1, K_2, \ldots$ as in the proof of Proposition 4.2. For each p in $K_i - \operatorname{Int}(K_{i-1})$ choose an open nbhd $V_p{}^i$ of p so that

(i) $V_p{}^i \subset \operatorname{Int}(K_{i+1} - K_{i-2}) \cap U^p$ where U^p is some open set in the covering $\{U_\alpha\}_{\alpha \in I}$ containing p, and

(ii) $V_p{}^i$ is the domain of a chart $\phi_p{}^i : V_p{}^i \to B_3$ which is onto and satisfies $\phi_p{}^i(p) = 0$. Let $W_p{}^i = (\phi_p{}^i)^{-1}(B_1)$. These sets cover $K_i - \operatorname{Int}(K_{i-1})$. Choose a finite subcover $W_{p_1}{}^i, \ldots, W_{p_{k_i}}{}^i$. Then the sets $\{W_{p_j}{}^i\}_{1 \le j < k_i}^{1 \le i \le \infty}$ give the required locally finite cover. □

Definition 4.4.

(1) *Let X be a topological space and let $f: X \to \mathbf{R}$ be continuous. Then the* support *of f denoted* $\operatorname{supp}(f)$ = *closure of the set* $\{x \in X \mid f(x) \neq 0\}$.

(2) *Let X be a C^k-differentiable manifold. Then a C^k-*partition of unity *on X is a collection $\{f_\alpha\}_{\alpha \in I}$ (I some index set) of C^k-differentiable functions mapping X into $\mathbf{R}$ such that*

(a) $\{\operatorname{supp}(f_\alpha)\}_{\alpha \in I}$ *is a locally finite covering of X,*

(b) $f_\alpha(p) \ge 0$ *for every $\alpha \in I$ and $p \in X$, and*

(c) $\sum_{\alpha \in I} f_\alpha(p) \equiv 1$ *for every $p \in X$. Note that condition* (a) *ensures that this is a finite sum.*

(3) *A partition of unity $\{f_\alpha\}_{\alpha \in I}$ on X is* subordinate to a covering $\{U_\beta\}_{\beta \in J}$ *if for every α in I, there exists a β in J for which* $\operatorname{supp}(f_\alpha) \subset U_\beta$.

Lemma 4.5. *Let B be an open ball of radius r centered at x_0 in $\mathbf{R}^n$. Then there exists a smooth function positive on B and zero off B.*

Proof. It is enough to show that there exists a smooth function $\gamma : \mathbf{R} \to \mathbf{R}$ such that $\gamma(s) = 0$ for $s \ge 1$ and $\gamma(s) > 0$ for $s < 1$. If γ exists, then consider $\rho : \mathbf{R}^n \to \mathbf{R}$ defined by $\rho(x) = \gamma(|x - x_0|^2/r^2)$. Clearly ρ is smooth and has the desired properties. Now just define

$$\gamma(s) = \begin{cases} 0 & \text{if } s \ge 1 \\ \exp\left(-\left(\dfrac{1}{s-1}\right)^2\right) & \text{if } s < 1 \end{cases}.$$

We leave it to the reader to check that γ is indeed smooth. □

Theorem 4.6. *Let X be a C^k differentiable manifold and let $\{U_\alpha\}_{\alpha \in I}$ be an open covering of X. Then there exists a countable partition of unity $\{\rho_\beta\}_{\beta \in J}$ on X subordinate to the covering $\{U_\alpha\}_{\alpha \in I}$. Moreover, if I is countable, then we may let $J = I$ and assume that $\operatorname{supp} \rho_\alpha \subset U_\alpha$ for all α in I.*

Proof. Let $\{V_\beta\}_{\beta \in J}$ be the locally finite refinement of $\{U_\alpha\}_{\alpha \in I}$ whose existence is guaranteed by Corollary 4.3. Define $g_\beta: X \to \mathbf{R}$ by

$$g_\beta(p) = \begin{cases} \gamma(\phi_\beta(p)) & \text{if } p \in V \\ 0 & \text{otherwise} \end{cases}$$

where $\gamma: \mathbf{R}^n \to \mathbf{R}$ is a smooth function which is positive on B_2 and zero off B_2 (using Lemma 4.5). Let $h(p) = \sum_{\beta \in J} g_\beta(p)$. Then h is well-defined (i.e., the sum is finite), and C^k since $\{V_\beta\}_{\beta \in J}$ is a locally finite covering for X. Also $h(p) > 0$ for all p. Let $\rho_\beta = (1/h) g_\beta$. Then $\{\rho_\beta\}_{\beta \in J}$ is a partition of unity subordinate to the cover $\{U_\alpha\}_{\alpha \in I}$. For the moreover part, let $U_1, U_2, \ldots$ be the covering and let $f_i = \sum_{\beta \in J_i} \rho_\beta$ where β is in J_i if $\operatorname{supp} \rho_\beta \subset U_i$ and $\operatorname{supp} \rho_\beta \not\subset U_j$ for $j < i$. ◻

Corollary 4.7. *Let X be a C^k differentiable manifold. Let U and V be open subsets of X with $\bar{U} \subset V$. Then there is a C^k differentiable function $f: X \to \mathbf{R}$ such that*

$$f(x) = \begin{cases} 1 & \textit{if } x \in U \\ 0 & \textit{if } x \notin V \\ 0 \le f(x) \le 1 & \textit{otherwise.} \end{cases}$$

Proof. Let $\{f_1, f_2\}$ be a partition of unity subordinate to the cover $\{V, X - \bar{U}\}$ given by Theorem 4.6. Take $f = f_1$. Certainly $\operatorname{supp} f \subset V$ and $f \equiv 1$ on U since $f_2 \equiv 0$ on U. ◻

We present the following Proposition just to indicate the great number of smooth functions which exist (as compared to, say, analytic functions).

Proposition 4.8. *Let C be a closed subset of $\mathbf{R}^n$; then there exists a smooth function $f: \mathbf{R}^n \to \mathbf{R}$ such that $f \ge 0$ everywhere and $C = f^{-1}(0)$.*

Proof. Cover $\mathbf{R}^n - C$ by a countable sequence of open balls $B_1, B_2, \ldots$ each contained in $\mathbf{R}^n - C$. Let f_i be a smooth function zero off B_i and positive on B_i. (Use Lemma 4.5.) Let

$$M_i = \sup_{|\alpha| \le i} \left(\frac{\partial^{|\alpha|} f_i}{\partial x^\alpha} \right).$$

(M_i is well-defined since each $\partial^{|\alpha|} f_i / \partial x^\alpha$ is compactly supported.) Let

$$f = \sum_{i=1}^{\infty} \frac{f_i}{2^i M_i}.$$

The choice of the M_i's guarantees that for each α the series

$$\sum_{i=1}^{\infty} \frac{1}{2^i M_i} \frac{\partial^{|\alpha|} f_i}{\partial x^\alpha}$$

converges uniformly. Using a standard theorem from advanced calculus (see Dieudonné, *Foundations of Modern Analysis*, 8.6.3, p. 157), $\partial^{|\alpha|}f/\partial x^\alpha$ exists and is continuous for each α, so that f is smooth. Thus f is the desired function. □

Exercise:

Let X be a smooth compact manifold. Show that there exists a 1:1 immersion of X into some Euclidean space, and thus conclude that any compact manifold can be realized as a submanifold of $\mathbf{R}^N$ for some large N.

§5. Vector Bundles

Definition 5.1.

(1) *Let E and X be smooth manifolds and $\pi: E \to X$ a submersion. Let $E_U = \pi^{-1}(U)$ for any subset U of X. Then E is a* family of vector spaces over X of dimension k *if for every p in X, E_p is a real vector space of dimension k whose operations (addition and scalar multiplication) are compatible with the topology on E_p induced from E. Let k be denoted by* $\dim_X E$.

(2) *A* section *of E is a smooth mapping $s: X \to E$ such that $\pi \cdot s = id_X$. $C^\infty(E)$ denotes the space of smooth sections of E.*

(3) *Let $\pi_E: E \to X$ and $\pi_F: F \to X$ be families of vector spaces over X. Then $\phi: E \to F$ is a* homomorphism from E to F *if*

(a) $\pi_F \cdot \phi = \pi_E$
(b) *ϕ is smooth*
(c) *For every $p \in X$, $\phi: E_p \to F_p$ is a linear map.*

ϕ is an isomorphism *if ϕ is a diffeomorphism and a homomorphism.*

Example. Let V be a vector space (finite dimensional), X a smooth manifold, and $E = X \times V$. Let $\pi: E \to X$ be a projection on the first factor. Then $E \xrightarrow{\pi} X$ is a family of vector spaces known as a *product family*. A family of vector spaces F over X is *trivial* if it is isomorphic to some product family.

Definition 5.2. *Let $E \xrightarrow{\pi} X$ be a family of vector spaces over X. E is a* vector bundle *over X if every point p in X has an open nbhd U_p so that the family of vector spaces E_{U_p} is trivial (i.e., a vector bundle is a locally trivial family of vector spaces). Note that* $\dim E = \dim X + \dim_X E$.

Example. Let X be a smooth manifold. Then TX (the tangent bundle over X) is a vector bundle with $\dim_X TX = \dim X$. The charts that were constructed in Proposition 3.8 to show that TX is a manifold also show that it is a locally trivial family of vector spaces.

When working with a vector space V, it is often useful to consider certain associated spaces such as the dual space V^*, the space $S^2(V^*)$ of symmetric bilinear forms on V, etc. In a similar fashion, when given a vector bundle E

over X, it is sometimes useful to construct associated vector bundles over X. For instance, one should be able to replace E_p by E_p^* (the dual space to E_p) for each p in X and make the new set into a vector bundle. One could also replace E_p by $S^2(E_p^*)$ (the space of symmetric bilinear forms on E_p), etc. The following will show how to formalize such a process to yield new vector bundles.

Let T be a covariant functor which takes (finite dimensional) vector spaces into (finite dimensional) vector spaces, (i.e., T: Vector spaces $\rightarrow$ Vector spaces and if V and W are vector spaces, then

$$T\colon \operatorname{Hom}(V, W) \rightarrow \operatorname{Hom}(T(V), T(W)).$$

This latter map has the property that if $f\colon V \rightarrow W$ and $g\colon W \rightarrow Z$, then $T(g \cdot f) = T(g) \cdot T(f)$. Note that $\operatorname{Hom}(V, W)$ denotes the set of linear mappings from V to W and is vector space isomorphic to $\mathbf{R}^{m \cdot n}$ where $n =$. dim V and $m =$ dim W.)

Definition 5.3. *T is* smooth *if for every pair of vector spaces V and W, the mapping*

$$T\colon \operatorname{Hom}(V, W) \rightarrow \operatorname{Hom}(T(V), T(W))$$

is smooth. (Note that the above isomorphism of $\operatorname{Hom}(V, W)$ *with* $\mathbf{R}^{m \cdot n}$ *gives* $\operatorname{Hom}(V, W)$ *the structure of a smooth manifold.)*

Proposition 5.4. *Let E be a vector bundle over X and T be a smooth covariant functor defined on (finite dimensional) vector spaces. Then $T(E) = \bigcup_{p \in X} T(E_p)$ (disjoint union) has the structure of a vector bundle over X.*

Proof. Let E be a set, X a smooth manifold, and $\pi\colon E \rightarrow X$ a map. Assume that E_p is a vector space for each p in X. To put a vector bundle structure on E is to make E into a smooth manifold so that E becomes a vector bundle over X with projection map π. Suppose F is a vector bundle and $\phi\colon E \rightarrow F$ is a bijection which is linear on the fibers and for which $\pi = \pi_F \cdot \phi$. Then there is a unique way to put a manifold structure on E so that E becomes a vector bundle and ϕ an isomorphism.

(1) We note that if $\phi\colon E \rightarrow F$ is a homomorphism, then there is a map $T(\phi)\colon T(E) \rightarrow T(F)$ which is linear on fibers. $T(\phi)(e) = T(\phi_p)(e)$ where $\phi_p = \phi | E_p\colon E_p \rightarrow F_p$ for $p \in X$ and $e \in E_p$. So $T(\phi)\colon \bigcup_{p \in X} T(E_p) \rightarrow \bigcup_{p \in X} T(F_p)$.

(2) Suppose that $E = X \times V$ is a product family where V is some vector space. Then $T(E) = \bigcup_{p \in X} T(V) = X \times T(V)$, the last equality being an obvious bijection. Give $T(E)$ a vector bundle structure by making this identification an isomorphism.

(3) Next assume that E is a trivial bundle. Then there exists an isomorphism $\phi\colon E \rightarrow X \times V = F$. As noted in (1), $T(\phi)\colon T(E) \rightarrow T(F)$. Since $T(F)$ is a vector bundle (by (2)), we can give $T(E)$ a vector bundle structure so that $T(\phi)$ is an isomorphism.

It is necessary to check that this vector bundle structure is independent of the choice of ϕ. So let $\psi\colon E \rightarrow G$ be an isomorphism where $G = X \times W$.

Then $\psi\cdot\phi^{-1}: X \times V \to X \times W$ is an isomorphism and can be identified with $\lambda: X \to \operatorname{Hom}(V, W)$ given by $\lambda(p) = \psi\cdot\phi^{-1}|(p \times V): V \to W$. Then $T(\psi\cdot\phi^{-1}): X \times T(V) \to X \times T(W)$ can be identified with $T\cdot\lambda$. Since $\psi\cdot\phi^{-1}$ is smooth, λ is smooth and since T is a smooth functor $T\cdot\lambda$ is smooth. Hence $T(\psi\cdot\phi^{-1})$ is smooth and an isomorphism. The diagram

$$\begin{array}{ccc} T(E) & \xrightarrow{\quad T(\phi)\quad} & T(F) \\ {\scriptstyle id}\big\downarrow & & \big\downarrow{\scriptstyle T(\psi\cdot\phi^{-1})} \\ T(E) & \xrightarrow{\quad T(\psi)\quad} & T(G) \end{array}$$

commutes and implies that the identity map on $T(E)$ is smooth as a map between the two possible vector bundle structures. Thus the two structures are the same.

(4) Let E be an arbitrary vector bundle. For each p in X, there is an open nbhd U_p so that E_{U_p} is a trivial bundle. By (3), $T(E_{U_p})$ has a unique structure as a vector bundle. Suppose $U_p \cap U_q \neq \varnothing$. Then $T(E_{U_p \cap U_p})$ has two structures as a vector bundle, namely $T(E_{U_p})_{U_p \cap U_q}$ and $T(E_{U_q})_{U_p \cap U_q}$. The uniqueness of the structures on $T(E_{U_p})$ and $T(E_{U_q})$ implies that these two structures are the same. So we have a unique way of making $T(E)$ into a vector bundle. □

Note. A similar proposition clearly holds when T is contravariant or when T is a functor of several variables, some covariant and some contravariant.

Examples.

(1) $T(V) = V^*$—the dual vector space to V. $T: \operatorname{Hom}(V, W) \to \operatorname{Hom}(W^*, V^*)$ is given by $A \mapsto A^*$—the adjoint of A. T is a continuous linear map and hence smooth. So $E^* = T(E)$ is a vector bundle. In particular, if $E = TX$, then $T(E)$ is denoted T^*X and is the *cotangent bundle of X*.

(2) $T(V_1, V_2) = V_1 \oplus V_2$.

$$T: \operatorname{Hom}(V_1, W_1) \times \operatorname{Hom}(V_2, W_2) \to \operatorname{Hom}(V_1 \oplus V_2, W_1 \oplus W_2)$$

is given by $T(f, g) \to f \oplus g$. T is continuous and bilinear, hence T is smooth. Given two vector bundles E and F over X, $T(E, F)$ is denoted by $E \oplus F$ and is called the *Whitney sum* of E and F. Note that $(E \oplus F)_p = E_p \oplus F_p$ for every $p \in X$. Hence $\dim_X (E \oplus F) = \dim_X E + \dim_X F$.

(3) $T(V) = S^2(V^*)$—the vector space of symmetric bilinear forms on V. $T: \operatorname{Hom}(V, W) \to \operatorname{Hom}(S^2(W^*), S^2(V^*))$ is defined as follows. Let $A \in \operatorname{Hom}(V, W)$, $B \in S^2(W^*)$, and $v_1, v_2 \in V$. Then $T(A)(B)$ is a symmetric bilinear form on V given by $T(A)(B)(v_1, v_2) = B(Av_1, Av_2)$. T is continuous and linear, hence T is smooth. If E is a vector bundle over X, then $T(E) = S^2(E^*)$ is a vector bundle over X. $\dim_X S^2(E^*) = n(n + 1)/2$ where $n = \dim_X E$.

(4) There is one more vector bundle which we shall need later. Let $G_{k,n}$ be the Grassmann manifold of k-planes in n-space. Let p be in $G_{k,n}$ and let E_p be the k-dimensional subspace of $\mathbf{R}^n$ associated with p. Then $E = \bigcup_{p \in G_{k,n}} E_p$ is a vector bundle over $G_{k,n}$ called the *canonical bundle*. Let ρ be the obvious projection. Recall from Example (3) after Definition 1.6 the chart nbhds W_p of $G_{k,n}$; i.e., $W_p = \{q \in G_{k,n} \mid \pi_{E_q,E_p}$ is a bijection onto $E_p\}$ where π_{E_q,E_p} is given by the restriction to E_q of orthogonal projection of $\mathbf{R}^n \to E_p$. The mappings $\psi_p : E_{W_p} \to W_p \times E_p$ defined by $v \mapsto (\rho(v), \pi_{E_{\rho(v)},E_p}(v))$ give the vector bundle structure to E. Check the details.

It is customary to give sections of certain vector bundles special names.

Definition 5.5. *Let X be a smooth manifold.*

(1) *A section of TX is called a* vector field.

(2) *A section of T^*X is called a* 1-form.

(3) *A section $s : X \to S^2(E^*)$ is a* metric on E *if $s(p)$ is a positive definite, symmetric, bilinear form for each p in X.*

(4) *A metric on TX is called a* Riemannian metric.

Locally the above sections have standard coordinate representations. Let U be a coordinate nbhd on X with $\phi_1, \ldots, \phi_n$ the system of coordinates and $\phi : U \to \mathbf{R}^n$ the corresponding chart. Equipping $\mathbf{R}^n$ with the standard coordinates $x_1, \ldots, x_n$, we may define

$$\left.\frac{\partial}{\partial \phi_i}\right|_p = (d\phi^{-1})_{\phi(p)}\left(\left.\frac{\partial}{\partial x_i}\right|_{\phi(p)}\right)$$

where $(\partial/\partial x_i)|_q$ is the unit vector in the x_i-direction based at q. (Thus $\partial/\partial x_i$ can be viewed as a vector field on $\mathbf{R}^n$.) Then $(\partial/\partial\phi_i)|_p \in T_pX$ and $\partial/\partial\phi_i : U \to T_UX$ is a locally defined section on X. The vectors

$$\left\{\left.\frac{\partial}{\partial \phi_i}\right|_p\right\}_{i=1}^n$$

are linearly independent at each p in U; so if s is a vector field on U, then

$$s(p) = \sum_{i=1}^n a_i(p) \left.\frac{\partial}{\partial \phi_i}\right|_p.$$

We note that $s : U \to TX$ is smooth iff $a_i : U \to \mathbf{R}$ is smooth for $1 \leq i \leq n$. So locally a vector field is a linear combination (over smooth functions) of the coordinate vector fields $\partial/\partial\phi_1, \ldots, \partial/\partial\phi_n$.

If $\{d\phi_1, \ldots, d\phi_n\}$ is the dual basis to $\partial/\partial\phi_1, \ldots, \partial/\partial\phi_n$ at each point of U, then every 1-form s can be written locally as $s = \sum_{i=1}^n a_i \, d\phi_i$. Also, $s : U \to T^*X$ is smooth iff $a_i : U \to R$ $(1 \leq i \leq n)$ is smooth. Finally, if s is a Riemannian metric, then locally $s = \sum_{i,j=1}^n a_{ij} \, d\phi_i \, d\phi_j$, i.e., if $\xi, \eta \in T_pX$, then

$$s_p(\xi, \eta) = \sum_{i,j=1}^n a_{ij}(p)(d\phi_i)_p(\xi)(d\phi_j)_p(\eta).$$

Here again $s : U \to S^2(T^*X)$ is smooth iff $a_{ij} : U \to \mathbf{R}$ is smooth for $1 \leq i,j \leq n$. Note that since s is symmetric $a_{ij} = a_{ji}$ for all i, j and that since s is positive $\det(a_{ij}(p)) \neq 0$ for each p in U.

Proposition 5.6. *Every vector bundle $\pi : E \to X$ has a metric.*

Proof. Let $E = X \times V$ be a product bundle. Then $S^2(E^*) = X \times S^2(V^*)$ is also a product bundle. Now let B be any positive definite symmetric bilinear form on V. Then define $s : X \to X \times S^2(V^*)$ by $s(p) = (p, B)$. s is smooth and a metric on $S^2(E^*)$. If E is a trivial bundle then there exists an isomorphism $\phi : E \to X \times V$ (for some product bundle $X \times V$). ϕ induces an isomorphism $\phi^{(2)} : S^2((X \times V)^*) \to S^2(E^*)$. If s is a metric on $X \times V$, then $\phi^{(2)} \cdot S$ is a metric on E.

Finally, let E be an arbitrary vector bundle. For each p in X, choose an open nbhd U_p of p so that E_{U_p} is trivial. Let $\{V_i\}_{i=1}^{\infty}$ be a countable locally finite refinement of $\{U_p\}_{p \in X}$. Let $\{\rho_i\}_{i=1}^{\infty}$ be a partition of unity subordinate to the cover $\{V_i\}_{i=1}^{\infty}$ of X. (See Theorem 4.6.) Let s_i be a metric on E_{V_i}. Define $\bar{s}_i : X \to S^2(E^*)$ by

$$\bar{s}_i(p) = \begin{cases} \rho_i(p)s_i(p) & \text{for all } p \in V_i \\ 0 & \text{otherwise;} \end{cases}$$

then $\bar{s}_i$ is a smooth section. Let $s = \sum_{i=1}^{\infty} \bar{s}_i$. This sum makes sense since for each p in X, only finitely many $\rho_i(p)$ are not zero. Let $v \in E_p$. Then

$$s(p)(v, v) = \sum \rho_i(p)s_i(p)(v, v) > \rho_i(p)s_i(p)(v, v)$$

where i is chosen so that $\rho_i(p) > 0$. Thus $s(p)$ is positive definite since $s_i(p)(v, v) \neq 0$, and s is a metric on E. □

Given a Riemannian metric $s : X \to S^2(T^*X)$ where X is a connected manifold, then there is a natural way to define a metric $d : X \times X \to \mathbf{R}$ so that (X, d) is a metric space. (There is, unfortunately, no way to change the fact that the word "metric" has two different though related meanings!) We show how to define d.

Let p and q be points in X and $c : \mathbf{R} \to X$ a (continuous, piecewise smooth) curve with $c(0) = p$ and $c(1) = q$. By piecewise smooth, we mean that the curve is infinitely differentiable except at a finite number of points. Let $(d/dt)|_{t_0}$ be the tangent vector in $T_{t_0}R$ defined by the curve $t \mapsto t_0 + t$ of $\mathbf{R} \to \mathbf{R}$. Then $t \to (d/dt)|_t$ is the canonical vector field on $\mathbf{R}$. Define $f : \mathbf{R} \to \mathbf{R}$ by

$$f(t) = \sqrt{s_{c(t)}\left((dc)_t\left(\frac{d}{dt}\bigg|_t\right), (dc)_t\left(\frac{d}{dt}\bigg|_t\right)\right)}$$

f is a piecewise smooth function and $\bar{c} = \int_0^1 f(t)\, dt$ makes sense. Note that $\bar{c}$ is just the *arc-length* of the curve c relative to the Riemannian metric s. Define $d(p, q)$ to be the infimum of $\bar{c}$ where c ranges over all piecewise smooth curves connecting p to q.

It should be noted that $d(p, q)$ is always defined and finite. Define an equivalence relation $\sim$ on X by $p \sim q$ if there exists a piecewise smooth curve of finite length connecting p to q. Since X is locally Euclidean, the equivalence classes are open. Since X is connected there is only one nonempty equivalence class. All steps in showing that d is a metric are easy except showing that if $d(p, q) = 0$, then $p = q$. This will be proved later. In any case, d is a pseudometric.

Example. Let $x_1, \ldots, x_n$ be the standard coordinates of $\mathbf{R}^n$. $s = \sum_{i=1}^n dx_i^2$ is a metric and induces the standard metric on $\mathbf{R}^n$. Let $c : \mathbf{R} \to \mathbf{R}^n$ be a curve with $c(t) = (c_1(t), \ldots, c_n(t))$. Note that

$$(dc)_t\left(\frac{d}{dt}\bigg|_t\right) = \sum_{i=1}^n \frac{\partial c_i}{\partial t}\frac{\partial}{\partial x_i}.$$

Hence

$$s_t\left((dc)_t\left(\frac{d}{dt}\bigg|_t\right), (dc)_t\left(\frac{d}{dt}\bigg|_t\right)\right) = \sum_{i=1}^n \left(\frac{\partial c_i}{\partial t}\right)^2.$$

So

$$\bar{c} = \int_0^1 \sqrt{\left(\frac{\partial c_1}{\partial t}\right)^2 + \cdots + \left(\frac{\partial c_n}{\partial t}\right)^2}\, dt$$

which is just the standard arc-length in $\mathbf{R}^n$. As is known from Euclidean geometry of $\mathbf{R}^n$ the shortest distance between two points is the straight line distance, so this metric on $\mathbf{R}^n$ is just the standard one.

Lemma 5.7. *Let $\bar{s} = \sum_{i,j=1}^n a_{ij}dx_idx_j$ be a Riemannian metric on $\mathbf{R}^n$. Let $\bar{d}$ be the induced pseudo-metric on $\mathbf{R}^n$. Then on a given compact set K there exist positive constants L and M so that $Md(p, q) \geq \bar{d}(p, q) \geq Ld(p, q)$ for every p and q in K where d is the standard metric on $\mathbf{R}^n$.*

$$(dc)_t\left(\frac{d}{dt}\bigg|_t\right) = \sum_{i=1}^n \frac{\partial c_i}{\partial t}\frac{\partial}{\partial x_i}.$$

Thus

$$\bar{s}_t\left((dc)_t\left(\frac{d}{dt}\bigg|_t\right), (dc)_t\left(\frac{d}{dt}\bigg|_t\right)\right) = \sum_{i,j=1}^n a_{ij}\frac{\partial c_i}{\partial t}\frac{\partial c_j}{\partial t}.$$

Let $v = (\partial c_1/\partial t, \ldots, \partial c_n/\partial t)$ and $A = (a_{ij})$. Then $s = v^tAv$ where v^t is the transpose of v. Now $t \mapsto |A(t)|$ is a continuous function and hence is bounded above by a positive constant M^2 on the compact set K.

Thus $s = v^t(Av) \leq |A|v^tv \leq M^2v^tv = M^2|v|^2$. Thus $\bar{c} \leq$ (length of c in the standard metric) $\times\, M$ which implies that $\bar{d}(p, q) \leq Md(p, q)$. Since A is a positive definite, symmetric matrix at each point we also have that $v^tAv \geq L^2|v|^2$ and the rest of the proof follows as above. □

We can now prove the following:

Proposition 5.8. *Let s be a Riemannian metric on a connected manifold X. Let d be the corresponding pseudo-metric on X. Then d is a metric and the topology induced by d on X is the same as the original topology on X.*

Proof. Fix p in X. For this proof we will call an open set in the topology induced by d on X, d-open. Then, it is sufficient to prove that every open nbhd of p contains a d-open nbhd of p and, conversely, that every d-open nbhd of p contains an open nbhd of p.

Let U be an open nbhd of p. Choose U', also an open nbhd of p, so that

(a) The closure of U' is compact and is contained in U, and
(b) $S^2(T^*X)|\bar{U}'$ is locally trivial.

Hence there exists a chart $\phi : \bar{U}' \to \mathbf{R}^n$ and a bundle homomorphism $\tilde{\phi}$ so that the diagram

$$\begin{array}{ccc} S^2(T^*X)|\bar{U}' & \xrightarrow{\tilde{\phi}} & \phi(\bar{U}') \times S^2(\mathbf{R}^n) \\ \downarrow & & \downarrow \\ \bar{U}' & \xrightarrow{\phi} & \phi(\bar{U}') \end{array}$$

commutes, where $n = \dim X$. Let B be an open ball in $\phi(\bar{U}')$ of radius r centered at $\phi(p)$. (Note B is open in $\mathbf{R}^n$.) Let $c : \mathbf{R} \to X$ be a curve centered at p, i.e., $c(0) = p$. Suppose that $c([0, 1])$ does not lie entirely within $\phi^{-1}(B)$. We claim that this curve has length at least N for some constant N not depending on c. If this last statement is true, then $B(p, N) = \{q \in X \mid d(p, q) < N\}$ is contained in U'. Also this statement completes the proof that d is a metric. For if $p, q \in X$ and $p \neq q$, then take U' small enough so that $p \in U'$ and $q \notin U'$. Then the length of any curve connecting p to q is greater than N and $d(p, q) \neq 0$.

To prove the claim, we note that $c^{-1}(\phi^{-1}(B))$ is open in $\mathbf{R}$, so that there exists a smallest t in $(0, 1)$ for which $c(t) \notin \phi^{-1}(B)$. $\bar{c}$ = length of c is $\geq$ length of $c([0, t])$. Let $s' = \tilde{\phi} \cdot s \cdot \phi^{-1}$. Then s' is a Riemannian metric on $\phi(\bar{U}') \subset \mathbf{R}^n$ and the length of $c([0, t])$ under the metric s is the same as the length of $(\phi \cdot c)([0, t])$ under the metric s'. Using Lemma 5.7, we see that for some constant L the length of $(\phi \cdot c)([0, t])$ under s' is $\geq L \times$ length of $(\phi \cdot c)([0, t])$ using the standard metric on $\mathbf{R}^n$, since $\phi(\bar{U}')$ is compact.

Now we note that since X is Hausdorff $c(t)$ is in U'. For if $c(t) \notin U'$, then there exists an open subset V of X such that $c(t) \in V$ and $V \cap \phi^{-1}(B) = \varnothing$. Also $c^{-1}(V)$ is open and contains t. Hence $c^{-1}(V) \cap c^{-1}(\phi^{-1}(B)) \neq \varnothing$, a contradiction. Thus $\phi \cdot c[0, t]$ is a curve connecting p with some point outside of B. Hence the length of $\phi \cdot c[0, t]$ is $\geq r$, but length of $\phi \cdot c[0, t) =$ length of $\phi \cdot c[0, t]$. Thus length of c is $\geq L \cdot r = N$.

For the converse we suppose that U is some d-open nbhd of p. Make the same construction as above for U', ϕ, $\tilde{\phi}$, s', r, and B. Let $\phi(q) \in B$. Then the straight line, c, from $\phi(p)$ to $\phi(q)$ has length $< r$. By Lemma 5.7, the length of c in the metric s' is $\leq Mr$ where M is a constant depending only on $\phi(\bar{U}')$. Thus the length of $\phi^{-1} \cdot c$, a curve connecting p to q in X is $< M \cdot r$ and $q \in B(p, Mr)$ = the ball in X of radius Mr about p. By choosing r small enough $B(p, Mr) \subset U$ since U is d-open and $B(p, Mr)$ is a basic d-open set. But the above says that $\phi^{-1}(B) \subset B(p, Mr)$ and $\phi^{-1}(B)$ is an open nbhd of p. □

Lemma 5.9. *Any differentiable manifold is metrizable.*

Proof. Let $X_1, X_2, \ldots$ be the components of X and let d_i be a metric on X_i making X_i into a metric space. (Use Proposition 5.8.) Define $d: X \times X \to \mathbf{R}$ by

$$d(x, y) = \begin{cases} \min\{d_m(x, y), 1\} & \text{if } x, y \in X_m \\ 1 & \text{if } x \in X_i,\ y \in X_j, \text{ and } i \neq j. \end{cases}$$

Then d is a metric on X compatible with the original topology. □

We need the following two results to show that X can be made into a complete metric space.

Lemma 5.10. *Let (X, d) be a metric space and $f: X \to \mathbf{R}$ be a continuous proper function. Then $d': X \times X \to \mathbf{R}$ defined by $d'(p, q) = d(p, q) + |f(p) - f(q)|$ for every p, q in X is a complete metric on X which is compatible with the given topology on X.*

Proof. That d' is a metric is clear. Let T be the topology on X induced by d and T' the topology induced by d'. Since d' is continuous in the topology T, we have that $T' \subset T$. Conversely, let U be in T and let p be in U. Choose ε so that $B'(p, \varepsilon) = \{x \in X \mid d'(x, p) < \varepsilon\} \subset B(p, \varepsilon) \subset U$. Thus U is in T' and $T = T'$. Finally we show that d' is complete. Let $\{x_n\}_{n=1}^{\infty}$ be a Cauchy sequence in the d' metric. Thus there exists a constant $L > 0$ so that $d'(x_1, x_n) < L$ for all positive integers n. Hence $|f(x_1) - f(x_n)| < L$ for all n, and

$$\{x_n\}_{n=1}^{\infty} \subset f^{-1}([f(x_1) - L, f(x_1) + L]).$$

Since f is proper this later set is compact and the sequence $\{x_n\}_{n=1}^{\infty}$ has a limit point in X and thus converges. □

Proposition 5.11. *There always exists a smooth proper function on a smooth manifold X. In particular, any differentiable manifold can be made into a complete metric space.*

Proof. Let $K_1, K_2, \ldots$ be a sequence of compact subsets of X such that $K_i \subset \operatorname{Int}(K_{i+1})$ for $i = 1, 2, \ldots$ and $X = \bigcup_{i=1}^{\infty} K_i$. Let $L_i = K_i - \operatorname{Int}(K_{i-1})$ with $L_1 = K_1$. Then $\bigcup_{i=1}^{\infty} L_i = \bigcup_{i=1}^{\infty} K_i = X$. Define smooth functions $\rho_i: X \to \mathbf{R}$ such that

$$\rho_i = \begin{cases} 1 & \text{on } L_i \\ 0 & \text{on } K_{i-2} \cup (X - K_{i+1}) \\ 0 \leq \rho \leq 1 & \text{on } X. \end{cases}$$

Then let $f = \sum_{i=1}^{\infty} i\rho_i$. This sum is locally finite and hence is a smooth function. We claim that f is, in fact, proper. First note that if $p \in L_i$, then $i \leq f(p) \leq 3i$ since $f(p) = (i - 1)\rho_{i-1}(p) + i\rho_i(p) + (i + 1)\rho_{i+1}(p)$. Then to show that f is proper we need only show that $f^{-1}([A, B])$ is compact where $A, B \in R$. Given a p in L_i such that $f(p) \in [A, B]$, we have the inequalities $i \leq B$ and $3i \geq A$, or that $i \in [A/3, B]$. Thus $f^{-1}[A, B] \subset \bigcup_i L_i$ where $i \in [A/3, B]$ and is then a closed subset of a compact set and hence compact.

The last assertion of the proposition follows immediately from Lemma 5.10. □

We now define what we mean by subbundles of a vector bundle and give one way to construct them.

Definition 5.12. *Let E be a vector bundle over X with projection π. F is a* subbundle *of E if F is a smooth submanifold of E and $\pi|F: F \to X$ is a vector bundle where, for each x in X, F_x has the vector space structure induced from E_x.*

Definition 5.13. *Let $E \to X$ and $F \to Y$ be vector bundles. $\phi: E \to F$ is a* homomorphism *if*

(1) *there exists a smooth function $f: X \to Y$ called the* base mapping, *so that*

$$\begin{array}{ccc} E & \xrightarrow{\phi} & F \\ \downarrow & & \downarrow \\ X & \xrightarrow{f} & Y \end{array}$$

commutes.

(2) *ϕ is smooth.*

(3) *$\phi_p: E_p \to F_{f(p)}$ is linear where $\phi_p = \phi|E_p$.*

Example. Let $f: X \to Y$ be a smooth mapping. The Jacobian of f is a map $(df)_p: T_pX \to T_{f(p)}Y$ for each $p \in X$. So $(df): TX \to TY$ defined by $(df)|T_pX = (df)_p$ is a mapping which is linear on the fibers. Locally, TX and TY are trivial and via trivializations are just $U \times \mathbf{R}^n$ and $V \times \mathbf{R}^m$ where $n = \dim X$, $m = \dim Y$ and $U \subset \mathbf{R}^n$, $V \subset \mathbf{R}^m$ are open. $(df): U \times \mathbf{R}^n \to V \times \mathbf{R}^m$ is given by $(p, v) \mapsto (f(p), (df_p)v)$ which is a smooth mapping. So (df) is smooth and with this extended definition of a homomorphism between vector bundles, (df) is a homomorphism.

Proposition 5.14. *Let $E \to X$ and $F \to Y$ be vector bundles and $\phi: E \to F$ be a homomorphism with base mapping f. Suppose ϕ_p has constant rank for all p in X. Then* $\operatorname{Ker} \phi = \bigcup_{p \in X} \operatorname{Ker} \phi_p$ *is a subbundle of E.*

Proof. The problem of showing that $\operatorname{Ker} \phi$ is a smooth manifold is a local one. By using trivializations we may assume that $U \subset \mathbf{R}^n$ and $V \subset \mathbf{R}^m$ are open subsets with $f: U \to V$, $\phi: U \times \mathbf{R}^s \to V \times \mathbf{R}^t$ and with the appropriate diagram commuting where $s = \dim_X E$ and $t = \dim_Y F$. Fix p in U and choose W, a vector space complement to $\operatorname{Ker} \phi_p$ in $\mathbf{R}^s$. Note that $\phi_p: W \to f(p) \times \mathbf{R}^t$ is 1:1. Since ϕ is continuous and $\dim(\operatorname{Ker} \phi_q)$ is constant throughout U, there exists an open nbhd U' of p on which $\phi_q: q \times W \to f(q) \times \mathbf{R}^t$ is 1:1 for all q in U', i.e., W is a vector space complement to $\operatorname{Ker} \phi_q$, for all q in U'. Let $\overline{W}$ be a vector space complement to $\phi_p(W) \subset f(p) \times \mathbf{R}^t$. We can then restrict U' to U'', also an open nbhd of p on which $\overline{W}$ is a vector space complement to $\phi_q(W)$ in $\mathbf{R}^t$ for all q in U''. Let $\sigma: V \times \mathbf{R}^t \to \mathbf{R}^t$ be projection on the second factor and $\tau: \mathbf{R}^t \to Z$

$= \mathbf{R}^t/\overline{W}$ be the natural projection. Then $g = \tau \cdot \sigma \cdot \phi : U' \times \mathbf{R}^s \to Z$ is a smooth mapping. Note that $\tau \cdot \sigma \cdot \phi(q, v) = \tau \cdot \sigma(f(q), \phi_q(v)) = \tau(\phi_q(v))$ and that $\tau(\phi_q(v)) = 0$ iff $\phi_q(v) = 0$ iff $(q, v) \in \operatorname{Ker} \phi_q$. So $g^{-1}(0) = \operatorname{Ker}(\phi | U'')$.

If we show that g is a submersion, then by Theorem 2.8 $\operatorname{Ker} \phi \cap (U'' \times \mathbf{R}^s)$ is a submanifold of $U'' \times \mathbf{R}^s$ and thus a submanifold of E. To show that g is a submersion, it is sufficient to show that if (q, v) in $U'' \times \mathbf{R}^s$, $r = g(q, v)$, and $c : \mathbf{R} \to Z$ is a (smooth) curve based at r, then there is a (smooth) curve $\bar{c} : \mathbf{R} \to U'' \times \mathbf{R}^s$ based at (q, v) with $g \cdot \bar{c} = c$. Let c be such a curve. Note that $\tau : \phi_q(W) \to Z$, $\sigma : f(q) \times \phi_q(W) \to \phi_q(W)$, and $\phi_q : W \to \phi_q(W)$ are isomorphisms so that $\phi_q^{-1} \cdot \sigma^{-1} \cdot \tau^{-1} \cdot c : \mathbf{R} \to q \times W$ is a smooth curve. Define $\bar{c}$ by

$$\bar{c}(t) = (\tau \cdot \sigma \cdot \phi_q)^{-1} \cdot c(t) + v - (\tau \cdot \sigma \cdot \phi_q)^{-1} \cdot c(0)$$

$\bar{c} : R \to q \times W \subset U'' \times W$ is a smooth curve based at (q, v) and $g \cdot \bar{c} = c$. □

Proposition 5.15. *Let F be a subbundle of E. Then there exists another subbundle G of E with $F \oplus G = E$. G is called a* complementary subbundle to *F.*

Proof. Choose a metric $s : X \to S^2(E^*)$ as given by Proposition 5.6. Let $\pi_s : E \to E$ be given by orthogonal projection onto F using s, i.e., on each fiber $(\pi_s)_p : E_p \to F_p$ is orthogonal projection. π_s is a homomorphism and $G = \operatorname{Ker} \pi_s$ is a subbundle of E by Proposition 5.14. At each p in X, $F_p \oplus G_p = E_p$ so $E = F \oplus G$. □

§6. Integration of Vector Fields

There is a close relationship between vector fields and smoothly parametrized families of curves which we shall explore now.

Definition 6.1. *A* one parameter group *on X is a smooth mapping $\phi : X \times \mathbf{R} \to X$ satisfying $\phi_0 = id_X$ and $\phi_{s+t} = \phi_s \cdot \phi_t$ for all s, t in $\mathbf{R}$ where $\phi_t(x) = \phi(x, t)$.*

Notes. (1) Let ϕ be a one-parameter group on X. Then ϕ_t is a diffeomorphism on X for each t. In fact, $\phi_{-t} = (\phi_t)^{-1}$.

(2) Let ζ_p be the tangent vector at $t = 0$ to the curve $p \mapsto \phi_t(p)$. Then the mapping $p \mapsto \zeta_p$ defines a vector field on X called the *infinitesimal generator of* ϕ. (The joint smoothness of ϕ in p and t guarantees that ζ is a smooth section.) We call a curve $c : I_\varepsilon \to X$ an *integral curve for* ζ if $(dc)_r((d/dt)|_r) = \zeta_{c(r)}$ for all r. The following lemma shows that the infinitesimal generator of ϕ is the vector field for which the curves $t \mapsto \phi_t(p)$ are integral curves. (Note $I_\varepsilon = (-\varepsilon, \varepsilon) \subset \mathbf{R}$.)

Lemma 6.2. *Let ζ be a vector field on a manifold X with p in X. Then there is a nbhd U of p in X, an $\varepsilon > 0$, and a unique smooth function $\phi : U \times I_\varepsilon \to X$ satisfying;*

(a) *The curves $t \mapsto \phi_t(q)$ are integral curves of ζ for all q in U;*

(b) $\phi_s \cdot \phi_t = \phi_{s+t}$ *on the domain* $\phi_t(U) \cap U$ *whenever* $|s|$, $|t|$, *and* $|s + t|$ *are* $< \varepsilon$; *and*

(c) $\phi_0 = id_U$.

Proof. This is, in reality, a theorem about first order systems of ordinary differential equations. First we transport the problem to $\mathbf{R}^n$. Let V be a chart nbhd of p with chart $h: V \to \mathbf{R}^n$. Choose another nbhd U of p with $\overline{U}$ compact and contained in V. Let $\eta = h_*(\zeta|V)$; i.e., $\eta_q = (dh)_{h^{-1}(q)}(\zeta_{h^{-1}(q)})$. η is a vector field defined on an open set in $\mathbf{R}^n$ and can be written in the form $\eta = \sum_{i=1}^n \eta_i(\partial/\partial x_i)$ where η_i are smooth functions on $h(V)$. Let $U' = h(\overline{U})$. Then for every x in U', we can consider the differential equations

$$(*) \qquad \frac{dy_i}{dt} = \eta_i(y) \qquad \text{with initial conditions}$$

$$y(0) = (y_1(0), \ldots, y_n(0)) = x.$$

By standard theorems on the existence and uniqueness of solutions to a system of o.d.e. [see, for example, Hurewicz, *Lectures on Differential Equations*, p. 28], there exists a smooth function $\psi: U' \times (-\varepsilon, \varepsilon) \to \mathbf{R}^n$ given by $\psi(x, t) = y(t)$ where y is the solution to (*) at x. If ε is chosen sufficiently small, we can assume that $\operatorname{Im} y \subset h(V)$. Note that $\psi(x, 0) = y(0) = x$ so that $\psi_0 = id_{U'}$. Next we claim that $\psi_t \cdot \psi_s = \psi_{t+s}$ when both sides are defined. For ψ_{t+s} and $\psi_t \cdot \psi_s$ are both solutions to $dy/dt = \eta(y)$ with initial values $\psi_s(x)$ when $t = 0$. By the uniqueness theorem for the initial value problem these must be identical.

Finally let $\phi_t = h^{-1} \cdot \psi_t \cdot h$. Then $\phi: U \times (-\varepsilon, \varepsilon) \to X$ is well-defined and satisfies (b) and (c). The uniqueness of ϕ follows from the local uniqueness of ψ once (a) has been satisfied. To prove that ζ is the infinitesimal generator of ϕ we apply the following lemma. □

Lemma 6.3. *Let* $h: X \to Y$ *be a diffeomorphism and* ϕ_t *a one parameter group on* X *with infinitesimal generator* ζ. *Then* $h_*\zeta$ *is the infinitesimal generator of the one parameter group* $\psi_t = h \cdot \phi_t \cdot h^{-1}$.

Proof. For each q in Y, $(h_*\zeta)_q = (dh)_{h^{-1}(q)}(\zeta_{h^{-1}(q)})$. Now $t \mapsto \phi_t(h^{-1}(q))$ is a curve representing $\zeta_{h^{-1}(q)}$ so that $t \mapsto h \cdot \phi_t(h^{-1}(q))$ is a curve representing $(dh)_{h^{-1}(q)}(\zeta_{h^{-1}(q)})$. □

Theorem 6.4. *Let* ζ *be a compactly supported vector field on a manifold* X; *i.e.,* ζ *is zero outside of some compact subset of* X. *Then there exists a unique one parameter group* ϕ *for which* ζ *is the infinitesimal generator.*

Proof. Let U be an open subset of X with $\overline{U}$ compact such that $\zeta \equiv 0$ off U. Applying Lemma 6.2 we can find for each point p in $\overline{U}$ an open nbhd U_p of p, a real number $\varepsilon_p > 0$, and a unique smooth function $\phi_p: U_p \times (-\varepsilon_p, \varepsilon_p) \to X$ satisfying (a), (b), and (c) of the last lemma. Choose a finite subcover of $\overline{U}$ by $U_{p_1}, \ldots, U_{p_k}$ and let $\varepsilon = \min_{1 \le i \le k} \varepsilon_{p_i}$. Define for $|t| < \varepsilon$, $\phi_t : X \to X$ by

$$\phi_t(p) = \begin{cases} \phi_{p_i}(p, t) & \text{for all } p \text{ in } U_{p_i} \\ p & \text{for all } p \text{ in } X - \bigcup_{i=1}^{k} U_{p_i} \end{cases}$$

ϕ_t is well-defined and unique by the uniqueness part of Lemma 6.2. Note that $\phi_0 = id_X$.

Finally, define $\phi : X \times \mathbf{R} \to X$ by $\phi_t = (\phi_{(t/n)})^n$ where n is an integer large enough so that $|t/n| < \varepsilon$. This is well-defined for if m is another such integer, then

$$(\phi_{t/m})^m = ((\phi_{t/mn})^n)^m = (\phi_{t/mn})^{n \cdot m} = ((\phi_{t/mn})^m)^n = (\phi_{t/n})^n$$

It is now easy to check that ϕ is a one parameter group whose infinitesimal generator is ζ. ◻

Notes. (1) On a compact manifold there is a 1:1 correspondence between vector fields and one parameter groups.

(2) For p in X, $\zeta_p = 0$ iff $\phi_t(p) = p$ for all t. Clearly if $\phi_t(p) = p$ for all t, then $\zeta_p = (d\phi_t/dt)(p)|_{t=0} = 0$. Conversely, assume that $\zeta_p = 0$. If we can show that $\phi_t(p) = p$ for all t in some nbhd of 0, then by the arguments in Theorem 6.4 we see that $\phi_t(p) = p$ for all t. Thus this question is a local one and we may assume that X is an open nbhd of 0 in R^n and that $p = 0$. Now as in Lemma 6.2 $\phi_t(0) = y(t)$ where $y(t) = (y_1(t), \ldots, y_n(t))$ is the solution to the system of ordinary differential equations $dy_i/dt = \eta_i(y)$ with initial condition $y_i(0) = 0$ where $\zeta = \sum_{i=1}^{n} \eta_i(\partial/\partial x_i)$. Since $\zeta_0 = 0$, $\eta_i(0) = 0$. Thus $y_i(t) \equiv 0$ is a solution to this system of equations. The uniqueness of such a solution guarantees that $\phi_t(0) = 0$ for small t.

Corollary 6.5. *Let X be a manifold and let ζ and η be two vector fields on X. Suppose that ζ is compactly supported and that η is the infinitesimal generator of a one parameter group. Then $\zeta + \eta$ is the infinitesimal generator of a one parameter group.*

Note. By taking $\eta = 0$ we see that this Corollary is a slight generalization of the last Theorem.

Proof. The proof is essentially the same as that of the last theorem. The only difference is in the definition of ϕ. Let ψ be the one parameter group associated with η and define

$$\phi_t(p) = \begin{cases} \phi_{p_i}(p, t) & \text{for all } p \text{ in } U_{p_i} \\ \psi(p, t) & \text{for } p \text{ in } X - \bigcup_{i=1}^{k} U_{p_i} \end{cases}$$

The rest of the proof proceeds as before. ◻

Chapter II

Transversality

§1. Sard's Theorem

In order to state and prove Sard's Theorem we need to know some elementary (Lebesgue) measure theory.

Let $a = (a_1, \ldots, a_n)$ and $b = (b_1, \ldots, b_n)$ be points of $\mathbf{R}^n$ with $a_i < b_i$ $(1 \le i \le n)$. Denote by $C(a, b)$ the open cube

$$\{(t_1, \ldots, t_n) \in \mathbf{R}^n \mid a_i < t_i < b_i, 1 \le i \le n\}.$$

Define the volume of $C(a, b)$ to be

$$\mathrm{vol}\,[C(a, b)] = (b_1 - a_1)\cdot \cdots \cdot(b_n - a_n)$$

Definition 1.1.

(1) *Let S be a subset of $\mathbf{R}^n$. Then S has* measure zero *if for every $\varepsilon > 0$, there is a covering of S by a countable number of open cubes $C_1, C_2, \ldots$ so that $\sum_{i=1}^{\infty} \mathrm{vol}\,[C_i] < \varepsilon$.*

(2) *Let X be a differentiable n-manifold and let S be a subset of X. Then S is of* measure zero *if there exists a countable open covering $U_1, U_2, \ldots$ of S and charts $\phi_i\colon U_i \to \mathbf{R}^n$ so that $\phi_i(U_i \cap S)$ is of measure zero in $\mathbf{R}^n$.*

To see that "measure zero" is well-defined on a manifold, we need the following two results:

Lemma 1.2. *A countable union of sets of measure zero in $\mathbf{R}^n$ is of measure zero.*

Proof. Let $S_1, S_2, \ldots$ be sets of measure zero in R^n. Given $\varepsilon < 0$, cover each S_i by open cubes whose total volume is less than $(\varepsilon/2^{i+1})$. Then the union of all of these cubes covers $S = \bigcup_{i=1}^{\infty} S_i$ and has total volume less than ε. □

Recall that if $A : \mathbf{R}^n \to \mathbf{R}^m$ is a linear map, then

$$|A| = \sup_{v \in \mathbf{R}^n - \{0\}} \frac{|Av|}{|v|}.$$

Also, if $l_{x,y}$ denotes the line between two points x and y in $\mathbf{R}^n$, then for any C^1-differentiable function $f\colon \mathbf{R}^n \to \mathbf{R}^m$

$$|f(x) - f(y)| \le |x - y| \sup_{p \in l_{x,y}} |(df)_p|.$$

(This is just a corollary to the Mean Value Theorem.)

Proposition 1.3. *Let $f\colon \mathbf{R}^n \to \mathbf{R}^n$ be C^1-differentiable and let S be a measure zero subset of $\mathbf{R}^n$. Then $f(S)$ has measure zero.*

Proof. Without loss of generality, S can be assumed to be contained in some large open cube. On this cube $|(df)_p|$ is bounded by some constant K, so that if $x, y \in S$, then $|f(x) - f(y)| < K|x - y|$. Given $\varepsilon > 0$, cover S by open cubes C_i whose total volume is less than $\varepsilon/(\sqrt{n}\,K)^n$. We note that $f(C_i)$ is contained in a cube whose volume is $(\sqrt{n}\,K)^n \operatorname{vol}(C_i)$ using the above inequality. (To see this assume C_i has equal length sides with length a. Let p be the center of C_i. Then $f(C_i)$ is contained in the sphere of radius $(K\sqrt{n}/2)a$ centered at $f(p)$ which is, in turn, contained in a cube centered at $f(p)$ all of whose sides have length $K\sqrt{n}\cdot a$.) Thus the total volume of cubes containing $f(S)$ is less than ε. □

This generalizes immediately to a statement of manifolds.

Corollary 1.4. *Let X and Y be differentiable n-manifolds, let $f\colon X \to Y$ be a C^1-differentiable, and let Z be a measure zero subset of X. Then $f(Z)$ has measure zero in Y.*

Proof. Let ψ be a chart on Y with domain V. Cover $f^{-1}(V)$ by a countable open covering $U_1, U_2, \ldots$ each of which is the domain for a chart $\phi_i : U_i \to \mathbf{R}^n$ and for which $f(U_i)$ is contained in V. Since Z is of measure zero in X, $\phi_i(Z \cap U_i)$ has measure zero in $\mathbf{R}^n$. Now $\psi \cdot f \cdot \phi_i^{-1}$ is C^1 on its domain in $\mathbf{R}^n$. By Proposition 1.3 $\psi \cdot f \cdot \phi_i^{-1} \cdot \phi_i(Z \cap U_i) = \psi(f(Z) \cap U_i)$ has measure zero in $\mathbf{R}^n$. Hence $\bigcup_{i=1}^{\infty} \psi(f(Z \cap U_i)) = \psi(f(Z \cap V))$ is of measure zero in $\mathbf{R}^n$. So $f(Z)$ has measure zero in Y. □

Lemma 1.5. *Let X be an n-dimensional submanifold of a differentiable m-manifold Y with $n < m$. Then X is of measure zero in Y.*

Proof. We first claim that an n-dimensional plane, $\mathbf{R}^n$, in $\mathbf{R}^m$ is of measure zero. $\mathbf{R}^n$ can be subdivided into a countable number of unit n-cubes so it is sufficient to show that the unit n-cube in $\mathbf{R}^m$ is of measure zero. Let $\varepsilon > 0$ be given. The unit n-cube can be covered by $(2/\varepsilon)^n$ cubes each of volume ε^m. Then the total volume of the cubes is $\varepsilon^m(2/\varepsilon)^n = 2^n\varepsilon^{m-n}$ which converges to zero as $\varepsilon \mapsto 0$ since $m > n$. Since X is a submanifold of Y, there exists a countable covering $U_1, U_2, \ldots$ of Y with charts $\psi_i : U_i \to \mathbf{R}^m$ such that $\psi_i(U_i \cap X)$ is contained in a fixed n-plane in $\mathbf{R}^m$. Hence $\psi_i(U_i \cap X)$ has measure zero in $\mathbf{R}^m$ and X has measure zero in Y. □

Proposition 1.6. *Let X and Y be differentiable manifolds of dimensions n and m respectively with $n < m$. Let $f\colon X \to Y$ be C^1-differentiable, then $f(X)$ has measure zero in Y.*

Proof. Let $s = m - n$. Define $\hat{f}\colon X \times \mathbf{R}^s \to Y$ by $\hat{f}(p, a) = f(p)$ for every p in X and a in $\mathbf{R}^s$. $X \times \{0\}$ is a submanifold of $X \times \mathbf{R}^s$ and, by Lemma 1.5, has measure zero in $X \times \mathbf{R}^s$. By Corollary 1.4 $\hat{f}(X \times \{0\}) = f(X)$ has measure zero in Y. □

We need one more result before coming to Sard's Theorem, namely Fubini's Theorem for measure zero sets.

Let $i_a: \mathbf{R}^{n-1} \to \mathbf{R} \times \mathbf{R}^{n-1} = \mathbf{R}^n$ be the embedding given by $i_a(x) = (a, x)$ where a is in $\mathbf{R}$.

Theorem 1.7. *Let A be a compact subset of $\mathbf{R}^n$. Suppose that for every $a \in \mathbf{R}$, $i_a^{-1}(A)$ has measure zero in $\mathbf{R}^{n-1}$. Then A is of measure zero in $\mathbf{R}^n$.*

Let I be a closed interval in $\mathbf{R}$. Suppose I is covered by subintervals $[a_1, b_1], \ldots, [a_m, b_m]$. Then the cover is minimal *if the covering minus any one element of the covering is no longer a covering.*

Lemma 1.8. Let $I = [a, b]$ be a closed interval in $\mathbf{R}$. Then the sum of the lengths of any minimal covering of I (by closed intervals in I) is less than $2(b\text{-}a)$.

Proof. Order the intervals of a minimal covering $[a_1, b_1], \ldots, [a_m, b_m]$ so that $a_1 \le a_2 \le \cdots \le a_m$. Then the minimality implies that $b_1 \le b_2 \le \cdots \le b_m$. Moreover, $[a_k, b_k] \cap [a_{k+2}, b_{k+2}] = \varnothing$ for $1 \le k \le m - 2$. Otherwise $a_{k+2} \le b_k$ and $[a_{k+1}, b_{k+1}] \subset [a_k, b_k] \cup [a_{k+2}, b_{k+2}]$ since $a_k \le a_{k+1}$ and $b_{k+1} \le b_{k+2}$. Hence the sum of the lengths of $[a_1, b_1], [a_3, b_3], [a_5, b_5], \ldots$ is less than $b - a$. Similarly for $[a_2, b_2], [a_4, b_4], \ldots$. □

Lemma 1.9. *Suppose the set $i_a^{-1}(A)$ is covered by open sets $\{U_1, \ldots, U_k\}$ of $\mathbf{R}^{n-1}$. Then there exists an open interval I_a about a such that $\{U_1, \ldots, U_k\}$ covers $i_t^{-1}(A)$ for every t in I_a.*

Proof. If there were no such interval, then there would exist a sequence $\{t_i\}_{i=1}^{\infty}$ of real numbers with $\text{Lim}_{i\to\infty} t_i = a$ and a point $x_i \in i_{t_i}^{-1}(A)$ such that x_i is in the complement of $U_1 \cup \cdots \cup U_k$. Since (t_i, x_i) is in A and A is compact, there exists a subsequence of the x_i's which converges to some point $\bar{x}$ in $\mathbf{R}^{n-1}$ and for which $(a, \bar{x})$ is in A. Since $\bigcup_{i=1}^{k} U_i$ is open, $\bar{x} \notin \bigcup_{i=1}^{k} U_i$. But $(a, \bar{x}) \in A$ implies that $\bar{x} \in i_a^{-1}(A)$ and the fact that $\{U_1, \ldots, U_k\}$ covers $i_a^{-1}(A)$ gives a contradiction. □

Proof of Theorem 1.7. Since A is compact and hence bounded, there is a closed interval I such that $A \subset I \times \mathbf{R}^{n-1}$. By hypothesis $i_a^{-1}(A)$ has measure zero for each a in I. Thus, given $\varepsilon > 0$, there is a cover of $i_a^{-1}(A)$ by open cubes in $\mathbf{R}^{n-1}$, $\{C_1^a, \ldots, C_{N_a}^a\}$ such that $\sum_{i=1}^{N_a} \text{vol}\,(C_i^a) < \varepsilon$. By Lemma 1.9, there exists an open interval I_a in I about a so that $C_1^a, \ldots, C_{N_a}^a$ covers $i_t^{-1}(A)$ for every t in I_a. Hence the collection of open sets $\{I_a \times C_i^a\}$ covers A. Thus there is a finite subcover $\{I_a \times C_i^a\}_{a\in B}^{1 \le i \le N_a}$ where B is some finite set.

Let $J_a = \bar{I}_a$. The finite collection $\{J_a\}_{a\in B}$ covers I and can be assumed to form a minimal covering of I. Then $\sum_{i=1}^{N_a} \text{vol}\,[J_a \times C_i^a] \le \varepsilon\,\text{vol}\,[J_a]$. Hence

$$\sum_{a\in B} \sum_{i=1}^{N_a} \text{vol}\,[J_a \times C_i^a] \le \varepsilon \sum_{a\in B} \text{length}\,(J_a) < 2\varepsilon\,\text{length}\,(I).$$

Since $\text{vol}\,[I_a \times C_i^a] = \text{vol}\,[J_a \times C_i^a]$, the total volume of the covering of A by $\{I_a \times C_i^a\}_{a\in B}^{1 \le i \le N_a}$ can be made arbitrarily small, A has measure zero in R^n. □

Since all of the results given in this section have been about measure zero sets, it is instructive, perhaps, to show at this time the following obvious but surprisingly complicated result.

Proposition 1.10. *Let S be a nonempty open subset of R^n. Then S is* not *of measure zero.*

Proof. Every open set S contains a nonempty open cube C whose closure is contained in S. Let $\{\tilde{C}_i\}_{i=1}^{\infty}$ be an open covering of S by open cubes. Since $\bar{C}$ is compact in $\mathbf{R}^n$, there is a finite subcover of $\bar{C}$ by $C_1, \ldots, C_m$. We claim that $\text{vol}\,[C] \le \sum_{\alpha=1}^{m} \text{vol}\,[C_\alpha]$. If this is true then we are done since $\sum_{i=1}^{\infty} \text{vol}\,[\tilde{C}_i] \ge \sum_{i=1}^{m} \text{vol}\,[C_i] \ge \text{vol}\,[C] > 0$. So the sums of the volumes of cubes in a covering of S are bounded away from zero and S does not have measure zero. To prove the claim, let N_α = number of integer lattice points of $\mathbf{R}^n$ (i.e., points of R^n all of whose coordinates are integers) which are contained in C_α. Now $C_\alpha = C(a^\alpha, b^\alpha)$ where $a^\alpha, b^\alpha \in R^n$. Let $a^\alpha = (a_1{}^\alpha, \ldots, a_n{}^\alpha)$ and $b^\alpha = (b_1{}^\alpha, \ldots, b_n{}^\alpha)$. Then for each j there are at most $b_j{}^\alpha - a_j{}^\alpha + 1$ and at least $l_j{}^\alpha = \max\{b_j{}^\alpha - a_j{}^\alpha - 1, 0\}$ integers in $[a_j{}^\alpha, b_j{}^\alpha]$. Hence

$$\prod_{j=1}^{n} l_j{}^\alpha \le N_\alpha \le \prod_{j=1}^{n} (b_j{}^\alpha - a_j{}^\alpha + 1).$$

Similarly let N = number of integer lattice points in $C = C(a, b)$ and obtain similar bounds on N. Certainly $N \le \sum_{i=1}^{m} N_i$ since $\{C_\alpha\}_{\alpha=1}^{m}$ covers C. Hence

$$\prod_{j=1}^{n} l_j \le \sum_{\alpha=1}^{m} \prod_{j=1}^{n} (b_j{}^\alpha - a_j{}^\alpha + 1).$$

For λ in $\mathbf{R}$ sufficiently large, let $C^\lambda = C(\lambda a, \lambda b)$ and $C_\alpha{}^\lambda(\lambda a^\alpha, \lambda b^\alpha)$. Apply the above argument to C^λ and $C_\alpha{}^\lambda$ to obtain

$$\prod_{j=1}^{n} (\lambda b_j - \lambda a_j - 1) \le \sum_{\alpha=1}^{m} \prod_{j=1}^{n} (\lambda b_j{}^\alpha - \lambda a_j{}^\alpha + 1)$$

Hence

$$\prod_{j=1}^{n} \left(b_j - a_j - \frac{1}{\lambda}\right) \le \sum_{\alpha=1}^{m} \prod_{j=1}^{n} \left(b_j{}^\alpha - a_j{}^\alpha + \frac{1}{\lambda}\right).$$

Taking the limits of both sides as $\lambda \to \infty$ we get

$$\text{vol}\,[C] = \prod_{j=1}^{n} (b_j - a_j) \le \sum_{\alpha=1}^{m} \prod_{j=1}^{n} (b_j{}^\alpha - a_j{}^\alpha) = \sum_{\alpha=1}^{m} \text{vol}\,[C_\alpha]. \qquad \square$$

Definition 1.11. *Let X and Y be differentiable manifolds and $f: X \to Y$ a C^1-mapping. Then*

(1) corank $(df)_p = \min(\dim X, \dim Y) - \text{rank}\,(df)_p$.
(2) *a point $p \in X$ is a* critical point of f *if corank $(df)_p > 0$.*
(3) *The set of critical points of f is denoted by $C[f]$.*

(3) *a point* $q \in Y$ *is a* critical value of f *if* $q \in f(C[f])$.

(4) *a point* $p \in X$ *is a* regular point of f *if* $p \notin C[f]$.

(5) *a point* $q \in Y$ *is a* regular value of f *if it is not a critical value of* f. *So, in particular, a point not in* Image f *is a regular value.*

Theorem 1.12. (*Sard's Theorem.*) *Let* X *and* Y *be smooth manifolds. Let* $f: X \to Y$ *be a smooth mapping. Then the set of critical values of* f *has measure zero in* Y.

Notes. (1) Sard's Theorem can be generalized as follows: Assume that $k > \max(\dim X - \dim Y, 0)$. If f is a C^k-differentiable mapping, then the measure of the set of critical values is zero. Since we will be using only smooth mappings in later chapters we will prove only the more restricted version here.

(2) If $\dim X < \dim Y$ then Sard's Theorem follows directly from Proposition 1.6 and the fact that a subset of a set of measure zero has measure zero (see [45]).

Sard's Theorem is in reality a local theorem and follows from:

Proposition 1.13. *Let* $f: U \to \mathbf{R}^m$ *be smooth where* U *is an open set in* $\mathbf{R}^n$. *Then the set of critical values of* f *is of measure zero in* $\mathbf{R}^m$.

The proof of Theorem 1.12 proceeds from Proposition 1.13 precisely as the proof of Corollary 1.4 proceeded from Proposition 1.3. The details are left for the reader.

The proof of Proposition 1.13 will be done by induction on n. Start the induction at $n = 0$. $\mathbf{R}^0$ is, by convention, just a point and the proposition is trivial in this case.

By induction, we assume that Sard's Theorem holds for all smooth mappings of $\mathbf{R}^{n-1} \to \mathbf{R}^m$, where m is arbitrary.

Lemma A. *Let* $f: U \to \mathbf{R}^m$ *be smooth, where* U *is an open subset of* $\mathbf{R}^n$. *Let* $f_1, \ldots, f_m: U \to \mathbf{R}$ *be the coordinate functions given by* f. *Assume that* $f_1(x_1, \ldots, x_n) = x_1$ *for all* $(x_1, \ldots, x_n) \in U$. *Let* $C =$ *critical point set of* f. *Then* $f(C)$ *has measure zero in* $\mathbf{R}^m$.

Proof. The proposition is trivial for $n = 1$, so we may assume $n > 1$. Given $a \in \mathbf{R}$, recall that $i_a: \mathbf{R}^{n-1} \to \mathbf{R}^n$ by $i_a(\bar{x}) = (a, \bar{x})$ where $\bar{x} = (x_2, \ldots, x_n)$. Define $g_a(\bar{x}) = (f_2(a, \bar{x}), \ldots, f_m(a, \bar{x}))$. Then the following diagram commutes.

$$\begin{array}{ccc} U_a & \xrightarrow{g_a} & \mathbf{R}^{m-1} \\ {\scriptstyle i_a}\downarrow & & \downarrow{\scriptstyle i_a} \\ U & \xrightarrow{f} & \mathbf{R}^m \end{array}$$

where $U_a = i_a^{-1}(U)$.

Note that

$$(df)_{(a,\bar{x})} = \left(\begin{array}{c|c} 1 & 0 \\ \hline * & (dg_a)_{\bar{x}} \end{array}\right).$$

Hence rank $(df)_{(a,\bar{x})} = \text{rank}\,(dg_a)_{\bar{x}} + 1$; i.e., $\bar{x}$ is a critical point of g_a iff $(a, \bar{x})$ is a critical point of f. So the critical point set of g_a is $i_a{}^{-1}(C)$.

By the induction hypothesis $g_a(i_a{}^{-1}(C))$ is of measure zero in $\mathbf{R}^{m-1}$. Since $i_a{}^{-1}(f(C)) = g_a(i_a{}^{-1}(C))$ we may conclude by Theorem 1.7 that $f(C)$ has measure zero. (Note that C is a closed set, which is a countable union of compact sets. Thus $f(C)$ is a countable union of compact sets so that 1.7 applies.) $\square$

Let $f: U \to \mathbf{R}^m$ be smooth. Let $C = C[f]$ be the critical point set of f. Denote by

$$C_i = \left\{ p \in C \;\middle|\; \frac{\partial^{|\alpha|}}{\partial x^\alpha} f_l(p) = 0 \quad \text{whenever} \quad 0 < |\alpha| \le i \quad \text{and} \quad 1 \le l \le m \right\}.$$

$$(i = 1, 2, \ldots)$$

The outline of the rest of the proof of Sard's Theorem is:

Lemma B. *$f(C - C_1)$ has measure zero.*

Lemma C. *$f(C_i - C_{i+1})$ has measure zero for $i \ge 1$.*

Lemma D. *For some i, $f(C_i)$ has measure zero.*

Proof of Lemma B. Let p be in $C - C_1$. Then there exists some partial derivative of f at p which is not zero. Assume that $(\partial f_1/\partial x_1)(p) \ne 0$. Let $h: U \to \mathbf{R}^n$ be defined by $h(x_1, \ldots, x_n) = (f_1(x_1, \ldots, x_n), x_2, \ldots, x_n)$. Then at p

$$(dh)_p = \left(\begin{array}{c|c} \dfrac{\partial f_1}{\partial x_1} & * \\ \hline 0 & I_{n-1} \end{array}\right)$$

which is invertible. By the Inverse Function Theorem, there exists open sets $U' \subset U$ and $V \subset \mathbf{R}^n$ with p in U' so that $h: U' \to V$ is a diffeomorphism. Let $g: V \to \mathbf{R}^m$ be given by $g = f \cdot h^{-1}$, then $f(C[f] \cap U') = g(C[g])$. Now $g_1(y_1, \ldots, y_n) = f_1 \cdot h^{-1}(y_1, \ldots, y_n) = y_1$. So we can apply Lemma A to g, and get that $g(C[g])$ has measure zero in $\mathbf{R}^m$. $\square$

Proof of Lemma C. On $C_i - C_{i+1}$ all ith partial derivatives vanish but not some $(i + 1)$st partial derivative. We may assume that g is an appropriate ith partial derivative so that $(\partial g/\partial x_1)(p) \ne 0$. Let $h: U \to \mathbf{R}^n$ be defined by $h(x) = (g(x), x_2, \ldots, x_n)$. Then $(dh)_p$ is non-singular, so that h restricted to

$U'_p \subset U$ is a diffeomorphism, where U'_p is an open nbhd of p. Let $V = h(U'_p)$. By definition $g(C_i) = 0$, so $h(C_i \cap U'_p) \subset \{0\} \times \mathbf{R}^{n-1}$ in $\mathbf{R}^n$. Let $k: \mathbf{R}^{n-1} \to \mathbf{R}^m$ be defined by $f \cdot h^{-1}$ restricted to $V \cap (\{0\} \times \mathbf{R}^{n-1})$.

Finally we note that $f(C_i \cap U'_p) \subset f(C[f] \cap U'_p) = k(C[k])$ and that by the induction hypothesis $k(C[k])$ has measure zero. Hence for each p in $C_i - C_{i+1}$, there is a nbhd U'_p of p for which $f(C_i \cap U'_p)$ has measure zero. We can choose a countable number of the U'_p's to cover $C_i - C_{i+1}$. So $f(C_i - C_{i+1})$ has measure zero in $\mathbf{R}^m$.

Proof of Lemma D. Without loss of generality, we may assume that U is an open cube with sides of length b, since U may be covered by a countable union of such sets, and that f is defined on a nbhd of $\bar{U}$. By Taylor's Theorem, if $x \in C_k$, and $y \in U$, then (*) $|f(y) - f(x)| \leq K|x - y|^{k+1}$ where K is some constant independent of y. Let r be a large integer. Subdivide U into subcubes with sides of length b/r denoted by $B_1, \ldots, B_N$ where $N = r^n$. Now $f(C_k \cap B_s)$ is contained in a ball D of radius $K(b/r)^{k+1}$ using (*), so the circumscribed cube has volume $(2K(b/r)^{k+1})^m$. Thus $f(C_k)$ is contained in the union of cubes whose volume is

$$N(2K)^m\left(\frac{b}{r}\right)^{m(k+1)} = \frac{(2K)^m b^{m(k+1)}}{r^{mk+m-n}}.$$

When $k > (n/m) - 1$, $mk + m - n > 0$. Therefore, as $r \to \infty$, the volume of the cubes containing $f(C_k) \to 0$. So $f(C_k)$ has measure zero in $\mathbf{R}^m$. □

Corollary 1.14. (*Brown*). *The set of regular values of a smooth mapping $f: X \to Y$ is dense in Y. (Recall from* 1.11 *that a point in Y which is not in* Im f *is a regular value of f.)*

Proof. Points of Y are either critical values or regular values for f. If the set of regular values is not dense, then there is a nonempty open set in Y consisting entirely of critical values. We have shown in Proposition 1.10 an open set of $\mathbf{R}^n$ does not have measure zero; this clearly extends to nonempty open subsets of Y, by using charts. Thus the set of critical values of f does not have measure zero, a contradiction to Sard's Theorem. Hence the regular values of f are dense in Y. □

Exercises

(1) Let $f: X \to \mathbf{R}^m$ be a 1:1 immersion and let $n = \dim X$. Let $v \neq 0$ be in $\mathbf{R}^m$ and let $\pi_v : \mathbf{R}^m \to \mathbf{R}^{m-1}$ be the orthogonal projection whose kernel is the subspace (v). Show that if $m > 2n + 1$, then there exists a vector v so that $\pi_v \cdot f: X \to \mathbf{R}^{m-1}$ is a 1:1 immersion. Hint: Consider the composite mapping g defined by

$$TX - \{0\text{-section}\} \xrightarrow{(df)} T\mathbf{R}^m = \mathbf{R}^m \times \mathbf{R}^m \xrightarrow{\pi_2} \mathbf{R}^m - \{0\} \xrightarrow{\psi} \mathbf{P}^{m-1}$$

where π_2 is projection on 2nd factor and ψ is the standard projection of $\mathbf{R}^m$ onto projective $(m - 1)$-space. Show that g is well-defined; i.e., $0 \notin \operatorname{Im} \pi_2 \cdot (df)$

and that $\pi_v \cdot f$ is an immersion iff $\tilde{v} \notin \operatorname{Im} g$ where $\tilde{v}$ is the point in $\mathbf{P}^{m-1}$ corresponding to the subspace (v) in $\mathbf{R}^m$. Next consider the composite mapping h defined by

$$X \times X - \Delta X \xrightarrow{\tilde{f}} \mathbf{R}^m - \{0\} \longrightarrow \mathbf{P}^{m-1}$$

where $\tilde{f}(p, q) = f(p) - f(q)$. Show that $\pi_v \circ f$ is 1:1 iff $\tilde{v} \notin \operatorname{Im} h$.

(2) Use Exercise (1) above and the Exercise of *I*, §4 to conclude that any compact n-manifold can be realized as a submanifold of $\mathbf{R}^{2n+1}$.

(3) Observe that the immersion part of the proof of Exercise 1 is valid when $m > 2n$. Thus show that there exists an immersion of any compact n-manifold into $\mathbf{R}^{2n}$.

(4) Does there exist a smooth function $f: \mathbf{R}^n \to \mathbf{R}^n$ such that $f^{-1}(a)$ is an uncountable set for each a in $\mathbf{R}^n$?

§2. Jet Bundles

Definition 2.1. *Let X and Y be smooth manifolds, and p in X. Suppose $f, g: X \to Y$ are smooth maps with $f(p) = g(p) = q$.*

(1) *f has* first order contact *with g at p if $(df)_p = (dg)_p$ as mappings of $T_pX \to T_qY$.*

(2) *f has* kth order contact *with g at p if $(df): TX \to TY$ has $(k-1)$st order contact with (dg) at every point in T_pX. This is written as $f \sim_k g$ at p. (k is a positive integer.)*

(3) *Let $J^k(X, Y)_{p,q}$ denote the set of equivalence classes under "$\sim_k$ at p" of mappings $f: X \to Y$ where $f(p) = q$.*

(4) *Let $J^k(X, Y) = \bigcup_{(p,q) \in X \times Y} J^k(X, Y)_{p,q}$ (disjoint union). An element σ in $J^k(X, Y)$ is called a k-*jet of mappings *(or just a k-*jet*) from X to Y.*

(5) *Let σ be a k-jet, then there exist p in X and q in Y for which σ is in $J^k(X, Y)_{p,q}$. p is called the* source *of σ and q is called the* target *of σ. The mapping $\alpha: J^k(X, Y) \to X$ given by $\sigma \mapsto$ (source of σ) is the* source map *and the mapping $\beta: J^k(X, Y) \to Y$ given by $\sigma \mapsto$ (target of σ) is the* target map.

Note that given a smooth mapping $f: X \to Y$ there is a canonically defined mapping $j^k f: X \to J^k(X, Y)$ called the *k-jet of f* defined by $j^k f(p) =$ equivalence class of f in $J^k(X, Y)_{p,f(p)}$ for every p in X. We will also show that $j^k f(p)$ is just an invariant way of describing the Taylor expansion of f at p up to order k and that $j^k f$ is a smooth mapping.

Note that $J^0(X, Y) = X \times Y$, so f has $\sim_0$ contact with g at p iff $f(p) = g(p)$, and $j^0 f(p) = (p, f(p))$ is just the graph of f.

Lemma 2.2 *Let U be an open subset of $\mathbf{R}^n$ and p be a point in U. Let $f, g: U \to \mathbf{R}^m$ be smooth mappings. Then $f \sim_k g$ at p iff*

$$\frac{\partial^{|\alpha|} f_i}{\partial x^\alpha}(p) = \frac{\partial^{|\alpha|} g_i}{\partial x^\alpha}(p)$$

for every multi-index α with $|\alpha| \le k$ and $1 \le i \le m$ where f_i and g_i are the coordinate functions determined by f and g, respectively and $x_1, \ldots, x_n$ are coordinates on U.

Proof. We proceed by induction on k. For $k = 1, f \sim_1 g$ iff $(df)_p = (dg)_p$ iff the first partial derivatives of f at p are identical with the first partial derivatives of g at p.

Assume the Lemma is true for $k - 1$. Let $y_1, \ldots, y_n$ be the coordinates of $\mathbf{R}^n$ in $U \times \mathbf{R}^n = TU$. Then $(df) : U \times \mathbf{R}^n \to \mathbf{R}^m \times \mathbf{R}^m = T\mathbf{R}^m$ is given by

$$(x, y) \mapsto (f(x), \bar{f}_1(y), \ldots, \bar{f}_m(y))$$

where

$$\bar{f}_i(x, y) = \sum_{j=1}^{n} \frac{\partial f_i}{\partial x_j}(x) y_j.$$

Similarly for (dg).

By assumption $(df) \sim_{k-1} (dg)$ at every point $(p, v) \in \{p\} \times \mathbf{R}^n$. By induction, the partial derivatives of (df) at points $(p, v) \in \{p\} \times \mathbf{R}^n$ are equal to the partial derivatives of (dg) at these same points. Let α be an n-tuple of non-negative integers with $|\alpha| \le k - 1$, then

$$\frac{\partial^{|\alpha|}\bar{f}_i}{\partial x^\alpha}(p, v) = \frac{\partial^{|\alpha|}\bar{g}_i}{\partial x^\alpha}(p, v).$$

Evaluate at $v = (0, \ldots, 1, \ldots, 0)$ with the 1 in the jth coordinate. Then we have that

$$\frac{\partial^{|\alpha|}}{\partial x^\alpha}\frac{\partial f_i}{\partial x_j}(p) = \frac{\partial^{|\alpha|}}{\partial x^\alpha}\frac{\partial g_i}{\partial x_j}(p).$$

Clearly all partial derivatives of f and g of order $\le k$ are obtained this way. To obtain the converse, just note that the partial derivatives of (df) of order $\le k - 1$ are determined by knowing the partial derivatives of f of order $\le k$. □

Corollary 2.3. *f and $g : U \to \mathbf{R}^m$ have kth order contact at p iff the Taylor expansions of f and g up to (and including) order k are identical at p.*

Lemma 2.4. *Let U be an open subset of $\mathbf{R}^n$ and V an open subset of $\mathbf{R}^m$. Let $f_1, f_2 : U \to V$ and $g_1, g_2 : V \to \mathbf{R}^l$ be smooth mappings so that $g_1 \cdot f_1$ and $g_2 \cdot f_2$ are defined. Let $p \in U$ and suppose that $f_1 \sim_k f_2$ at p and $g_1 \sim_k g_2$ at $q = f_1(p) = f_2(p)$. Then $g_1 \cdot f_1 \sim_k g_2 \cdot f_2$ at p.*

Proof. Again proceed by induction. For $k = 1$, this is just the chain rule, i.e.,

$$d(g_1 \cdot f_1)_p = (dg_1)_q(df_1)_p = (dg_2)_q(df_2)_p = d(g_2 \cdot f_2)_p.$$

Assume true for $k - 1$. Then again apply the chain rule, using the inductive assumption that

$$(dg_1) \cdot (df_1) \underset{k-1}{\sim} (dg_2) \cdot (df_2) \qquad \text{at all } (p, v) \text{ in } \{p\} \times \mathbf{R}^n. \quad \square$$

Proposition 2.5. *Let X, Y, Z, and W be smooth manifolds.*

(1) *Let $h: Y \to Z$ be smooth; then h induces a mapping $h_*: J^k(X, Y) \to J^k(X, Z)$ defined as follows; Let σ be in $J^k(X, Y)_{p,q}$ and let $f: X \to Y$ represent σ. Then $h_*(\sigma) =$ the equivalence class of $h \cdot f$ in $J^k(X, Z)_{p,h(q)}$.*

(2) *Let $a: Z \to W$ be smooth. Then $a_* \cdot h_* = (a \cdot h)_*$ as mappings of $J^k(X, Y) \to J^k(X, W)$ and $(id_Y)_* = id_{J^k(X,Y)}$. Thus if h is a diffeomorphism, h_* is a bijection.*

(3) *Let $g: Z \to X$ be a smooth diffeomorphism; then g induces a mapping $g^*: J^k(X, Y) \to J^k(Z, Y)$ defined as follows: let τ be in $J^k(X, Y)_{p,q}$ and let $f: X \to Y$ represent τ. Then $g^*(\tau) =$ equivalence class of $f \cdot g$ in $J^k(Z, Y)_{g^{-1}(p),q}$.*

(4) *Let $a: W \to Z$ be a smooth diffeomorphism. Then $a^*g^* = (g \cdot a)^*$ as mappings of $J^k(X, Y) \to J^k(W, Y)$ and $(id_X)^* = id_{J^k(X,Y)}$ so that g^* is a bijection.*

Proof. A simple application of Lemma 2.4 shows that h_* and g^* are well-defined mappings. The rest of the proposition is equally easy. □

Let $A_n{}^k$ be the vector space of polynomials in n-variables of degree $\leq k$ which have their constant term equal to zero. Choose as coordinates for $A_n{}^k$ the coefficients of the polynomials. Then $A_n{}^k$ is isomorphic to some Euclidean space and is, in this way, a smooth manifold. Let $B^k_{n,m} = \bigoplus_{i=1}^m A_n{}^k$. $B^k_{n,m}$ is also a smooth manifold.

Let U be an open set in $\mathbf{R}^n$ and $f: U \to \mathbf{R}$ be smooth. Define $T_k f: U \to A_n{}^k$ by $T_k(f)(x_0)$ is the polynomial in x of degree k given by the first k terms of the Taylor series of f at x_0 after the constant term.

Let V be an open subset of $\mathbf{R}^m$. Then there is a canonical bijection $T_{U,V}: J^k(U, V) \to U \times V \times B^k_{n,m}$ given by

$$T_{U,V}(\sigma) = (x_0, y_0, T_k f_1(x_0), \ldots, T_k f_m(x_0))$$

where

$$x_0 = \alpha(\sigma) = \text{source of } \sigma,$$

$$y_0 = \beta(\sigma) = \text{target of } \sigma,$$

$$f: U \to V \text{ is smooth and represents } \sigma,$$

and

$$f_i: U \to \mathbf{R} \ (1 \leq i \leq m) \text{ are the coordinate functions associated to } f.$$

By Corollary 2.3, $T_{U,V}$ is well-defined; i.e., independent of the choice of f, and injective. That $T_{U,V}$ is onto is clear. □

Lemma 2.6. *Let U and U' be open subsets of $\mathbf{R}^n$ and let V and V' be open subsets of $\mathbf{R}^m$. Suppose $h: V \to V'$ and $g: U \to U'$ are smooth mappings with g a diffeomorphism. Then*

$$T_{U',V'}(g^{-1})^* h_* T_{U,V}^{-1}: U \times V \times B^k_{n,m} \to U' \times V' \times B^k_{n,m}$$

is a smooth mapping.

Proof. Let $D = (x_0, y_0, f_1(x), \ldots, f_m(x))$ with $f_i \in A_n{}^k$ $(1 \leq i \leq m)$. Define $f: U \to \mathbf{R}^m$ by $f(x) = y_0 + (f_1(x - x_0), \ldots, f_m(x - x_0))$. Then $f(x_0) =$

y_0 and let σ = equivalence class of f in $J^k(U, V)_{(x_0,y_0)}$. Thus $T_{U,V}(\sigma) = D$. Now $(g^{-1})^* h_*(\sigma) = j^k(h \cdot f \cdot g^{-1})(g(x_0))$.

So

$$T_{U',V'}(g^{-1})^* h_* T_{U,V}^{-1}(D) = T_{U',V'}(j^k(h \cdot f \cdot g^{-1})(g(x_0)))$$
$$= (g(x_0), h(y_0), T_k((h \cdot f \cdot g^{-1})_1)(g(x_0)), \ldots, T_k((h \cdot f \cdot g^{-1})_m)(g(x_0)))$$

where $(h \cdot f \cdot g^{-1})_i : U' \to \mathbf{R}$ are the coordinate functions of $h \cdot f \cdot g^{-1} : U' \to \mathbf{R}^m$. To show that this mapping is smooth we need only show that the mapping of $U \times V \times B^k_{n,m} \to A_n{}^k$ given by $D \mapsto T_k((h \cdot f \cdot g^{-1})_i)(g(x_0))$ is smooth.

Let $\phi = h \cdot f \cdot g^{-1}$. Then

$$T_k(\phi_i)(g(x_0)) = \sum_{1 \le |\alpha| \le k} \frac{\partial^{|\alpha|} \phi_i}{\partial x^\alpha} (g(x_0))(x - g(x_0))^\alpha$$

To show that $D \mapsto T_k(\phi_i)(g(x_0))$ is smooth it is enough to show that $D \mapsto (\partial^{|\alpha|}\phi_i/\partial x^\alpha)(g(x_0))$ mapping $U \times V \times B^k_{n,m} \to \mathbf{R}$ is smooth for each multi-index α for which $|\alpha| \le k$. This is done by the chain rule and induction on $l = |\alpha|$.

In fact, one can show by induction that $(\partial^{|\alpha|}\phi_i/\partial x^\alpha)(g(x_0))$ is sums and products of terms of the form

$$\frac{\partial^{|\beta|}}{\partial y^\beta} \frac{\partial h_i}{\partial y_j}(y_0), \quad \frac{\partial^{|\beta|}}{\partial x^\beta} \frac{\partial f_i}{\partial x_j}(0), \quad \frac{\partial^{|\gamma|}}{\partial x^\gamma} \frac{\partial g_i{}^{-1}}{\partial x_j}(g(x_0))$$

where $|\gamma| \le |\beta| - 1$ and $|\beta| \le |\alpha|$ and where $y_1, \ldots, y_m$ are coordinates on $\mathbf{R}^m$ and h_i, g_i are the coordinate functions determined by h and g respectively. Each of these terms vary smoothly with D; hence $(\partial^{|\alpha|}\phi_i/\partial x^\alpha)(g(x_0))$ varies smoothly with D. □

Theorem 2.7. *Let X and Y be smooth manifolds with $n = \dim X$ and $m = \dim Y$. Then*

(1) *$J^k(X, Y)$ is a smooth manifold with*

$$\dim J^k(X, Y) = m + n + \dim (B^k_{n,m}).$$

(2) *$\alpha : J^k(X, Y) \to X$, $\beta : J^k(X, Y) \to Y$, and $\alpha \times \beta : J^k(X, Y) \to X \times Y$ are submersions.*

(3) *If $h : Y \to Z$ is smooth, then $h_* : J^k(X, Y) \to J^k(X, Z)$ is smooth. If $g : X \to Y$ is a diffeomorphism, then $g^* : J^k(Y, Z) \to J^k(X, Z)$ is a diffeomorphism.*

(4) *If $g : X \to Y$ is smooth, then $j^k g : X \to J^k(X, Y)$ is smooth.*

Proof.

(1) Let U be the domain for a chart ϕ on X and V be the domain for a chart ψ on Y. Let $U' = \phi(U)$ and $V' = \psi(V)$. Then $(\phi^{-1})^* \psi_* : J^k(U, V) \to J^k(U', V')$ and $\tau_{U,V} \equiv T_{U',V'} \cdot (\phi^{-1})^* \psi_* : J^k(U, V) \to U' \times V' \times B^k_{n,m}$. Give $J^k(X, Y)$ the manifold structure induced by declaring that $\tau_{U,V}$ is a chart. To see that this structure is well-defined we need only check to see what hap-

pens on overlaps. Let ϕ_1, ψ_1, U_1, V_1, U_1', V_1' be the data for another chart τ_{U_1,V_1}. Then note that

$$\begin{aligned}\tau_{U_1,V_1}\cdot(\tau_{U,V})^{-1} &= T_{U_1',V_1'}(\phi_1{}^{-1})^*(\psi_1)_*(\psi_*)^{-1}\phi^* T_{U',V'}^{-1}\\ &= T_{U_1',V_1'}(\phi_1{}^{-1}\cdot\phi)^*\cdot(\psi_1\cdot\psi^{-1})_* T_{U',V'}^{-1}\end{aligned}$$

since lower *'s and upper *'s commute. This last mapping is smooth by Lemma 2.6.

(2) In local coordinates α has the form

$$\phi\cdot\alpha\cdot\tau_{U,V}^{-1}(D) = \phi\cdot\alpha\cdot(\psi_*)^{-1}\cdot\phi^* T_{U',V'}^{-1}(D) = \phi\cdot\alpha\cdot j^k(\psi^{-1}\cdot f\cdot\phi)\cdot\phi^{-1}(x_0)$$

where f is defined in Lemma 2.6. Thus $\phi\cdot\alpha\cdot\tau_{U,V}^{-1}(D) = x_0$ since $\alpha\cdot j^k g = id_X$ for any mapping g. So α is a smooth mapping and a submersion. Similarly

$$\begin{aligned}\psi\cdot\beta\cdot\tau_{U,V}^{-1}(D) &= \psi\cdot\beta\cdot j^k(\psi^{-1}\cdot f\cdot\phi)\cdot\phi^{-1}(x_0)\\ &= \psi\cdot\psi^{-1}\cdot f\cdot\phi\cdot\phi^{-1}(x_0) = f(x_0) = y_0\end{aligned}$$

since $\beta\cdot j^k g = g$ for any mapping g. Thus β is also a smooth mapping and a submersion. Since $T_{(p,q)}(X \times Y) \cong T_pX \oplus T_qY$, $\alpha \times \beta : J^k(X, Y) \to X \times Y$ is a submersion.

(3) is obvious from the calculations in (1) and (2).

(4) $j^k g$: $X \to J^k(X, Y)$. Suppose g: $\mathbf{R}^n \to \mathbf{R}^m$. Then $j^k g$: $\mathbf{R}^n \to J^k(\mathbf{R}^n, \mathbf{R}^m) = \mathbf{R}^n \times \mathbf{R}^m \times B^k_{n,m}$ and is given by

$$j^k g(x_0) = (x_0, g(x_0), (T_k g_1)(x_0), \ldots, (T_k g_m)(x_0))$$

where $g_1, \ldots, g_m$ are the coordinate functions of g. Now $T_k g_i$ is a smooth function being only the sum of partial derivatives of the g_i's. So in the local situation $j^k g$ is a smooth function. With the standard use of the charts given above, one can see that $j^k g$ is smooth as a mapping of $X \to J^k(X, Y)$. □

Remarks.

(1) $J^k(X, Y)$ is, in general, not a vector bundle since there is no natural addition in $J^k(X, Y)_{p,q}$. However, if $Y = \mathbf{R}^m$, then $J^k(X, Y)$ is a vector bundle over $X \times \mathbf{R}^m$ where the addition of jets in $J^k(X, \mathbf{R}^m)_{p,q}$ is given by the addition of functions representing these jets.

(2) $J^1(X, Y)$ is canonically isomorphic to Hom (TX, TY) where the isomorphism ψ is given as follows: Let σ be a 1-jet with source p and target q, and let f: $X \to Y$ represent σ. Then $\psi(\sigma) = (df)_p$ in Hom (T_pX, T_qY). As an exercise show that ψ is well-defined and a diffeomorphism. Also note that $\alpha \times \beta = \pi\cdot\psi$ where π is the projection which makes Hom (TX, TY) into a vector bundle over $X \times Y$. Using this identification we can think of $J^1(X, Y)$ as a vector bundle over $X \times Y$.

(3) Although $J^k(X, Y)$ is not a vector bundle, it does have more structure than just the fact that it is a manifold would indicate. We isolate that structure with the following Definition.

Definition 2.8. *Let E, X, and F be smooth manifolds and let $\pi : E \to X$ be a submersion. Then E is a* fiber bundle over X with fiber F and projection π

if for every p in X, there exists a nbhd U of p and a diffeomorphism $\phi_U : E_U \to U \times F$ *where* $E_U = \pi^{-1}(U)$ *such that the diagram commutes*

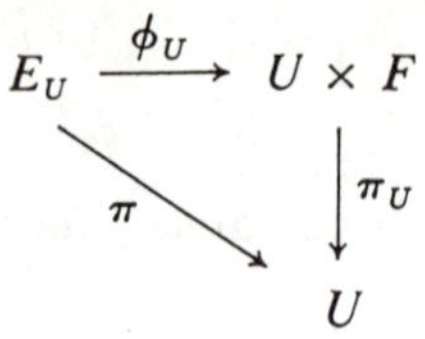

where π_U *is the obvious projection.*

Notes. (1) E_p is diffeomorphic (under ϕ_U) with F for all p in X.

(2) Clearly every vector bundle of dimension n over X is a fiber bundle with fiber $F = \mathbf{R}^n$. But not every fiber bundle with a Euclidean space as fiber is a vector bundle. (Consider $J^k(X, Y) \to X \times Y$!)

Exercises

(1) There is an obvious canonical projection $\pi_{k,l} : J^k(X, Y) \to J^l(X, Y)$ for $k > l$ defined by forgetting the jet information of order $> l$. Show that $J^k(X, Y)$ is a fiber bundle over $J^l(X, Y)$ with projection $\pi_{k,l}$ and identify the fiber.

(2) Let $J^1(X, \mathbf{R})_{X,0}$ be the set of all 1-jets whose target is 0.

(a) Show that $J^1(X, \mathbf{R})_{X,0}$ is a vector bundle over X whose projection is the source mapping.

(b) Show that $J^1(X, \mathbf{R})_{X,0}$ is canonically isomorphic (as vector bundles) with T^*X.

§3. The Whitney C^∞ Topology

Definition 3.1. *Let X and Y be smooth manifolds.*

(i) *Denote by* $C^\infty(X, Y)$, *the set of smooth mappings from X to Y.*

(ii) *Fix a non-negative integer k. Let U be a subset of* $J^k(X, Y)$. *Then denote by* $M(U)$ *the set*

$$\{f \in C^\infty(X, Y) \mid j^k f(X) \subset U\}.$$

Note that $M(U) \cap M(V) = M(U \cap V)$.

(iii) *The family of sets* $\{M(U)\}$ *where U is an open subset of* $J^k(X, Y)$ *form a basis for a topology on* $C^\infty(X, Y)$. *This topology is called the* Whitney C^k topology. *Denote by* W_k *the set of open subsets of* $C^\infty(X, Y)$ *in the Whitney* C^k *topology.*

(iv) *The* Whitney C^∞ topology *on* $C^\infty(X, Y)$ *is the topology whose basis is* $W = \bigcup_{k=0}^{\infty} W_k$. *This is a well-defined basis since* $W_k \subset W_l$ *whenever* $k \le l$. *To see this use the canonical mapping* $\pi_k^l : J^l(X, Y) \to J^k(X, Y)$ *which assigns to* σ *in* $J^l(X, Y)$ *the equivalence class of f in* $J^k(X, Y)$ *where f represents* σ. *Then* $M(U) = M((\pi_k^l)^{-1}(U))$ *for every open set U in* $J^k(X, Y)$.

In order to develop a feeling for these topologies we will describe a nbhd basis in the Whitney C^k topology for a function f in $C^\infty(X, Y)$. Choose a metric d on $J^k(X, Y)$ compatible with its topology. This is possible since all manifolds are metrizable by (I,5.9). Define

$$B_\delta(f) \equiv \{g \in C^\infty(X, Y) \mid \forall x \in X, d(j^k f(x), j^k g(x)) < \delta(x)\}$$

where $\delta : X \to \mathbf{R}^+$ is a continuous mapping. We claim that $B_\delta(f)$ is an open set for every such δ. For consider the continuous mapping $\Delta : J^k(X, Y) \to R$ defined by $\sigma \mapsto \delta(\alpha(\sigma)) - d(j^k f(\alpha(\sigma)), \sigma)$. Let $U = \Delta^{-1}(0, \infty)$. Then U is open in $J^k(X, Y)$ and $B_\delta(f) = M(U)$. Now let W be an open nbhd of f in $C^\infty(X, Y)$, let V be an open set in $J^k(X, Y)$ so that $f \in M(V) \subset W$, and let $m(x) = \inf \{d(\sigma, j^k f(x)) \mid \sigma \in \alpha^{-1}(x) \cap (J^k(X, Y) - V)\}$. Note that $m(x) = \infty$ if $\alpha^{-1}(x) \subset V$. Let $\delta : X \to \mathbf{R}^+$ be any continuous function such that $\delta(x) < m(x)$ for every x in X. It is possible to choose such a δ since m is bounded below on any compact subset in X by a positive constant. Then, by using a partition of unity argument, one may construct a δ globally. With this δ, $B_\delta(f) \subset W$. Finally, let γ and δ be continuous functions mapping X into $\mathbf{R}^+$. Define $\eta(x) = \min \{\gamma(x), \delta(x)\}$ and note that $\eta: X \to \mathbf{R}^+$ is continuous and that $B_\eta(f) = B_\delta(f) \cap B_\gamma(f)$. Thus the collection $\{B_\delta(f)\}$ forms a nbhd basis of f in the Whitney C^k topology on $C^\infty(X, Y)$. We may think of $B_\delta(f)$ as those smooth mappings of $X \to Y$ all of whose first k partial derivatives are δ-close to f.

On a compact manifold, we may find a countable nbhd basis of f by taking $B_n(f) = B_{\delta_n}(f)$ where $\delta_n(x) = 1/n$ for all x in X. Clearly this is a nbhd basis since if $\delta : X \to \mathbf{R}^+$ is continuous and X is compact, then δ is bounded below by $1/n$ for some large n. So $C^\infty(X, Y)$ satisfies the first axiom of countability *if* X is compact. From the above, one may prove easily that a sequence of functions f_n in $C^\infty(X, Y)$ converges to f (in the Whitney C^k topology) iff $j^k f_n$ converges uniformly to $j^k f$. Thus, in the local situation, f_n and all of the partial derivatives of f_n of order $\leq k$ converge uniformly to f.

On noncompact manifolds, convergence of $f_n \mapsto f$ is a concept stronger even than uniform convergence, since one has as much "control at infinity" as is wanted. Said precisely, the sequence of mappings f_n converge to f (in the Whitney C^k topology) iff there is a compact subset K of X such that $j^k f_n$ converges uniformly to $j^k f$ on K and all but a finite number of the f_n's equal f off K. The "if" part is clear, and we shall prove the "only if" part by contradiction. Assume f_n converges to f and that there does not exist a compact set K with the above property. Let $K_1, K_2, \ldots$ be a sequence of compact subsets of X such that $K_i \subset \operatorname{Int}(K_{i+1})$ and $X = \bigcup_{i=1}^\infty K_i$. We now define $\delta : X \to \mathbf{R}^+$ so that infinitely many f_n are not in $B_\delta(f)$. There is a function f_{l_1} in the sequence such that $f_{l_1} \neq f$. Thus there is an x_1 such that $d(j^k f_{l_1}(x_1), j^k f(x_1)) = a_1 > 0$. Choose m_1 so that x_1 is in K_{m_1} and let $\delta = a_1/2$ on K_{m_1}. Assume inductively that we have chosen functions $f_{l_1}, \ldots, f_{l_s}$ with $l_1 < \cdots < l_s$; a compact set K_{m_s}; a continuous positive-valued function δ defined on K_{m_s}; and points $x_1, \ldots, x_s$ in K_{m_s} so that for every $i \leq s$

$$d\left(j^k f_{l_i}(x_i), j^k f(x_i)\right) \geq \delta(x_i).$$

Now choose $f_{l_{s+1}}$ where $l_{s+1} > l_s$ so that $f_{l_{s+1}} \neq f$ off K_{m_s+1}. Let x_{s+1} be a point not in K_{m_s+1} where $d(j^k f_{l_{s+1}}(x_{s+1}), j^k f(x_{s+1})) = a_{s+1} > 0$. Then choose m_{s+1} so that x_{s+1} is in $K_{m_{s+1}}$. Extend δ to be a continuous positive-valued function on $K_{m_{s+1}}$ which is $\equiv a_{s+1}$ on $K_{m_{s+1}} - K_{m_s+1}$. In this way we construct a subsequence $f_{l_1}, f_{l_2}, \ldots$ and a continuous positive-valued function δ defined on X so that for every $j, f_{l_j} \notin B_\delta(f)$. Thus f_n does not converge to f and we have a contradiction. Finally we note that for a noncompact manifold X, $C^\infty(X, Y)$ in the Whitney C^k topology does not satisfy the first axiom of countability. To see this, let $W_1, W_2, \ldots$ be a countable nbhd basis of f in $C^\infty(X, Y)$. Then choose for each m a continuous function $\delta_m : X \to \mathbf{R}^+$ so that $B_{\delta_m}(f) \subset W_m$ and a sequence of points $x_1, x_2, \ldots$ with no limit point. Now construct a continuous function δ so that $\delta(x_m) < \delta_m(x_m)$ for every m. Since $W_1, W_2, \ldots$ is a nbhd basis of f, there is an m such that $W_m \subset B_\delta(f)$ which implies that $B_{\delta_m}(f) \subset B_\delta(f)$ which is a contradiction.

Thus we see that there is a great qualitative difference in the Whitney C^k topology on $C^\infty(X, Y)$ depending on whether or not the domain X is compact. If X is compact then we get a standard type of topology. If, on the other hand, X is not compact we have defined a very fine topology on $C^\infty(X, Y)$, one with many open sets. In either case, though, a theorem which asserts that a given set is dense in $C^\infty(X, Y)$ is saying that this set is indeed quite large and is a rather strong result.

Definition 3.2. *Let F be a topological space. Then*

(a) *A subset G of F is* residual *if it is the countable intersection of open dense subsets of F.*

(b) *F is a* Baire space *if every residual set is dense.*

Proposition 3.3. *Let X and Y be smooth manifolds. Then $C^\infty(X, Y)$ is a Baire space in the Whitney C^∞ topology.*

Proof. For each integer k choose a metric d_k on $J^k(X, Y)$ which makes $J^k(X, Y)$ into a complete metric space. Use Proposition I, 5.11.

Let $U_1, U_2, \ldots$ be a countable sequence of open dense subsets of $C^\infty(X, Y)$ and let V be another open subset of $C^\infty(X, Y)$. We must show that $V \cap \bigcap_{i=1}^\infty U_i \neq \varnothing$. Since V is open in the Whitney C^∞ topology, there is an open subset W in $J^{k_0}(X, Y)$ such that $M(\overline{W}) \subset V$ and $M(W) \neq \varnothing$. It is clearly enough to show that $M(\overline{W}) \cap \bigcap_{i=1}^\infty U_i \neq \varnothing$.

To do this we inductively choose a sequence of functions $f_1, f_2, \ldots$ in $C^\infty(X, Y)$; a sequence of integers $k_1, k_2, \ldots$; and for each i an open subset W_i in $J^{k_i}(X, Y)$ satisfying:

(A_i) $f_i \in M(W) \cap \bigcap_{j=1}^{i-1} M(W_j) \cap U_i$.

(B_i) $M(\overline{W}_i) \subseteq U_i$ and $f_i \in M(W_i)$

(C_i) $d_s(j^s f_i(x); j^s f_{i-1}(x)) < 1/2^i$ for all x in X, $i > 1$ and $0 \leq s \leq i$.

We first show that by choosing the above data we can prove the theorem. Define $g^s(x) = \operatorname{Lim}_{i \to \infty} j^s f_i(x)$. This makes sense since d_s is a complete metric and for each x the sequence $j^s f_1(x), j^s f_2(x), \ldots$ is a Cauchy sequence in

$J^s(X, Y)$ by (C). Note that $j^0 f_i(x) = (x, f_i(x))$, so we can define $g: X \to Y$ with $g^0(x) = (x, g(x))$. We claim that g is smooth. If so, we are done. Indeed, each f_i is in $M(W)$ by (A) and thus $g = \text{Lim}_{i\to\infty} f_i$ is in $M(\overline{W})$. Now, by (B), W_s was chosen so that $M(\overline{W}_s) \subset U_s$ and by (A) each f_i for $i > s$ was chosen to be in $M(W_s)$. Thus $g = \text{Lim}_{i\to\infty} f_i$ is in $M(\overline{W}_s)$. Since s is arbitrary $g \in M(\overline{W}) \cap \bigcap_{s=1}^{\infty} U_s$ and we are done.

We will now show that g is smooth. This is a local question, so choose x in X and compact nbhds K of x and L of $g(x)$ with $g(K) \subset L$. By choosing K and L small enough we may assume that they are contained in chart nbhds and then via these charts that K and L are subsets of $\mathbf{R}^n$ and $\mathbf{R}^m$ respectively. Since the metric d_s is compatible with the topology on $J^s(X, Y)$, conditions (C) translate, in the local situation, to the fact that $j^s f_i$ converge uniformly to g^s on K. Using local coordinates we see that the coordinate functions of $j^s f_i$ are just $\partial^{|\beta|} f_i/\partial x^\beta$ for $|\beta| \leq s$. Thus locally $\partial^{|\beta|} f_i/\partial x^\beta$ converges uniformly on K. Using a classical theorem (Dieudonné 8.6.3, p. 157),

$$\lim_{i\to\infty} \frac{\partial^{|\beta|} f_i}{\partial x^\beta} = \frac{\partial^{|\beta|} g}{\partial x^\beta} \text{ on } K \text{ for all } |\beta| \leq s.$$

Since s is arbitrary all partial derivatives of g exist and are continuous at x, in fact $g^s(x) = j^s g(x)$ and g is smooth.

Finally we will show that one can choose the f_i, k_i, and W_i inductively satisfying (A_i), (B_i), and (C_i). Choose f_1 in $M(W) \cap U_1$. This is possible since $M(W)$ is open and nonempty while U_1 is dense. Thus (A_1) is satisfied. Since U_1 is open and f is in U_1 we may choose k_1 and an open set W_1 in $J^{k_1}(X, Y)$ so that $f_1 \in M(W_1)$ and $M(\overline{W}_1) \subset U_1$. Thus (B_1) is satisfied. (C_1) is vacuous. Now assume inductively that the data is chosen for all $j \leq i - 1$. We will choose f_i satisfying (A_i) and (C_i) and then we can easily choose W_i and k_i so that (B_i) holds. Consider the set

$$D_i = \left\{ g \in C^\infty(X, Y) \mid d_s(j^s g(x), j^s f_{i-1}(x)) < \frac{1}{2^i} \right.$$
$$\left. \text{for } 0 \leq s \leq i \text{ and for all } x \text{ in } X \right\}.$$

If D_i is open, then $E_i = M(W) \cap \bigcap_{j=1}^{i-1} M(W_j) \cap D_i$ is open. It is easy to check that f_{i-1} is in E_i using the inductive hypotheses (A_{i-1}) and (B_{i-1}) and the definition of D_i. Since U_i is dense and E_i is open and nonempty we may choose f_i in $U_i \cap E_i$. By the definition of E_i, (A_i) is satisfied, and by the definition of D_i, (C_i) is satisfied. So the proof of the Theorem reduces to showing that D_i is open in $C^\infty(X, Y)$. Let

$$F_s = \left\{ g \in C^\infty(X, Y) \mid d_s(j^s g(x), j^s f_{i-1}(x)) < \frac{1}{2^i} \; \forall x \in X \right\}.$$

Since $D_i = F_1 \cap \cdots \cap F_i$, it is enough to show that F_s is open in $C^\infty(X, Y)$. Now define $B_x = \alpha^{-1}(x) \cap B(1/2^i, j^s f_{i-1}(x))$ where $\alpha: J^s(X, Y) \to X$ is the source mapping and

$$B\left(\frac{1}{2^i}, j^s f_{i-1}(x)\right) = \left\{ \sigma \in J^s(X, Y) \mid d_s(\sigma, j^s f_{i-1}(x)) < \frac{1}{2^i} \right\}.$$

Let $G = \bigcup_{x \in X} B_x$. It is easy to see that $F_s = M(G)$, so that we need only show that G is an open subset of $J^s(X, Y)$. Let σ be a point in G and $x = \alpha(\sigma)$. Note that the mapping $\Psi: X \to \mathbf{R}$ defined by $q \mapsto d_s(j^s f_{i-1}(q), j^s f_{i-1}(x))$ is continuous. Thus $H = \alpha^{-1}\Psi^{-1}(-\delta/2, \delta/2)$ is an open subset of $J^s(X, Y)$ where $\delta = 1/2^i - d_s(\sigma, j^s f_{i-1}(x))$. (Note that $\delta > 0$ since σ is in G.) Clearly $H \cap B(\delta/2, \sigma)$ is open and contains σ so that if $H \cap B(\delta/2, \sigma) \subset G$, we are done. Let $\tau \in H \cap B(\delta/2, \sigma)$. To show that $\tau \in G$, we need to show that $d_s(\tau, j^s f_{i-1}(\alpha(\tau))) < 1/2^i$. But

$$d_s(\tau, j^s f_{i-1}(\alpha(\tau))) \le d_s(\tau, \sigma) + d_s(\sigma, j^s f_{i-1}(x))$$
$$+ d_s(j^s f_{i-1}(x), j^s f_{i-1}(\alpha(\tau))) < \frac{\delta}{2} + \left(\frac{1}{2^i} - \delta\right) + \frac{\delta}{2} = \frac{1}{2^i}. \quad \square$$

Proposition 3.4. *Let X and Y be smooth manifolds. The mapping $j^k: C^\infty(X, Y) \to C^\infty(X, J^k(X, Y))$ defined by $f \mapsto j^k f$ is continuous in the Whitney C^∞ topology.*

Proof. Let U be an open subset of $J^l(X, J^k(X, Y))$. Then $M(U)$ is a basic open set in $C^\infty(X, J^k(X, Y))$. It is sufficient to show that $(j^k)^{-1}(M(U))$ is an open subset of $C^\infty(X, Y)$. First we define a mapping

$$\alpha_{k,l}: J^{k+l}(X, Y) \to J^l(X, J^k(X, Y))$$

as follows: let σ be a $(k + l)$-jet in $J^{k+l}(X, Y)$ with source x and let $f: X \to Y$ represent σ. By Theorem 2.7 (4), $j^k f: X \to J^k(X, y)$ is a smooth mapping. Define $\alpha_{k,l}(\sigma) = j^l(j^k f)(x)$. That $\alpha_{k,l}(\sigma)$ is well-defined, i.e., does not depend on the choice of representative f, can be seen from Corollary 2.3, and the fact that $j^l(j^k f)(x)$ depends only on the partial derivative of f at x of order $\le k + l$. For the same reasons, it is clear that $\alpha_{k,l}$ is a smooth mapping. (In fact, it is an embedding.)

Thus $\alpha_{k,l}^{-1}(U)$ is an open subset of $J^{k+1}(X, Y)$. We claim that $M(\alpha_{k,l}^{-1}(U)) = (j^k)^{-1}(M(U))$ and thus $(j^k)^{-1}(M(U))$ is an open subset of $C^\infty(X, Y)$. The claim follows trivially from the fact that $\alpha_{k,l} \cdot j^{k+l} f = j^l \cdot j^k f$ as mappings of $X \to J^l(X, J^k(X, Y))$. $\square$

Proposition 3.5. *Let X, Y, and Z be smooth manifolds. Let $\phi: Y \to Z$ be smooth. Then the mapping $\phi_*: C^\infty(X, Y) \to C^\infty(X, Z)$ given by $f \mapsto \phi \cdot f$ is a continuous mapping in the Whitney C^∞ topology.*

Proof. Let U be an open set in $J^k(X, Z)$. $M(U)$ is then a basic open set of $C^\infty(X, Z)$. Recall from Theorem 2.7 that there is a differentiable mapping $\hat{\phi}_*: J^k(X, Y) \to J^k(X, Z)$ defined by $\sigma = j^k f(x) \mapsto j^k(\phi \cdot f)(x)$. Thus $\hat{\phi}_*^{-1}(U)$ is an open set in $J^k(X, Y)$. It is easy to check that $\phi_*^{-1}(M(U)) = M(\hat{\phi}_*^{-1}(U))$ so that ϕ_* is continuous. $\square$

We shall now investigate the properties of $C^\infty(X, \mathbf{R})$ which is a vector space over $\mathbf{R}$. It would be nice if $C^\infty(X, \mathbf{R})$ were a topological vector space, but alas, scalar multiplication viewed as a mapping of $\mathbf{R} \times C^\infty(X, \mathbf{R}) \to C^\infty(X, \mathbf{R})$ is not continuous unless X is compact. For if X is not compact and $f: X \to \mathbf{R}$ is a smooth function with noncompact support, then $\gamma_f: \mathbf{R} \to$

$C^\infty(X, \mathbf{R})$ defined by $r \mapsto rf$ would be continuous. Thus $\operatorname{Lim}_{n\to\infty} f/n = 0$. But this cannot happen unless there is some compact set K off of which each function f/n is zero which contradicts the assumption on the support of f. Addition and multiplication of functions fare better.

Proposition 3.6. *Let X, Y, and Z be smooth manifolds. Then $C^\infty(X, Y) \times C^\infty(X, Z)$ is homeomorphic (in the C^∞ topology) with $C^\infty(X, Y \times Z)$ by using the standard identification $(f, g) \mapsto f \times g$ where $(f \times g)(x) = (f(x), g(x))$.*

To prove this proposition we need the following Lemma on topological spaces.

Lemma 3.7. *Let A, B, and P be Hausdorff spaces. Suppose that P is locally compact and paracompact. Let $\pi_A: A \to P$ and $\pi_B: B \to P$ be continuous. Set*

$$A \times_P B = \{(a, b) \in A \times B \mid \pi_A(a) = \pi_B(b)\}$$

and give $A \times_P B$ the topology induced from $A \times B$. Let $K \subset A$ and $L \subset B$ be subsets such that $\pi_A|K$ and $\pi_B|L$ are proper. Let U be an open nbhd of $K \times_P L$ in $A \times_P B$. Then there exists a nbhd V of K in A and a nbhd W of L in B such that $V \times_P W \subseteq U$.

Proof. First note that if X and Y are Hausdorff spaces with Y locally compact and if $f: X \to Y$ is continuous and proper, then f is a closed mapping. For let Z be a closed subset of X and y be a point in $\overline{f(Z)}$. Let $y_1, y_2, \ldots$ be a sequence of points in $f(Z)$ with $\operatorname{Lim}_{i\to\infty} y_i = y$. Since Y is locally compact, there is a compact nbhd V of y. We may assume that y_i in V for all i. Choose $x_1, x_2, \ldots$ so that $f(x_i) = y_i$. Since f is proper $f^{-1}(V)$ is compact. Thus by restriction to a subsequence we may assume that the sequence $x_1, x_2, \ldots$ converges. Suppose $\operatorname{Lim}_{i\to\infty} x_i = x$. Then x is in Z since Z is closed and by the continuity of f, $f(x) = y$. So y is in $f(Z)$ and $f(Z)$ is closed.

Now consider $\pi_A \times \pi_B: A \times B \to P \times P$. Note that Δ_P, the diagonal of $P \times P$, is closed and that $A \times_P B = (\pi_A \times \pi_B)^{-1}(\Delta_P)$. So $E = A \times B - A \times_P B$ is open. For each p in P, let $K_p = K \cap (\pi_A)^{-1}(p)$ and $L_p = L \cap (\pi_B)^{-1}(p)$. Note that $K_p \times L_p = K_p \times_P L_p \subset U$ and that $U \cup E$ is open in $A \times B$. Since $\pi_A|K$ and $\pi_B|L$ are proper, K_p and L_p are compact. Thus there is an open nbhd V_p of K_p in A and W_p of L_p in B such that $V_p \times W_p \subset U \cup E$. To see this, choose for each (k, l) in $K_p \times L_p$ open nbhds $V^{k,l}$ of k in A and $W^{k,l}$ of l in B such that $V^{k,l} \times W^{k,l} \subset U \cup E$. For a fixed k, the collection $\{W^{k,l}\}_{l \in L_p}$ is an open covering of L_p. Since L_p is compact, there is a finite subcovering $W^{k,l_1}, \ldots, W^{k,l_m}$. Set $V^k = V^{k,l_1} \cap \cdots \cap V^{k,l_m}$ and $W^k = W^{k,l_1} \cup \cdots \cup W^{k,l_m}$ and note that $V^k \times W^k \subset U \cup E$. The collection $\{V^k\}_{k \in K_p}$ is an open covering of K_p. So, by compactness, there is a finite subcovering $V^{k_1}, \ldots, V^{k_n}$. Set $W_p = W^{k_1} \cap \cdots \cap W^{k_n}$ and $V_p = V^{k_1} \cup \cdots \cup V^{k_n}$.

To continue with the proof of the lemma, note that $\pi_A|K$ and $\pi_B|L$ satisfy the hypotheses of the first paragraph; thus $\pi_A(K - V_p)$ and $\pi_B(L - W_p)$ are closed in P and $P_p = P - \pi_A(K - V_p) - \pi_B(L - W_p)$ is open. Moreover, p is in P_p. Thus the collection $\{P_p\}_{p \in P}$ is an open covering of P and since P is paracompact there is a locally finite refinement $\{P_\alpha\}$. For each α choose $\alpha(p)$, a point in P, so that $P_\alpha \subset P_{\alpha(p)}$. Let $V_\alpha = V_{\alpha(p)} \cup \pi_A{}^{-1}(P - P_\alpha)$ and $W_\alpha = W_{\alpha(p)} \cup \pi_B{}^{-1}(P - P_\alpha)$. Let $V = \bigcap_\alpha V_\alpha$ and $W = \bigcap_\alpha W_\alpha$. To complete the proof, we need to show that $K \subset V$, $L \subset W$, V and W are open, and $V \times_P W \subset U$.

(a) $K \subset V$. It is clearly enough to show that $K \subset V_\alpha$ for each α. So assuming that k is in K, we must show that $k \in V_{\alpha(p)}$ or $k \in \pi_A{}^{-1}(P - P_\alpha)$. But this is trivial, since if $k \notin V_{\alpha(p)}$, then $\pi_A(k) \notin P_{\alpha(p)}$.

(b) $L \subset W$. Just the same as (a).

(c) V is open. Let v be a point in V. Since $\{P_\alpha\}$ is locally finite, there exists a nbhd U of $\pi_A(v)$ and finitely many P_α's; namely, $P_{\alpha_1}, \ldots, P_{\alpha_r}$, for which $U \cap P_{\alpha_i} \neq \varnothing$. Let $U' = \pi_A{}^{-1}(U) \cap V_{\alpha_1} \cap \cdots \cap V_{\alpha_r}$ and note that U' is an open nbhd of v. If $U \cap P_\alpha = \varnothing$, then $U' \subset \pi_A{}^{-1}(P - P_\alpha) \subset V_\alpha$. If $U \cap P_\alpha \neq \varnothing$, then $\alpha = \alpha_i$ for some i and $U' \subset V_\alpha$. So $U' \subset V$ and V is open.

(d) W is open. Just the same as (c).

(e) $V \times_P W \subset U$. Choose $(v, w) \in V \times_P W$ and set $p = \pi_A(v) = \pi_B(w)$. Choose an α such that p is in P_α. Thus $v \in V_\alpha - \pi_A{}^{-1}(P - P_\alpha) \subset V_{\alpha(p)}$. Similarly w is in $W_{\alpha(p)}$. Hence $(v, w) \in V_{\alpha(p)} \times W_{\alpha(p)} \subset U \cup E$. Since $\pi_A(v) = \pi_B(w)$ $(u, v) \notin E$, thus $V \times {}_P W \subset U$. □

Proof of Proposition 3.6. The projections $\pi_Y : Y \times Z \to Y$ and $\pi_Z : Y \times Z \to Z$ induce continuous mappings $(\pi_Y)_* : C^\infty(X, Y \times Z) \to C^\infty(X, Y)$ and $(\pi_Z)_* : C^\infty(X, Y \times Z) \to C^\infty(X, Z)$ by Proposition 3.5. Since the identification of $C^\infty(X, Y \times Z)$ with $C^\infty(X, Y) \times C^\infty(X, Z)$ is given by $(\pi_Y)_* \times (\pi_Z)_*$, it is continuous. To show that the identification is a homeomorphism we need only show that it is an open mapping.

To do this we let (f, g) be in $C^\infty(X, Y \times Z)$. Choose an open set W in $J^k(X, Y \times Z)$ so that (f, g) is in $M(W)$. Now notice that

$$J^k(X, Y \times Z) = J^k(X, Y) \times_X J^k(X, Z)$$

where $A = J^k(X, Y)$, $B = J^k(X, Z)$, $P = X$, and $\pi_A : J^k(X, Y) \to X$ and $\pi_B : J^k(X, Z) \to X$ are the respective source mappings. Applying Lemma 3.7, there are open sets U in $J^k(X, Y)$ and V in $J^k(X, Z)$ such that $U \times {}_X V \subset W$. Finally we note that $M(U) \times M(V) \subset (\pi_Y)_* \times (\pi_Z)_*(M(W))$, so that this identification is an open mapping. □

Corollary 3.8. *Addition and multiplication of smooth functions are continuous operations in the C^∞ topology, i.e., $C^\infty(X, \mathbf{R}) \times C^\infty(X, \mathbf{R}) \to C^\infty(X, \mathbf{R})$ given by $(f, g) \mapsto f + g$ or $(f, g) \to f \cdot g$ is continuous.*

Proof. $+ : \mathbf{R} \times \mathbf{R} \to \mathbf{R}$ given by $(x, y) \mapsto x + y$ is smooth so $(+)_* : C^\infty(X, \mathbf{R} \times \mathbf{R}) \to C^\infty(X, \mathbf{R})$ is continuous by Proposition 3.5. Thus via the homeomorphism of $C^\infty(X, \mathbf{R} \times \mathbf{R})$ with $C^\infty(X, \mathbf{R}) \times C^\infty(X, \mathbf{R})$ given by Proposition 3.6 $(f, g) \mapsto f + g$ is continuous. The proof for multiplication is similar. □

For completeness sake, we make some further remarks about the Whitney C^∞ topology on $C^\infty(X, Y)$.

Proposition 3.9. *Let X, Y, and Z be smooth manifolds with X compact. Then the mapping of $C^\infty(X, Y) \times C^\infty(Y, Z) \to C^\infty(X, Z)$ given by $(f, g) \mapsto g \cdot f$ is continuous.*

Remark. This proposition is not true if X is not compact although if we replace $C^\infty(X, Y)$ by the open subset of proper mappings of X into Y, then the conclusion is still valid.

Proof. Let D be the fiber product $J^k(X, Y) \times_Y J^k(Y, Z)$ described in Lemma 3.7, where $A = J^k(X, Y)$, $B = J^k(Y, Z)$, $P = Y$, $\pi_A = \beta$ (the target mapping), and $\pi_B = \alpha$ (the source mapping). The mapping $\gamma: D \to J^k(X, Z)$ defined by $(\sigma, \tau) \mapsto \tau \cdot \sigma$ is continuous. (Note $\tau \cdot \sigma = j^k(g \cdot f)(\alpha(\sigma))$ where f represents σ in $J^k(X, Y)$ and g represents τ in $J^k(Y, Z)$). To prove the proposition it is enough to show that if f is in $C^\infty(X, Y)$, if g is in $C^\infty(Y, Z)$, and if $S \subseteq J^k(X, Z)$ is open with $g \cdot f$ in $M(S)$, then there are open sets $V \subset J^k(X, Y)$ and $W \subset J^k(Y, Z)$ with $\gamma(V \times_Y W) \subset S$. Then if f' is in $M(V)$ and if g' is in $M(W)$, $g' \cdot f'$ will be in $M(S)$. Thus composition will be a continuous mapping in the Whitney C^k topology for arbitrary k and thus continuous in the C^∞ topology.

First we note that $j^k(g \cdot f)(X) = \gamma(j^k f(X) \times_Y j^k g(Y))$ i.e.,

$$j^k f(X) \times_Y j^k g(Y) \subset \gamma^{-1}(S).$$

We now apply Lemma 3.7 with $K = j^k f(X)$, $L = j^k g(Y)$, and $U = \gamma^{-1}(S)$ to show the existence of the desired V and W. That Lemma 3.7 is applicable follows from the facts that U is open (since γ is continuous), $\pi_A | K$ is compact (since X and thus K are compact), and $\pi_B | L$ is proper (since $\pi_B \cdot j^k g = id_Y$). □

Notes. (1) Let $f: X \to Y$ be smooth; then π induces $f^*: C^\infty(Y, Z) \to C^\infty(X, Z)$ given by $g \mapsto g \cdot f$. The Remark after this last proposition shows that if f is not a proper mapping then this "nice" functorially defined mapping is not continuous. In particular, if X is an open subset of Y and f is just given by inclusion then π^* is not continuous; i.e., the restriction mapping of $C^\infty(Y, Z) \to C^\infty(X, Z)$ given by $g \mapsto g | X$ is not continuous.

(2) An easy consequence of the proof of Proposition 3.9 is that f^* is continuous if f is proper. The only use that was made of the compactness of X was to show that $\pi_A | K$ is proper; this statement is still true if f is proper.

For future reference we make one last comment about the continuity of these types of functorial mappings. If X is a set, let $X^l = X \times \cdots \times X$.

Proposition 3.10. *Let X and Y be smooth manifolds. The mappings $\delta_l: C^\infty(X, Y)^l \to C^\infty(X^l, Y^l)$ given by $(f_1, \ldots, f_l) \mapsto f_1 \times \cdots \times f_l$ where $(f_1 \times \cdots \times f_l)(x_1, \ldots, x_l) = (f_1(x_1), \ldots, f_l(x_l))$ is continuous.*

Proof. We assume that $l = 2$ as the proof for general l is essentially the same. First we claim that the mapping $\gamma: J^k(X, Y) \times J^k(X, Y) \to J^k(X^2, Y^2)$ given by $(\sigma, \tau) \to \sigma \times \tau$ (where $\alpha(\sigma \times \tau) = (\alpha(\sigma), \alpha(\tau))$, $\beta(\sigma \times \tau) = (\beta(\sigma), \beta(\tau))$, and if f represents σ and if g represents τ, then $f \times g$ represents $\sigma \times \tau$) is continuous. To see that this claim is sufficient, let W be an open nbhd in $J^k(X^2, Y^2)$

with $f \times g$ in $M(W)$. Since $\gamma^{-1}(W)$ is open in $J^k(X, Y)^2$ and contains $j^kf(X) \times j^kg(X)$, there exist open sets U and V in $J^k(X, Y)$ so that $j^kf(X) \times j^kg(X) \subset U \times V \subset \gamma^{-1}(W)$. Thus $f \times g \in M(U) \times M(V) \subset \delta_2^{-1}(M(W))$ and δ_2 is continuous.

To see that γ is continuous (in fact, a smooth embedding), look in local coordinates. In these coordinates for fixed sources and targets γ is just a linear injection which varies smoothly with σ, τ. □

§4. Transversality

Definition 4.1. *Let X and Y be smooth manifolds and $f: X \to Y$ be a smooth mapping. Let W be a submanifold of Y and x a point in X. Then f intersects W* transversely *at x (denoted by $f \pitchfork W$ at x) if either*

(a) *$f(x) \notin W$, or*

(b) *$f(x) \in W$ and $T_{f(x)}Y = T_{f(x)}W + (df)_x(T_xX)$. If A is a subset of X, then f intersects W* transversely *on A (denoted by $f \pitchfork W$ on A) if $f \pitchfork W$ at x for all $x \in A$. Finally, f* intersects *W* transversely *(denoted by $f \pitchfork W$) if $f \pitchfork W$ on X.*

(c) *If $A \subset W$ we say that f* intersects *W* transversely *on A if $f \pitchfork W$ at x for all x for which $f(x) \in A$.*

Examples.

(1) Let $X = \mathbf{R} = W$, $Y = \mathbf{R}^2$, and $f(x) = (x, x^2)$. Then $f \pitchfork W$ at all nonzero x.

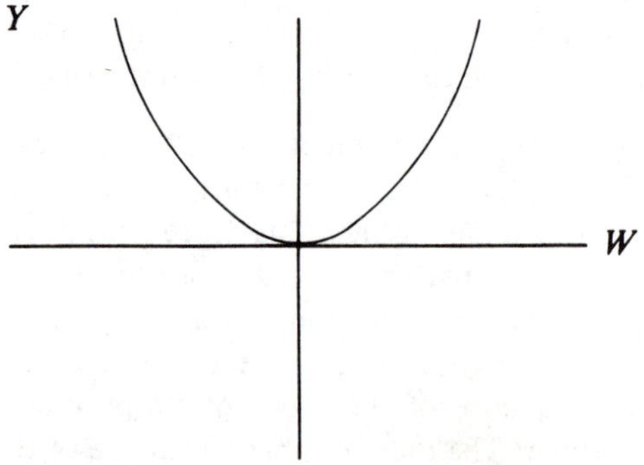

Notice that f can be perturbed ever so slightly to be transversal to W; e.g.,

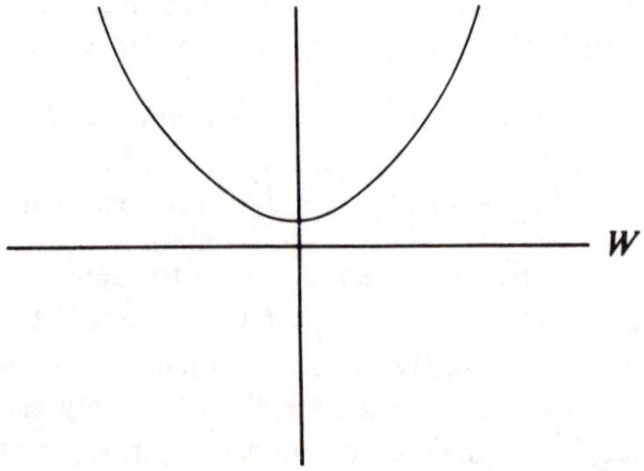

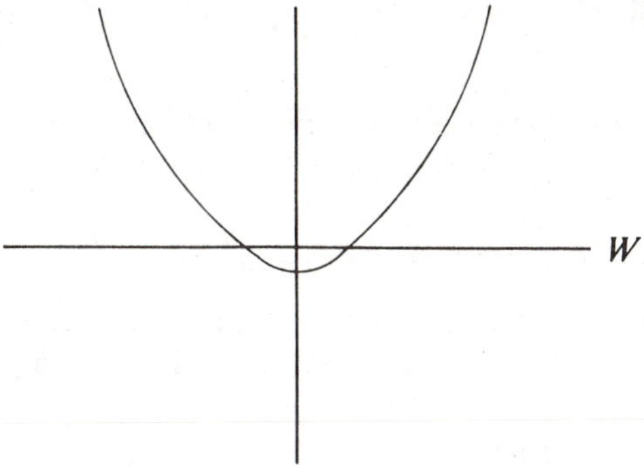

(2) Suppose that X, Y, W are as in (1) and that f is given by the graph

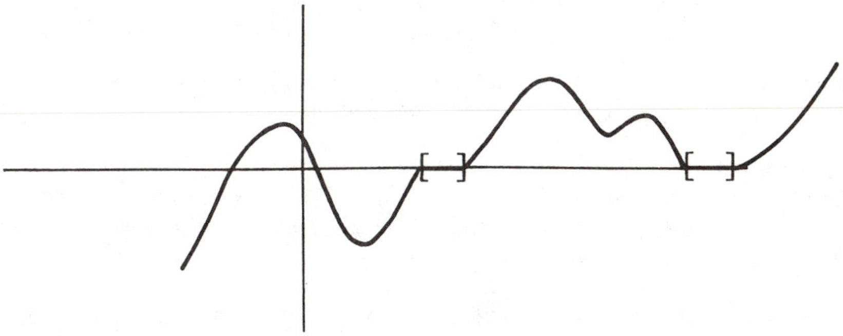

Then f does not intersect W transversally on the segments within the brackets and does elsewhere.

(3) If $X = \mathbf{R} = W$ and $Y = \mathbf{R}^3$, then if f is any mapping of $X \to Y$, it is transversal to W only if $f(X) \cap W = \varnothing$. Notice that here too a nontransversal mapping can be approximated closely by a transversal mapping, since in 3-space f can avoid W by just "going around" it. Moreover, f doesn't have to move far to accomplish this task. This will be made precise shortly.

In any case, it becomes apparent quickly that the relative dimensions of X, Y, and W play an important part in determining what transversality means in a particular instance. Also, for any trio X, Y, and W, the set of transversal mappings is quite large. In fact, the Thom Transversality Theorem is just this observation formalized.

Before discussing this theorem, we will give some consequences of the property that a mapping is transversal.

Proposition 4.2. *Let X and Y be smooth manifolds, $W \subset Y$ a submanifold. Suppose* $\dim W + \dim X < \dim Y$ (*i.e.*, $\dim X < \operatorname{codim} W$). *Let $f: X \to Y$ be smooth and suppose that $f \pitchfork W$. Then $f(X) \cap W = \varnothing$.*

Proof. Suppose $f(x) \in W$. Then

$$\begin{aligned}\dim (T_{f(x)}W + (df)_x(T_xX)) &\leq \dim T_{f(x)}W + \dim T_xX \\ &= \dim W + \dim X < \dim Y = \dim T_{f(x)}Y.\end{aligned}$$

So it is impossible for $T_{f(x)}W + (df)_x(T_xX) = T_{f(x)}Y$. Hence if $f \pitchfork W$ at x, $f(x) \notin W$. ☐

Lemma 4.3. *Let X, Y be smooth manifolds, $W \subset Y$ a submanifold, and $f: X \to Y$ smooth. Let $p \in X$ and $f(p) \in W$. Suppose there is a nbhd U of $f(p)$ in Y and a submersion $\phi: U \to \mathbf{R}^k (k = \text{codim } W)$ such that $W \cap U = \phi^{-1}(0)$. Then $f \pitchfork W$ at p iff $\phi \cdot f$ is a submersion at p.*

Remark. Such a nbhd U always exists. For there exists a chart nbhd U of $f(p)$, a chart $\alpha: U \to \mathbf{R}^m$ ($m = \dim Y$) and a decomposition of $\mathbf{R}^m = \mathbf{R}^k \times \mathbf{R}^{m-k}$ so that $W \cap U = \alpha^{-1}(0 \times \mathbf{R}^{m-k})$. Let $\pi: \mathbf{R}^m \to \mathbf{R}^k$ be projection on the first factor, then let $\phi = \pi \cdot \alpha$.

Proof. One can show easily that $\text{Ker } (d\phi)_{f(p)} = T_{f(p)}W$. So $f \pitchfork W$ at p

$$\text{iff} \quad T_{f(p)}Y = T_{f(p)}W + (df)_p(T_pX)$$
$$\text{iff} \quad T_{f(p)}Y = \text{Ker } (d\phi)_{f(p)} + (df)_p(T_pX)$$

Since $(d\phi)_{f(p)}$ is onto we see that $(d\phi \cdot f)_p$ is onto iff this last equality holds. Thus $\phi \cdot f$ is a submersion at p iff $f \pitchfork W$ at p. ☐

Theorem 4.4. *Let X and Y be smooth manifolds, W a submanifold of Y. Let $f: X \to Y$ be smooth and assume that $f \pitchfork W$. Then $f^{-1}(W)$ is a submanifold of X. Also* $\text{codim} f^{-1}(W) = \text{codim } W$. *In particular, if* $\dim X = \text{codim } W$, *then $f^{-1}(W)$ consists only of isolated points.*

Proof. It is sufficient to show that for every point $p \in f^{-1}(W)$, there exists an open nbhd V of p in X so that $V \cap f^{-1}(W)$ is a submanifold. Choose U and ϕ as in Lemma 4.3. Choose V a nbhd of p so that $f(V) \subset U$. By Lemma 4.3 $\phi \cdot f$ is a submersion at p. Thus, by contracting V if necessary, we assume that $\phi \cdot f$ is a submersion on V. Thus $f^{-1}(W) \cap V = (\phi \cdot (f|V))^{-1}(0)$ is a submanifold, by (I,2.8). ☐

Proposition 4.5. *Let X and Y be smooth manifolds with W a submanifold of Y. Let $T_W = \{f \in C^\infty(X, Y) \mid f \pitchfork W\}$. Then T_W is an open subset of $C^\infty(X, Y)$ (in the Whitney C^1, and thus, C^∞, topology) if W is a closed submanifold of Y.*

Proof. Define a subset U of $J^1(X, Y)$ as follows: let σ be a 1-jet with source x and target y and let $f: X \to Y$ represent σ. Then $\sigma \in U$ iff either (i) $y \notin W$ or (ii) $y \in W$ and $T_yY = T_yW + (df)_x(T_xX)$. Recall that $M(U) = \{f \in C^\infty(X, Y) \mid (j^1f)(X) \subset U\}$. It is clear that $T_W = M(U)$, so that if we can show that U is open, then so is T_W.

We show that $V = J^1(X, Y) - U$ is closed. Let $\sigma_1, \sigma_2, \ldots$ be a convergent sequence of 1-jets with σ_i in V for all i. Let $\sigma = \text{Lim}_{i \to \infty} \sigma_i$, we will show that σ is in V. Let p = source σ and q = target of σ. Since the targets of σ_i are in W for all i and W is closed, q is in W. Let $f: X \to Y$ represent σ. Choose coordinate nbhds U' of p in X and V' of q in Y so that $f(U') \subset V'$. Assume that the chart defined on V' takes W onto a subspace of dimension k. Via these charts we may assume that $X = \mathbf{R}^n$, $x = 0$,

$Y = \mathbf{R}^m$, and $W = \mathbf{R}^k \subset \mathbf{R}^m$. Let $\phi : \mathbf{R}^m \to \mathbf{R}^m/\mathbf{R}^k = \mathbf{R}^{m-k}$ be projection. By applying Lemma 4.3 we see that $f \pitchfork W$ at 0 iff $\phi \cdot f$ is a submersion at 0 iff $\phi\,(df)_0 \notin F$ where

$$F = \{A \in \operatorname{Hom}(\mathbf{R}^n, \mathbf{R}^{m-k}) \mid \operatorname{rank} A < m - k\}.$$

Consider the mapping

$$\mathbf{R}^n \times W \times \operatorname{Hom}(\mathbf{R}^n, \mathbf{R}^m) \subset J^1(\mathbf{R}^n, \mathbf{R}^m) \xrightarrow{\eta} \operatorname{Hom}(\mathbf{R}^n, \mathbf{R}^{m-k})$$

given by $(x, w, B) \mapsto \phi \cdot B$. Since F is closed and η is continuous $\eta^{-1}(F)$ is closed in $\mathbf{R}^n \times W \times \operatorname{Hom}(\mathbf{R}^n, \mathbf{R}^m)$ which, in turn, is a closed subset of $J^1(\mathbf{R}^n, \mathbf{R}^m)$. Now $V = \eta^{-1}(F)$ since $\tau = (x, y, (dg)_x)$ is in V iff y is in W and g does not intersect W transversely at x iff $\eta(\tau)$ is in F. Since V is closed in this local situation σ is in V. □

Note that Proposition 4.5 does not hold if W is not assumed to be closed. As an example take $X = S^1$, $Y = \mathbf{R}^3$, and $W = \{(t, 0, 0) \mid 0 < t < 2\}$. Transversality in these dimensions means that $f(X) \cap W = \varnothing$. Let $f : S^1 \to \mathbf{R}^3$ be given by $f(x, y) = (x, y, 0)$ where $S^1 \subset R^2$ is thought of as the unit circle centered at the origin. Then $f \pitchfork W$ but arbitrarily small perturbations of f given by $f_\varepsilon(x, y) = (x - \varepsilon, y, 0)$ intersect W and are thus not transverse to W.

Lemma 4.6. *Let X, B, and Y be smooth manifolds with W a submanifold of Y. Let $j: B \to C^\infty(X, Y)$ be a mapping (not necessarily continuous) and define $\Phi : X \times B \to Y$ by $\Phi(x, b) = j(b)(x)$. Assume that Φ is smooth and that $\Phi \pitchfork W$. Then the set $\{b \in B \mid j(b) \pitchfork W\}$ is dense in B.*

Proof. Let $W_\Phi = \Phi^{-1}(W)$. Since $\Phi \pitchfork W$, W_Φ is a submanifold of $X \times B$. (Apply Theorem 4.4.) Let π be the restriction to W_Φ of the projection of $X \times B \to B$. First note that if $b \notin \operatorname{Im} \pi$, then $j(b)(X) \cap W = \varnothing$ so $j(b) \pitchfork W$. Now if $\dim W_\Phi < \dim B$, then $\pi(W_\Phi)$ has measure zero in B by Proposition 1.6 and for a dense set of b in B; namely $B - \operatorname{Im} \pi$, $j(b) \pitchfork W$. Thus, in this case, the Lemma is true and we may assume that $\dim W_\Phi \geq \dim B$. We claim that if b is a regular value for π, then $j(b) \pitchfork W$. If this claim is true, then the lemma is proved for we may apply Sard's Theorem (actually Corollary 1.14) to π.

To prove the claim let b be a regular value for π and let x be in X. If $(x, b) \notin W_\Phi$, then $j(b)(x) \notin W$ and $j(b) \pitchfork W$ at x. So we may assume that (x, b) is in W_Φ. Since b is a regular value for π and $\dim W_\Phi \geq \dim B$, we have that $T_{(x,b)}(X \times B) = T_{(x,b)}W_\Phi + T_{(x,b)}(X \times \{b\})$. Apply $(d\Phi)_{(x,b)}$ to both sides and obtain

$$(d\Phi)_{(x,b)}T_{(x,b)}(X \times B) = T_{j(b)(x)}W + (dj(b))_x(T_x X).$$

Now we assumed that $\Phi \pitchfork W$ so

$$T_{\Phi(x,b)}Y = T_{\Phi(x,b)}W + (d\Phi)_{(x,b)}(T_{(x,b)}(X \times B)).$$

Combining these two equalities we have that

$$T_{j(b)(x)}Y = T_{j(b)(x)}W + (dj(b))_x(T_x X).$$

Thus $j(b) \pitchfork W$ at x. □

Remark. If we let $G\colon X \times B \to Y$ be a B-parameter family of mappings of $X \to Y$ where $G_b(x) = G(x, b)$ and we let $j\colon B \to C^\infty(X, Y)$ be given by $j(b) = G_b$, then $\Phi = G$. Assume $G \pitchfork W$, then the set $\{b \in B \mid G_b \pitchfork W\}$ is dense in B. This remark is *the basic fact* about transversality; that is, if a parametrized family of mappings intersects a given submanifold transversely, then for a dense set of parameters the individual mappings also intersect this submanifold transversely.

In the same spirit, we have the following:

Corollary 4.7. *Let $G\colon X \times B \to Y$ be a smooth mapping. Let $\Phi(x, b) = j^k G_b(x)$ Assume that $\Phi \pitchfork W$ where W is a submanifold of $J^k(X, Y)$. Then the set $\{b \in B \mid j^k G_b \pitchfork W\}$ is dense in B.*

Proof. Define $j\colon B \to C^\infty(X, J^k(X, Y))$ by $b \mapsto j^k G_b$ and apply Lemma 4.6. □

Definition 4.8. *Let X and Y be smooth manifolds with $f\colon X \to Y$ a smooth mapping. Let W be a submanifold of Y with W' a subset of W. Then $f \pitchfork W$ on W' if for every x in X with $f(x)$ in W', $f \pitchfork W$ at x.*

Theorem 4.9. (*Thom Transversality Theorem*). *Let X and Y be smooth manifolds and W a submanifold of $J^k(X, Y)$. Let*

$$T_W = \{f \in C^\infty(X, Y) \mid j^k f \pitchfork W\}.$$

Then T_W is a residual subset of $C^\infty(X, Y)$ in the C^∞ topology. Moreover, if W is closed, then T_W is open.

Proof. We need to show that T_W is the countable intersection of open dense subsets. To construct the sets which will go into this countable intersection, we first choose a countable covering of W by open subsets $W_1, W_2, \ldots$ in W such that each W_r satisfies

(a) the closure of W_r in $J^k(X, Y)$ is contained in W,

(b) $\overline{W}_r$ is compact,

(c) there exist coordinate nbhds U_r in X and V_r in Y such that $\pi(\overline{W}_r) \subset U_r \times V_r$ where $\pi\colon J^k(X, Y) \to X \times Y$ is the projection mapping, and

(d) $\overline{U}_r$ is compact.

This choice is possible since around each point w in W, we may choose an open set W_w satisfying (a), (b), (c), and (d), since W is a submanifold of $J^k(X, Y)$. Since W is second countable we may extract a countable subcovering from $\{W_w\}_{w \in W}$. Let

$$T_{W_r} = \{f \in C^\infty(X, Y) \mid j^k(f) \pitchfork W \text{ on } \overline{W}_r\}.$$

It is clear that $T_W = \bigcap_{r=1}^\infty T_{W_r}$. Thus the proof reduces to showing that each T_{W_r} is open and dense in $C^\infty(X, Y)$.

Define $T_r = \{g \in C^\infty(X, J^k(X, Y)) \mid g \pitchfork W \text{ on } \overline{W}_r\}$. The proof of Proposition 4.5 can be easily adapted to show that T_r is open since $\overline{W}_r$ is closed and contained in W. Since $j^k\colon C^\infty(X, Y) \to C^\infty(X, J^k(X, Y))$ is continuous (by Proposition 3.4), $T_{W_r} = (j^k)^{-1}(T_r)$ is open.

We now continue with the harder part of the Theorem, that is, to show that T_{W_r} is dense. Choose charts $\psi: U_r \to \mathbf{R}^n$ and $\eta: V_r \to \mathbf{R}^m$ and smooth functions $\rho: \mathbf{R}^n \to [0, 1] \subset \mathbf{R}$ and $\rho': \mathbf{R}^m \to [0, 1] \subset \mathbf{R}$ such that

$$\rho = \begin{cases} 1 & \text{on a nbhd of } \psi \cdot \alpha(\overline{W}_r) \\ 0 & \text{off } \psi(U_r) \end{cases}$$

and

$$\rho' = \begin{cases} 1 & \text{on a nbhd of } \eta \cdot \beta(\overline{W}_r) \\ 0 & \text{off } \eta(V_r) \end{cases}$$

where $n = \dim X$, $m = \dim Y$, α is the source map, and β is the target mapping. The choice of ρ and ρ' are possible since $\overline{W}_r$ is compact.

We will use Corollary 4.7 to show that f can be perturbed slightly so that $j^k f$ is transverse to W on $\overline{W}_r$. The perturbation will be accomplished locally using the data defined in the previous paragraph. Let B' be the space of polynomial mappings of $\mathbf{R}^n \to \mathbf{R}^m$ of degree k. For b in B', define $g_b: X \to Y$ by

$$g_b(x) = \begin{cases} f(x) & \text{if } x \notin U_r \text{ or } f(x) \notin V_r \\ \eta^{-1}(\rho(\psi(x))\rho'(\eta f(x))b(\psi(x)) + \eta f(x)) & \text{otherwise.} \end{cases}$$

The choice of ρ and ρ' guarantees that g_b is a smooth function from X to Y and is just a polynomial perturbation of f done locally and smoothed out so that it is equal to f off the domain of interest. Define $G: X \times B' \to Y$ by $G(x, b) = g_b(x)$. By inspection of the formula defining g_b, one sees that G is smooth.

Define $\Phi(x, b) = j^k g_b(x)$. In order to apply Corollary 4.7, we need to know that $\Phi \pitchfork W$ on $\overline{W}_r$. Now it is not necessarily true that the transversality condition holds on all of $X \times B'$, but we will find an open nbhd B of 0 in B' so that $\Phi: X \times B \to J^k(X, Y)$ will $\pitchfork W$ on some nbhd of $\overline{W}_r$. We can then apply the Corollary on $X \times B$ rather than $X \times B'$. Assuming that this transversality condition holds, then given $f: X \to Y$ we can find a sequence $b_1, b_2, \ldots$ in B converging uniformly to 0 so that $j^k g_{b_i} \pitchfork W$ on $\overline{W}_r$. Since $g_0 = f$ and $g_b = f$ off U_r, $\operatorname{Lim}_{i \to \infty} g_{b_i} = f$ in $C^\infty(X, Y)$ and T_{W_r} is dense in $C^\infty(X, Y)$.

Let $\varepsilon = \frac{1}{2}\min\{d(\operatorname{supp} \rho', \mathbf{R}^m - \eta(V_r)), d(\eta\beta(\overline{W}_r), (\rho')^{-1}[0, 1))$. Set $B = \{b \in B' \mid b(x) < \varepsilon \, \forall x \in \operatorname{supp} \rho\}$. So B is an open nbhd of 0 in B'. Suppose that (x, b) is in $X \times B$ and that $\Phi(x, b)$ is in $\overline{W}_r$, then we will show that $\Phi: X \times B \to J^k(X, Y)$ is locally a diffeomorphism. If true, Φ would satisfy any transversality condition. Since $\Phi(x, b) \in \overline{W}_r$, we have that $x \in \alpha(\overline{W}_r)$ and $g_b(x) \in \beta(\overline{W}_r)$. Then $s = d(\eta f(x), \eta g_b(x)) < \varepsilon$ since

$$\eta g_b(x) = \rho(\psi(x))\rho'(\eta f(x))b(\psi(x)) + \eta f(x).$$

So

$$s = |\rho\psi(x)\rho'\eta f(x) b\psi(x)| \begin{cases} \le |b\psi(x)| < \varepsilon & \text{if } \psi(x) \in \operatorname{supp} \rho \\ = 0 & \text{if } \psi(x) \notin \operatorname{supp} \rho. \end{cases}$$

Using the definition of ε we observe that $\eta f(x)$ is in $\operatorname{Int}(\rho')^{-1}(1)$ since $g_b(x)$ is in $\beta(\overline{W}_r)$. Recall that $\rho \equiv 1$ on a nbhd of $\psi_\alpha(\overline{W}_r)$ so that $\eta g_b(x) = b\psi(x) + \eta f(x)$ and that $g_b(x') = \eta^{-1}(b\psi + \eta f)(x')$ for all x' in a nbhd of x. Clearly this argument also holds for all b' in some nbhd of b in B. It now follows that

$\Phi: X \times B \to J^k(X, Y)$ is locally a diffeomorphism near (x, b). For let σ be in $J^k(X, Y)$ near $\Phi(x, b)$, let $x' = \alpha(\sigma)$, and let b' be the unique polynomial mapping of degree $\leq k$ such that $\sigma = j^k(\eta^{-1}(b' \cdot \psi + \eta \cdot f))(x')$. Then $\sigma \mapsto (x', b')$ is a smooth mapping and is the inverse of Φ. □

Corollary 4.10. *For each integer i, let W_i be a submanifold of $J^{k_i}(X, Y)$. Then the set of smooth mappings $f: X \to Y$ for which $j^{k_i}f \pitchfork W_i$ is dense in $C^\infty(X, Y)$. If the number of W_i's is finite and each W_i is closed, then this set is open as well.*

Proof. Follows immediately from Theorem 4.9 and the fact that $C^\infty(X, Y)$ is a Baire space. □

Corollary 4.11. *Let X and Y be smooth manifolds and W a submanifold of $J^k(X, Y)$ such that $\alpha(\overline{W})$ is contained in an open subset U of X. Let $f: X \to Y$ be a smooth mapping and V an open nbhd of f in $C^\infty(X, Y)$. Then there exists a smooth mapping g in V such that $j^k g \pitchfork W$ and $g = f$ off U.*

Proof. This is really a corollary to the proof of the Thom Transversality Theorem. Under the assumptions of the corollary it is clear that for each W_r, $\alpha(\overline{W}_r) \subset U$. Thus we can choose U_r so that $U_r \subset U$. Then note that the perturbation $g_b = f$ off U_r, and thus off U, for each b in B. So the constructed transversal mapping does in fact equal f off U. □

Corollary 4.12. (*Elementary Transversality Theorem*). *Let X and Y be smooth manifolds with W a submanifold of Y. Then*

(a) *the set of smooth mappings of X to Y which intersect W transversally is dense in $C^\infty(X, Y)$ and if W is closed, then this set is also open.*

(b) *let U_1 and U_2 be open subsets of X with $\overline{U}_1 \subset U_2$. Let f be in $C^\infty(X, Y)$ and V be an open nbhd of f in $C^\infty(X, Y)$. Then there is a smooth mapping $g: X \to Y$ in V such that $g = f$ on U_1 and $g \pitchfork W$ off U_2.*

Proof.

(a) Note that $J^0(X, Y) = X \times Y$ and $j^0 f(x) = (x, f(x))$. The projection $\beta: X \times Y \to Y$ is a submersion so $\beta^{-1}(W)$ is a submanifold of $X \times Y$. If $j^0 f \pitchfork \beta^{-1}(W)$ at x, then $f \pitchfork W$ at x. For either $j^0 f(x) \notin \beta^{-1}(W)$ in which case $f(x) \notin W$ or $j^0 f(x) \in \beta^{-1}(W)$ and

$$T_{(x,f(x))}\beta^{-1}(W) + (dj^0 f)_x T_x X = T_{(x,f(x))}(X \times Y).$$

Apply $(d\beta)_{(x,f(x))}$ to each side to obtain

$$T_{f(x)}W + (df)_x(T_x X) = T_{f(x)}Y.$$

Thus $f \pitchfork W$ at x. Since the set of transversal mappings to W contains the set $\{f \in C^\infty(X, Y) \mid j^0 f \pitchfork \beta^{-1}(W)\}$ which is dense by Theorem 4.9, we are done. Note that the last part of (a) is just Proposition 4.5.

For (b) note that $W' = \beta^{-1}(W) \cap (X \times Y - \alpha^{-1}(\overline{U}_2))$ is a submanifold of $X \times Y$ since $X \times Y - \alpha^{-1}(\overline{U}_2)$ is an open subset of $X \times Y$. Also $\alpha(W')$ is contained in the open set $X - \overline{U}_1$ so by Corollary 4.11, there exists $g: X \to Y$ in V such that $g = f$ on U_1 and $j^0 g \pitchfork W'$. This latter condition is the same as $j^0 g \pitchfork \beta^{-1}(W)$ off U_2. Thus $g \pitchfork W$ off U_2 as in (a). □

We now present a generalization of transversality in jet spaces to transversality in multijet spaces which is useful for studying the injectivity, or alternately, the self-intersections of smooth mappings.

Let X and Y be smooth manifolds. Define

$$X^s = X \times \cdots \times X \quad (s\text{-times}) \text{ and}$$
$$X^{(s)} = \{(x_1, \ldots, x_s) \in X^s \mid x_i \neq x_j \quad \text{for} \quad 1 \leq i < j \leq s\}.$$

Let $\alpha : J^k(X, Y) \to X$ be the source map. Define $\alpha^s : J^k(X, Y)^s \to X^s$ in the obvious fashion. Then $J_s^k(X, Y) = (\alpha^s)^{-1}(X^{(s)})$ is the *s-fold k-jet bundle*. A *multijet bundle* is some s-fold k-jet bundle. $X^{(s)}$ is a manifold since it is an open subset of X^s. Thus $J_s^k(X, Y)$ is an open subset of $J^k(X, Y)^s$ and is also a smooth manifold. Now let $f: X \to Y$ be smooth. Then we can define $j_s^k f: X^{(s)} \to J_s^k(X, Y)$ in the natural way; i.e.,

$$j_s^k f(x_1, \ldots, x_s) = (j^k f(x_1), \ldots, j^k f(x_s)).$$

Theorem 4.13. (*Multijet Transversality Theorem*). *Let X and Y be smooth manifolds with W a submanifold of $J_s^k(X, Y)$. Let*

$$T_W = \{f \in C^\infty(X, Y) \mid j_s^k f \pitchfork W\}.$$

Then T_W is a residual subset of $C^\infty(X, Y)$. Moreover if W is compact, then T_W is open.

First we need a lemma.

Lemma 4.14. *Let W be a submanifold of $J_s^k(X, Y)$ with W' a compact subset of W. Then $T_{W'} = \{f \in C^\infty(X, Y) \mid j_s^k f \pitchfork W \text{ on } W'\}$ is open in $C^\infty(X, Y)$.*

Proof. Let $x = (x_1, \ldots, x_s)$ be in $X^{(s)}$. Choose disjoint open nbhds U_i of x_i in X. Then choose open nbhds V_i of x_i with $\overline{V}_i \subset U_i$. Let $U^x = \times_{i=1}^s U_i$ and $V^x = \times_{i=1}^s \overline{V}_i$. Note that V^x is a closed subset of X^s. Let $T_x = \{f \in C^\infty(X, Y) \mid j_s^k f \pitchfork W \text{ on } W' \cap (\alpha^s)^{-1}(V^x)\}$ where $\alpha : J^k(X, Y) \to X$ is the source map. Suppose T_x is an open subset of $C^\infty(X, Y)$. Since the collection $\{\text{Int } V^x\}$ where x is in $\alpha^{(s)}(W')$ is an open covering of $\alpha^{(s)}(W')$ and $\alpha^{(s)}(W')$ is compact, we may extract a finite subcover indexed by $x^1, \ldots, x^l$. Noting that $T_{W'} = \bigcap_{i=1}^l T_{x^i}$ we see that $T_{W'}$ is open.

To show that T_x is open we consider the mapping $A : C^\infty(X, Y) \to C^\infty(X^s, J^k(X, Y)^s)$ defined by $(j^k f)^s$ and the set

$$T = \{g \in C^\infty(X^s, J^k(X, Y)^s) \mid g \pitchfork W \text{ on } W' \cap (\alpha^s)^{-1}(V^x)\}.$$

Since V^x is bounded away from the generalized diagonal in X^s by U^x we see that $j_s^k f \pitchfork W$ on $W' \cap (\alpha^{(s)})^{-1}(V^x)$ iff $(j^k f)^s \pitchfork W$ on $W' \cap (\alpha^s)^{-1}(V^x)$. Thus $T_x = A^{-1}(T)$ and it is sufficient to show that T is open and that A is continuous. Since $W' = (\alpha^s)^{-1}(V^x)$ is closed we may apply Proposition 4.5 (or more precisely the adaptation of this Proposition mentioned in Theorem 4.9) to conclude that T is open. Next we observe that $A = \delta_s \cdot j^k$ where $\delta_s : C^\infty(X, Z) \to C^\infty(X^s, Z^s)$ is given by $f \mapsto f^s$. Now j^k is continuous by Proposition 3.4 and δ_s is continuous by Proposition 3.10, so A is continuous. □

Remark. If W is a submanifold of $J_s^k(X, Y)$ such that $\alpha^{(s)}(W)$ is a compact subset of $X^{(s)}$, then T_W is open in $C^\infty(X, Y)$.

Proof of Theorem 4.13. The moreover part follows immediately from Lemma 4.14. The idea of the main part of this Theorem is the same as that for the Thom Transversality Theorem. Thus we shall just indicate what changes need to be made in the proof of Theorem 4.9 in order to prove this Theorem.

Choose open sets W_r in W satisfying (a), (b), and in place of (c) and (d)
(c′) there exist coordinate patches $U_{r,1}, \ldots, U_{r,s}$ in X and $V_{r,1}, \ldots, V_{r,s}$ in Y such that $\{\overline{U}_{r,i}\}_{i=1}^s$ are mutually disjoint and $\pi_s(\overline{W}_r) \subset U_{r,1} \times \cdots \times U_{r,s} \times V_{r,1} \times \cdots \times V_{r,s}$ where $\pi_s : J_s^k(X, Y) \to X^{(s)} \times Y^s$ (not $Y^{(s)}$) is the obvious projection, and
(d′) $\overline{U}_{r,i}$ is compact for $1 \leq i \leq s$.

Let

$$T_{W_r} = \{f \in C^\infty(X, Y) \mid f \pitchfork W \text{ on } \overline{W}_r\}.$$

Since $T_W = \bigcap_{r=1}^\infty T_{W_r}$, the proof reduces to showing that each T_{W_r} is open and dense. Since $\overline{W}_r$ is compact we apply Lemma 4.14 to show that T_{W_r} is open.

To prove the density of T_{W_r}, we wish to make a polynomial perturbation on each $U_{r,i}$ which is smoothed to equal f off of $U_{r,i}$. The only technical point is that these perturbations be done simultaneously. Choose charts $\psi_i : U_{r,i} \to \mathbf{R}^n$ and $\eta_i : V_{r,i} \to \mathbf{R}^m$ and smooth functions $\rho_i : R^n \to [0, 1]$ and $\rho_i' : R^m \to [0, 1]$ such that

$$\rho_i = \begin{cases} 1 & \text{on a nbhd of } \psi_i \cdot \alpha_i(\overline{W}_r) \\ 0 & \text{off } \psi_i(U_{r,i}) \end{cases}$$

and

$$\rho_i' = \begin{cases} 1 & \text{on a nbhd of } \eta_i \cdot \beta_i(\overline{W}_r) \\ 0 & \text{off } \eta_i(V_{r,i}) \end{cases}$$

where α_i and β_i are the source and target maps onto the ith component of X and Y in $X^{(s)}$ and Y^s, respectively. Let $b = (b_1, \ldots, b_s) \in (B')^s$. Then define $g_b : X \to Y$ by

$$g_b(x) = \begin{cases} \eta_i^{-1}(\rho_i \cdot \psi(x) \rho_i' \cdot \eta \cdot f(x) b_i \cdot \psi(x) + \eta_i f(x)) & \text{if } x \in U_{r,i} \text{ and } f(x) \in V_{r,i} \\ f(x) & \text{otherwise} \end{cases}$$

where $b = (b_1, \ldots, b_s)$. Again g_b is a smooth function. Let

$$\varepsilon_i = \tfrac{1}{2} \min\{ d(\operatorname{supp} \rho_i', \mathbf{R}^m - \eta_i(V_{r,i})), d(\eta_i \beta_i(\overline{W}_r), (\rho_i')^{-1}[0,1))$$

Set $B_i = \{b_i \in B' : |b_i(\psi(x))| < \varepsilon_i \; \forall x \in \operatorname{supp} \rho_i\}$ and set $B = B_1 \times \cdots \times B_s$. Define $G(x, b) = g_b(x)$. Then $G : X \times B \to Y$ is a smooth mapping. Since $U_{r,i}$ was chosen so that $\{\overline{U}_{r,i}\}_{i=1}^s$ is disjoint we can show that $H : X^{(s)} \times B \to J_s^k(X, Y)$ is locally a diffeomorphism where $H(\bar{x}, b) = j_s^k g_b(\bar{x})$ where $\bar{x} = (x_1, \ldots, x_s)$. This is similar to the proof in Theorem 4.9.

To complete the proof we invoke the following Lemma.

Lemma 4.15. *Let X, Y, and B be smooth manifolds. Let W be a submanifold of $J_s^k(X, Y)$. Let W' be an open subset of W such that $\overline{W}' \subset W$. Let $G: X \times B \to Y$ be a smooth mapping. Define $H: X^{(s)} \times B \to J_s^k(X, Y)$ by $H(x, b) = j_s^k(G_b)(x)$ and assume that $H \pitchfork W$ on $\overline{W}'$. Then the set*

$$\{b \in B \mid j_s^k G_b \pitchfork W \text{ on } \overline{W}'\}$$

is dense in B.

Proof. Let $X = X^{(s)}$, $Z = J_s^k(X, Y)$ and $j: B \to C^\infty(X^{(s)}, J_s^k(X, Y))$ be given by $j(b) = j_s^k G_b$. Apply Lemma 4.6. □

Exercises

(1) (a) Let X and Y be smooth manifolds with W a submanifold of Y. Assume that $\dim X = \operatorname{codim} W$. Let p be in X and let $f: X \to Y$ be smooth. Assume that $f(p)$ in W and $f \pitchfork W$ at p. Then there exists a nbhd N of f in $C^\infty(X, Y)$ (in the C^1 topology) and an open nbhd U of p in X such that if g is in N, then $g^{-1}(W) \cap U$ consists of one point q and $g \pitchfork W$ at q. (Hint: Use Lemma 4.3 and note the similarity with Theorem 4.4.)

(b) Assume that X is compact. Let $f': X \to Y \pitchfork W$. Show that there is an open nbhd N of f in $C^\infty(X, Y)$ such that the number of points in $f^{-1}(W)$ is equal to the number of points in $g^{-1}(W)$ for any g in N.

(2) Let $f: X \to X$ be smooth with p in X a fixed point for f.

Definition: p is a *non-degenerate fixed point* iff $(df)_p: T_pX \to T_pX$ does not have 1 as an eigenvalue.

Let Diff (X) be the group of smooth diffeomorphisms on X and give Diff (X) the relative topology as a subset of $C^\infty(X, X)$.

(a) Show that $\{f \in \text{Diff}\,(X) \mid \text{fixed points of } f \text{ are nondegenerate}\}$ is open and dense in Diff (X), and

(b) Show that nondegenerate fixed points are isolated.

Hint: Consider what it means for $j^0 f \pitchfork \Delta X$ at p in X where ΔX is the diagonal in $X \times X = J^0(X, X)$.

(3) Let $\{W_\alpha\}_{\alpha \in I}$ be a family of submanifolds of $J^k(X, Y)$ where I is some index set. Let

$$T_W = \{f \in C^\infty(X, Y) \mid \forall \alpha \in I, j^k f \pitchfork W_\alpha\}.$$

Suppose that $\bigcup_{\alpha \in I} W_\alpha$ is closed in $J^k(X, Y)$. Then show that T_W is an open subset of $C^\infty(X, Y)$.

§5. The Whitney Embedding Theorem

Let X and Y be smooth manifolds. We want to show that if $\dim Y$ is large enough relative to $\dim X$, then the set of immersions of X into Y is dense in $C^\infty(X, Y)$. The idea of this proof of the Whitney Immersion Theorem will be to translate the fact that f is an immersion into (a finite number of) transversality conditions.

Let σ be in $J^1(X, Y)$; then σ defines a unique linear mapping of $T_pX \to T_qY$ where p is the source of σ and q is the target of σ. Let f be a representative of σ in $C^\infty(X, Y)$. Then $(df)_p$ is that linear mapping. Define rank σ = rank $(df)_p$ and corank $\sigma = l -$ rank σ where $l = \min(\dim X, \dim Y)$. Let

$$S_r = \{\sigma \in J^1(X, Y) \mid \operatorname{corank} \sigma = r\}.$$

We will show that S_r is a submanifold of $J^1(X, Y)$. The significance of the submanifolds S_r is illustrated by the following obvious lemma.

Lemma 5.1. *$f: X \to Y$ is an immersion iff $j^1f(X) \cap (\bigcup_{r \neq 0} S_r) = \varnothing$.*

One also observes the following:

Lemma 5.2. *Let S be an $m \times n$ matrix where $S = \begin{pmatrix} A & B \\ C & D \end{pmatrix}$ where A is a $k \times k$ invertible square matrix. Then* rank $S = k$ *iff* $D - CA^{-1}B = 0$.

Proof. The matrix

$$T = \begin{pmatrix} I_k & 0 \\ -CA^{-1} & I_{m-k} \end{pmatrix}$$

is an $m \times m$ invertible matrix. So

$$\operatorname{rank} S = \operatorname{rank} TS = \operatorname{rank} \begin{pmatrix} A & B \\ 0 & D - CA^{-1}B \end{pmatrix}.$$

Clearly this latter matrix has rank $= k$ iff $D - CA^{-1}B = 0$. □

Let V and W be vector spaces of dimension n and m respectively. Let $q = \min\{n, m\}$. Let $S: V \to W$ be linear, then define corank $(S) = q -$ rank (S). Define $L^r(V, W) = \{S \in \operatorname{Hom}(V, W) \mid \operatorname{corank} S = r\}$.

Proposition 5.3. *$L^r(V, W)$ is a submanifold of* Hom (V, W) *with* codim $L^r(V, W) = (m - q + r)(n - q + r)$.

Proof. Let S be in $L^r(V, W)$ and let $k = q - r = \operatorname{rank}(S)$. Choose bases of V and W so that the matrix of $S = \begin{pmatrix} A & B \\ C & D \end{pmatrix}$ where A is a $k \times k$ invertible matrix. Choose an open nbhd U of S in Hom (V, W) so that if S' is in U and $S' = \begin{pmatrix} A' & B' \\ C' & D' \end{pmatrix}$ then A' is a $k \times k$ invertible matrix. Consider the smooth mapping $f: U \to \operatorname{Hom}(\mathbf{R}^{n-k}, \mathbf{R}^{m-k})$ given by $f(S') = D' - C'(A')^{-1}B'$. f is a submersion since if we fix A, B, C, then

$$g: \operatorname{Hom}(\mathbf{R}^{n-k}, \mathbf{R}^{m-k}) \to \operatorname{Hom}(\mathbf{R}^{n-k}, \mathbf{R}^{m-k})$$

given by $g(D) = f\begin{pmatrix} A & B \\ C & D \end{pmatrix} = D - CA^{-1}B$ is a diffeomorphism. (In particular $(dg)_D$ = identity, so $(df)_S$ is surjective.) By Lemma 5.2 $f^{-1}(0) = L^r(V, W) \cap U$ which is a submanifold since f is a submersion. Moreover

$$\operatorname{codim} L^r(V, W) = \dim \operatorname{Hom}(\mathbf{R}^{n-k}, \mathbf{R}^{m-k}) = (n - k)(m - k). \quad \square$$

Theorem 5.4. *S_r is a submanifold of $J^1(X, Y)$ with* codim $S_r = (n - q + r)(m - q + r)$. *In fact, S_r is a subfiber-bundle of $J^1(X, Y)$ with fiber $L^r(\mathbf{R}^n, \mathbf{R}^m)$.*

Proof. Let $U \subset X$ and $V \subset Y$ be coordinate charts. Then $J^1(X, Y)_{U \times V} \cong U \times V \times \mathrm{Hom}(\mathbf{R}^n, \mathbf{R}^m)$ and under this isomorphism $S_r' \cong U \times V \times L^r(\mathbf{R}^n, \mathbf{R}^m)$ where $S_r' = S_r \cap J^1(X, Y)_{U \times V}$. Applying Proposition 5.3, S_r is then a submanifold. ▯

Let $\mathrm{Im}(X, Y)$ be the subset of immersions in $C^\infty(X, Y)$.

Lemma 5.5. *$\mathrm{Im}(X, Y)$ is an open subset of $C^\infty(X, Y)$.*

Proof. S_0 is an open subset of $J^1(X, Y)$. $M(S_0) = \mathrm{Im}(X, Y)$ by Lemma 5.1. ▯

Theorem 5.6. (*Whitney Immersion Theorem*). *Let X and Y be smooth manifolds with* $\dim Y \geq 2 \cdot \dim X$. *Then* $\mathrm{Im}(X, Y)$ *is an open dense subset of $C^\infty(X, Y)$ (in the Whitney C^∞ topology).*

Proof. Codim $S_r = (n - q + r)(m - q + r)$ where $m = \dim Y$, $n = \dim X$, and $q = \min(m, n) = n$. Thus if $r \geq 1$

$$\operatorname{codim} S_r = r(m - n + r) \geq r(n + r) \geq n + 1.$$

For these relative dimensions $j^1 f \pitchfork S_r$ iff $j^1 f(X) \cap S_r = \varnothing$ when $r \geq 1$. Thus the Thom Transversality Theorem (4.9, 4.10) and Lemma 5.1 imply the result. ▯

Theorem 5.7. (*Whitney* 1:1 *Immersion Theorem*). *Let X and Y be smooth manifolds. Assume that* $\dim Y \geq 2 \dim X + 1$. *Then the set of* 1:1 *immersions of X into Y is a residual set and hence dense in $C^\infty(X, Y)$.*

Proof. Since the set of immersions is open and dense, we need only show that the set of 1:1 mappings of X into Y is a residual set. First note that $f: X \to Y$ is 1:1 iff $j_2{}^0 f: X^{(2)} \to J_2{}^0(X, Y)$ does not intersect W where $W = (\beta^2)^{-1}(\Delta Y)$. Note that W is a submanifold since $\beta^2 : J_2{}^0(X, Y) \to Y^2$ is a submersion. Since codim W = codim $\Delta Y = \dim Y > 2 \cdot \dim X = \dim X^{(2)}$ we have that $j_2{}^0 f \pitchfork W$ iff $j_2{}^0 f(X^{(2)}) \cap W = \varnothing$. So f is 1:1 iff $j_2{}^0 f \pitchfork W$ and the result follows from the multijet transversality theorem. ▯

Proposition 5.8. *Let X and Y be smooth manifolds with X compact. Then the set of* 1:1 *immersions is open in $C^\infty(X, Y)$.*

Lemma A. *Let $f: X \to Y$ be smooth and an immersion at p in X. Then there is a nbhd U_p of p in X and a nbhd of W_f of f in $C^\infty(X, Y)$ so that if g is in W_f, then $g|U_p$ is a* 1:1 *immersion.*

Proof. Let U be a nbhd of p and $\phi : U \to \mathbf{R}^n$ a chart. Assume that U is small enough so that there exists a chart nbhd V of $f(p)$ with chart $\psi : V \to \mathbf{R}^m$ such that $f(U) \subset V$ and $\psi \cdot f \cdot \phi^{-1}$ is a linear injection. This is possible since f is an immersion at p. Choose an open nbhd U_p of p so that $\overline{U}_p$ is compact and contained in U. Suppose that $g: X \to Y$ so that $g(\overline{U}_p) \subset V$. Then define $\tilde{g} = \psi \cdot g \cdot \phi^{-1} : \phi(\overline{U}_p) \to \mathbf{R}^m$. Let $p' = \phi(p)$ and let $M = \inf_{|x|=1} |(d\tilde{f})_{p'}(x)|$. Note that $M > 0$ since $(d\tilde{f})_{p'}$ is 1:1. Let

$$W_f = \left\{ g \in C^\infty(X, Y) \mid g(\overline{U}_p) \subset V \text{ and } |(d\tilde{f})_{p'} - (d\tilde{g})_x| < \frac{M}{2} \, \forall x \in \phi(\overline{U}_p) \right\}.$$

W_f is an open nbhd in $C^\infty(X, Y)$ since the first condition is a C^0 open one and the second one is C^1 open. Using the triangle inequality in R^m we have that

$$|(d\tilde{f})_{p'}(x_1) - (d\tilde{f})_{p'}(x_2)| \le |\tilde{g}(x_1) - \tilde{g}(x_2)| + |((d\tilde{f})_{p'} - \tilde{g})(x_1) - ((d\tilde{f})_{p'} - \tilde{g})(x_2)|.$$

Now if g is in W_f, then

$$|((d\tilde{f})_{p'} - \tilde{g})(x_1) - ((d\tilde{f})_p - \tilde{g})(x_2)| \le (d((d\tilde{f})_{p'} - \tilde{g}))_x| \, |x_1 - x_2| \le \frac{M}{2} |x_1 - x_2|$$

where the first inequality is given by the Mean Value Theorem for some x in $\phi(\overline{U}_p)$. Thus for g in W_f,

$$|\tilde{g}(x_1) - \tilde{g}(x_2)| \ge M|x_1 - x_2| - \frac{M}{2}|x_1 - x_2| = \frac{M}{2}|x_1 - x_2|$$

and g is 1:1. Since the set of immersions of X into Y is open we are done. $\square$

Proof of Proposition 5.8. Let $f: X \to Y$ be a 1:1 immersion. First we show that there is a nbhd Z of ΔX in $X \times X$ and an open nbhd W' of f in $C^\infty(X, Y)$ such that for every g in W' g is an immersion and $g^2(Z - \Delta X) \cap \Delta Y = \varnothing$ where $g^2 : X^2 \to Y^2$ is given by $g \times g$. For every p in X, choose U_p and $W_f{}^p$ as guaranteed by Lemma A. Since X is compact we can choose $p_1, \ldots, p_s$ so that $U_{p_1}, \ldots, U_{p_s}$ cover X. Let $W' = \bigcap_{i=1}^s W_f{}^{p_i}$ and $Z = \bigcup_{i=1}^s (U_{p_i} \times U_{p_i})$. Let g be in W'. Then $g|U_{p_i}$ is an immersion for each i, so g is an immersion. Also if (p, q) is in $Z - \Delta X$ then $(p, q) \in U_{p_i} \times U_{p_i}$ for some i and thus $g(p) \neq g(q)$ since $g|U_{p_i}$ is 1:1.

Now suppose that there does not exist an open nbhd of f consisting entirely of 1:1 immersions. Since the set of immersions is open, there does not exist a nbhd of f consisting of 1:1 mappings. Since X is compact and $C^\infty(X, Y)$ satisfies the first axiom of countability, there is a sequence of functions f_n converging to f each of which is not 1:1. We may assume that each f_n is in W'. Let (p_n, q_n) be in $X \times X - \Delta X$ such that $f_n(p_n) = f_n(q_n)$. The pair (p_n, q_n) exists since f_n is not 1:1. By the choice of W', (p_n, q_n) is not in Z. $X \times X - Z$ is compact so we may assume that $\operatorname{Lim}_{n\to\infty} (p_n, q_n)$ exists and is (p, q). Since $(p, q) \notin Z$, $p \neq q$. Also $\operatorname{Lim}_{n\to\infty} f_n(p_n) = f(p)$ and $\operatorname{Lim}_{n\to\infty} f_n(q_n) = f(q)$. Since with an appropriate choice of metric on Y f_n converges uniformly to f, $f(p) = f(q)$ contradicting the fact that f is 1:1. $\square$

Proposition 5.9. (*Whitney Embedding Theorem*). *Let X be a smooth manifold of dimension n. Then there exists an embedding f of X into $\mathbf{R}^{2n+1}$.*

Proof. By Theorem 5.7 the set of 1:1 immersions of $X \to \mathbf{R}^{2n+1}$ is dense. We showed in the first paragraph of Lemma 3.7 that the image of a closed set under a continuous proper mapping is closed. Thus if there is a 1:1 proper immersion of $X \to \mathbf{R}^{2n+1}$, the result is proved. The following lemma is thus sufficient.

Lemma 5.10. *Let X be a smooth manifold. Then the space of smooth proper mappings of $X \to \mathbf{R}^m$ is nonempty and open in $C^\infty(X, \mathbf{R}^m)$ (in the C^0 topology).*

Proof. By I,5.11 there exists a proper mapping of $X \to \mathbf{R}$. Compose this mapping with any linear injection of $\mathbf{R} \to \mathbf{R}^m$ to obtain a proper mapping of $X \to \mathbf{R}^m$. To show that the set of proper mappings is open, let $f: X \to \mathbf{R}^m$ be proper and let $V_x = \{y \in \mathbf{R}^m \mid d(y, f(x)) < 1\}$. Let $V = \bigcup_{x \in X} \{x\} \times V_x$ in $J^0(X, \mathbf{R}^m) = X \times \mathbf{R}^m$. The continuity of f guarantees that V is open and clearly f is in $M(V)$. Now let g be in $M(V)$, then g is proper for $g^{-1}(\bar{B}_r(y)) \subset f^{-1}(\bar{B}_{r+1}(y))$. Since f is proper $g^{-1}(\bar{B}_r(y))$ is a closed subset of a compact set and thus compact. $\square$

§6. Morse Theory

In the last section we used transversality to analyze what "most" mappings of a manifold X into some high dimensional manifold Y look like. We now use the same technique to analyze the other extreme: that is, to determine the structure of most mappings of X into $\mathbf{R}$. In particular, the Whitney Theorem shows that a generic mapping is as nice as possible in the differentiable sense, namely, the Jacobian always has the maximum rank possible. With Morse Theory we will show that, in general, the Jacobian is not of maximal rank; in other words, the mapping has singularities, but these singularities must be of a particular type.

Since the dimension of $\mathbf{R}$ is 1, the only non-empty submanifolds of $J^1(X, R)$ of the type S_r (see §5) are S_0 and S_1. Thus p in X is a singularity (or critical point) for $f: X \to R$ iff $j^1f(p)$ is in S_1.

Definition 6.1.
(a) *Let p be a singularity of $f: X \to \mathbf{R}$. p is* nondegenerate *if $j^1f \pitchfork S_1$ at p.*
(b) *f is a* Morse function *if all of the singularities of f are nondegenerate.*

Theorem 6.2. *Let X be a smooth manifold. Then the set of Morse functions is an open dense subset of $C^\infty(X, R)$.*

Proof. Apply the Thom Transversality Theorem. $\square$

Proposition 6.3. *Let $f: X \to \mathbf{R}$ be a smooth function with a nondegenerate critical point p. Then there exists a nbhd V of p in X such that no other critical points of f are in V, i.e., nondegenerate critical points are isolated.*

Proof. Note that codim S_1 = dim X (see 5.4) and apply Theorem 4.4 (or, more precisely, Exercise (1)(a) of §4). $\square$

Now that we know that nondegenerate critical points are isolated and that they represent a generic situation for functions, we wish to analyze their character locally.

Proposition 6.4. *Let U be an open subset of $\mathbf{R}^n$ and $f: U \to \mathbf{R}$ be smooth. Then f has a nondegenerate critical point at p iff the* Hessian of f at p = $((\partial^2 f/\partial x_i\, \partial x_j)(p))$ *is nonsingular.*

Proof. $J^1(U, R) \cong U \times \mathbf{R} \times \operatorname{Hom}(\mathbf{R}^n, \mathbf{R})$. Note that the projection $\pi: J^1(U, \mathbf{R}) \to \operatorname{Hom}(\mathbf{R}^n, \mathbf{R})$ is a submersion and $\pi^{-1}(0) = S_1$. Now apply Lemma 4.3; that is, $j^1 f \pitchfork S_1$ at p iff $\pi \cdot j^1 f$ is a submersion at p. But $\pi \cdot j^1 f$ at x is

$$(df)_x = \left(\frac{\partial f}{\partial x_1}(x), \ldots, \frac{\partial f}{\partial x_n}(x)\right)$$

in the standard coordinates on $\operatorname{Hom}(\mathbf{R}^n, \mathbf{R})$. Thus $\pi \cdot j^1 f$ is a submersion at p iff the mapping of $\mathbf{R}^n \to \mathbf{R}^n$ given by

$$x \to \left(\frac{\partial f}{\partial x_1}(x), \ldots, \frac{\partial f}{\partial x_n}(x)\right)$$

is a submersion at p iff $((\partial^2 f/\partial x_i\, \partial x_j)(p))$ is nonsingular. □

We shall now give an invariant definition for the Hessian of a function at a critical point. To do so we need the concept of intrinsic derivative.

Let E be a vector bundle over X. E always has a distinguished global section called the zero section $i_0: X \to E$ which is defined by $i_0(p) = 0$ in E_p. Let $\pi: E \to X$ be the projection mapping. Then it is clear that

$$\operatorname{Ker}(d\pi)_{i_0(p)} \cap (di_0)_p(T_p X) = \{0\}$$

and that $T_{i_0(p)} E_p = \operatorname{Ker}(d\pi)_{i_0(p)}$ so that $T_{i_0(p)} E \cong T_{i_0(p)} E_p \oplus T_p X$. Now $T_{i_0(p)} E_p \cong E_p$ in a canonical way since E_p is a vector space. Finally let $\sigma: T_{i_0(p)} E \to E_p$ be the obvious projection.

Definition 6.5.

(a) *Let $\psi: X \to E$ be a section such that $\psi(p) = 0$. Then define $(D\psi)_p: T_p X \to E_p$ by $(D\psi)_p(v) = \sigma \cdot (d\psi)_p(v)$. ($D\psi_p$ is called the* intrinsic derivative of ψ at the zero p.

(b) *Let $f: X \to \mathbf{R}$ be smooth with a critical point at p. The 1-form $df: X \to T^*X$ has a zero at p. Define $(d^2 f)_p(v, w) = \langle (D(df))_p v, w\rangle$ where $v, w \in T_p X$. Then $(d^2 f)_p$ is a bilinear form on $T_p X$ called the* Hessian of f at the critical point p.

We leave the following two Lemmas as exercises.

(A) Let U be an open subset of $\mathbf{R}^n$ and $f: U \to \mathbf{R}$ be smooth with a critical point at p. Then with the standard identifications

$$(d^2 f)_p\left(\left.\frac{\partial}{\partial x_i}\right|_p, \left.\frac{\partial}{\partial x_j}\right|_p\right) = \frac{\partial^2 f}{\partial x_i\, \partial x_j}(p).$$

(B) Let $\phi : X \to Y$ and $f\colon Y \to R$ be smooth. Suppose that q is a critical point for f and $p \in \phi^{-1}(q)$. Then p is a critical point for $\phi * f$ and $(d^2\phi * f)_p = (\phi * (d^2f))_p$ where $\phi * (d^2f)_p(v, w) = (d^2f)_q((d\phi)_p v, (d\phi)_p w)\ \forall v, w \in T_pX$.
(Hint: Compute both lemmas by using local coordinates.)

Proposition 6.6. *Let $f\colon X \to \mathbf{R}$ be smooth with a critical point at p. Then $(d^2f)_p$ is a symmetric bilinear form on T_pX and is nondegenerate iff p is a nondegenerate critical point of f.*

Proof. Let U be a nbhd of p in X and $\phi : U \to \mathbf{R}^n$ a chart centered at p. Let $U' = \phi(U)$ and $g = f \cdot \phi^{-1}$. By Lemma (B) above 0 is a critical point of. g and $(d^2f)_p = (d^2\phi * g)_p = \phi * (d^2g)_0$. By Lemma (A), $(d^2g)_0$ is symmetric. Hence $(d^2f)_p$ is symmetric. By Proposition 6.4, 0 is a nondegenerate critical point of g iff $(d^2g)_0$ is a nondegenerate bilinear form iff $(d^2f)_p$ is a nondegenerate bilinear form since $(d\phi)_p : T_pX \to T_0\mathbf{R}^n$ is an isomorphism. Finally the diagram

$$\begin{array}{ccc} J^1(U, \mathbf{R}) & \xrightarrow{(\phi^*)^{-1}} & J^1(U', \mathbf{R}) \\ {\scriptstyle j^1f}\uparrow & & \uparrow{\scriptstyle j^1g} \\ U & \xrightarrow{\ \phi\ } & U' \end{array}$$

commutes and $\phi * (S_1 \cap J^1(U', \mathbf{R})) = S_1 \cap J^1(U, \mathbf{R})$ so that $j^1g \pitchfork S_1$ at 0 iff $j^1f \pitchfork S_1$ at p. Thus 0 is a nondegenerate critical point of g iff p is a nondegenerate critical point of f. $\square$

We recall the following proposition from Linear Algebra.

Proposition 6.7. *Let B be a symmetric, nondegenerate, bilinear form on a real vector space V of dimension n. Then there is an integer $k \leq n$ and a basis $v_1, \ldots, v_n$ of V such that $B(v_i, v_j) = s_i\delta_{ij}$ where*

$$s_i = \begin{cases} -1 & \text{if } i \leq k \\ 1 & \text{if } i > k. \end{cases}$$

k is called the index *of B.*

Definition 6.8. *Let p be a nondegenerate critical point of $f\colon X \to R$. Then the* index *of f at p is the index of $(d^2f)_p$.*

Theorem 6.9. (*Morse Theorem*). *Let $g : \mathbf{R}^n \to \mathbf{R}$ be given by*

$$g(x_1, \ldots, x_n) = c - (x_1{}^2 + \cdots + x_k{}^2) + x_{k+1}^2 + \cdots + x_n{}^2$$

where c is some constant. Then
(a) *g has a nondegenerate critical point of index k at the origin and has no other singularities.*
(b) *Let $f\colon X \to \mathbf{R}$ be smooth with a nondegenerate critical point of index k*

at p. Then there is a nbhd U of p and a chart $\alpha : U \to \mathbf{R}^n$ centered at p such that

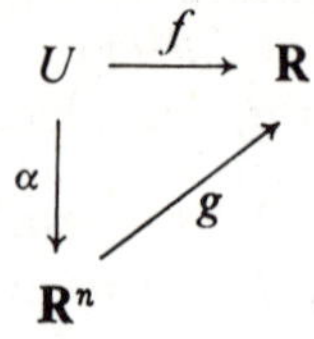

commutes where $c = f(p)$.

A simple calculation gives the proof of (a). We shall need a sequence of Lemmas to prove (b). Note that in the coordinates $\alpha_1, \ldots, \alpha_n$ defined by α, f has the "normal form" $f(x) = f(p) - (\alpha_1{}^2(x) + \cdots + \alpha_k{}^2(x)) + \alpha_{k+1}^2(x) + \cdots + \alpha_n{}^2(x)$ for all x in a nbhd of p. Thus the behavior of a function in the nbhd of a non-degenerate critical point is determined.

Lemma 6.10. *Let U be an open convex subset of $\mathbf{R}^n$, $a \in U$, and $f: U \to \mathbf{R}$ smooth. Then there exist $g_1, \ldots, g_n : U \to \mathbf{R}$ all smooth so that for every x in U*

$$f(x) = f(a) + \sum_{i=1}^{n} g_i(x)(x_i - a_i).$$

Moreover $g_i(a) = (\partial f/\partial x_i)(a)$.

Proof. Fix x in U and let $\phi(t) = f(a + t(x - a))$. This is well-defined for t in $[0,1]$ by convexity. Then

$$f(x) - f(a) = \phi(1) - \phi(0) = \int_0^1 \frac{d\phi}{dt}\,dt$$

and

$$\frac{d\phi}{dt}(t) = \sum_{i=1}^{n} \frac{\partial f}{\partial x_i}(a + t(x - a))(x_i - a_i)$$

by the chain rule. Let

$$g_i(x) = \int_0^1 \frac{\partial f}{\partial x_i}(a + t(x - a))\,dt. \qquad \square$$

Lemma 6.11. *Let f be a smooth function on an open convex subset U of $\mathbf{R}^n$. Let a be a critical point for f. Then there are smooth functions $g_{ij} : U \to \mathbf{R}$ $(1 \le i, j \le n)$ such that*

(a) $g_{ij} = g_{ji}$;

(b) $f(x) = f(a) + \sum_{i,j=1}^{n} g_{ij}(x)(x_i - a_i)(x_j - a_j)\ \forall x \in U$;

and

(c) $g_{ij}(a) = \dfrac{1}{2}\dfrac{\partial^2 f}{\partial x_i\,\partial x_j}(a)$.

Proof. From Lemma 6.10 there are functions $g_1, \ldots, g_n$ such that $f(x) = f(a) + \sum_{i=1}^{n} g_i(x)(x_i - a_i)$ and $g_i(a) = (\partial f/\partial x_i)(a) = 0$ since a is a critical point of f. Now apply Lemma 6.10 to each $g_i(x)$ to insure the existence of smooth functions $h_{ij}: U \to \mathbf{R}$ so that $g_i(x) = \sum_{j=1}^{n} h_{ij}(x)(x_j - a_j)$. This is possible since $g_i(a) = 0$. Let $g_{ij} = \frac{1}{2}(h_{ij} + h_{ji})$. Then we have written

$$f(x) = f(a) + \sum_{i,j=1}^{n} g_{ij}(x)(x_i - a_i)(x_j - a_j) \ \forall x \in U$$

and $g_{ij} = g_{ji}$. To see that $g_{ij}(a) = \frac{1}{2}(\partial^2 f/\partial x_i \, \partial x_j)(a)$ differentiate both sides of the above equation with $\partial^2/\partial x_i \, \partial x_j$ and observe that the only terms on the right-hand side which do not disappear when evaluated at a are $g_{ij}(a)$ and $g_{ji}(a)$. ☐

Lemma 6.12. *Let U be an open convex nbhd of 0 in $\mathbf{R}^n$. Let $f: U \to \mathbf{R}$ be a smooth function with a non-degenerate critical point at 0. Assume that $(\partial^2 f/\partial x_i \, \partial x_j)(0) = \pm\delta_{ij}$. Then there exists a nbhd V of 0 contained in U and smooth functions $h_i: V \to \mathbf{R}$ $(1 \le i \le n)$ satisfying*

(a) $h_i(0) = 0$ and $\dfrac{\partial h_i}{\partial x_j}(0) = \pm\,\delta_{ij}$;

(b) $f(x) = f(a) + (\pm h_1^2(x) \pm \cdots \pm h_n^2(x)) \ \forall x \in V$.

Proof. The Lemma will be proved by induction on r. The induction hypothesis for each r with $0 \le r \le n$ is that there exist smooth functions $h_i: V_r \to \mathbf{R}$ $(1 \le i \le n)$ and $g_{ij}: V_r \to \mathbf{R}$ $(r + 1 \le i, j \le n)$ where V_r is a nbhd of 0 contained in V, satisfying

(a′) $g_{ij} = g_{ji}$;

(b′) $g_{ij}(0) = \dfrac{1}{2}\dfrac{\partial^2 f}{\partial x_i \, \partial x_j}(0)$

(c′) $h_i(0) = 0$ and $\dfrac{\partial h_i}{\partial x_j}(0) = \pm\delta_{ij}$;

and

(d′) $f(x) = f(0) + (\pm h_1^2(x) \pm \cdots \pm h_r^2(x)) \pm \displaystyle\sum_{i,j=r+1}^{n} g_{ij}(x)h_i(x)h_j(x)$.

The Lemma is proved by taking $r = n$.

For $r = 0$, let $h_i(x) = x_i$ and use Lemma 6.11 to obtain the g_{ij}'s which satisfy (a′) through (d′) on $V_0 = U$.

Assume that the induction hypothesis is true for $r - 1$, giving the existence of smooth functions u_i $(1 \le i \le n)$ and v_{ij} $(r \le i, j \le n)$ of V_{r-1} into R. By (b′), $v_{rr}(0) = \frac{1}{2}(\partial^2 f/\partial x_r^2)(0) \neq 0$. Thus there is a nbhd V_r of 0 contained in V_{r-1} on which v_{rr} is nowhere zero. Let $h_i = u_i | V_r$ for $i \neq r$ and define

$$h_r(x) = \sqrt{|v_{rr}(x)|}\left(u_r(x) + \frac{1}{v_{rr}(x)} \sum_{j=r+1}^{n} v_{rj}(x)u_j(x)\right).$$

h_r is well-defined and smooth on V_r. A straight-forward calculation will show that (c′) holds. Now let $\delta = \text{sign}\,(v_{rr}(0))$. From the induction hypothesis (d′)

we know that

$$f(x) = f(0) + (\pm h_1{}^2(x) \pm \cdots \pm h_{r-1}{}^2(x)) + s(x)$$

where $s(x) = \sum_{i,j=r}^{n} v_{ij}(x)u_i(x)u_j(x)$. Let $g_{ij} = v_{ij} - v_{rj}v_{ri}/v_{rr}$ for $r + 1 \leq i, j \leq n$. Then (a′) and (b′) hold. Finally to see that (d′) is true, compute

$$\text{(i)}\ \ \delta h_r{}^2 = v_{rr}u_r{}^2 + 2\sum_{j=r+1}^{n} v_{rj}u_ru_j + \sum_{i,j=r+1}^{n} \frac{v_{ri}v_{rj}}{v_{rr}} u_iu_j$$

and

$$\text{(ii)}\ \ \sum_{i,j=r+1}^{n} g_{ij}h_ih_j = \delta h_r{}^2 + s.$$

Then

$$f(x) = f(a) + (\pm h_1{}^2(x) \pm \cdots \pm h_{r-1}^2(x)) - \delta h_r{}^2(x) + \sum_{i,j=r+1}^{n} g_{ij}h_i(x)h_j(x). \quad \square$$

Proof of Theorem 6.9 (b). Let U be a coordinate nbhd of p and $\psi : U \to \mathbf{R}^n$ a chart. Then $(d^2(f\cdot\psi^{-1}))$ is a symmetric, nondegenerate, bilinear form on $T_0\mathbf{R}^n$. Choose a matrix A which diagonalizes this form, i.e.

$$d^2(f\cdot\psi^{-1}\cdot A) = \begin{pmatrix} -I_k & 0 \\ 0 & I_{n-k} \end{pmatrix}$$

where I_s is the $s \times s$ identity matrix. Let $\eta = A^{-1}\cdot\psi$. η is also a chart on U. Now $g(x) = f(\eta^{-1}(x)) - f(p)$ satisfies the conditions of Lemma 6.12 and $g(0) = 0$. Thus there exist smooth functions $h_1, \ldots, h_n$ defined on a nbhd V of 0 in $\mathbf{R}^n$ so that $g(x) = \pm h_1{}^2(x) \pm \cdots \pm h_n{}^2(x)$ where $h_i(0) = 0$ and $(\partial h_i/\partial x_j)(0) = \pm\delta_{ij}$. Now define $H : V \to \mathbf{R}^n$ by $H(x) = (h_1(x), \ldots, h_n(x))$. H is a diffeomorphism on a nbhd of 0. Let $\alpha = H\cdot\eta$ be a chart defined on a nbhd U' of p in X. Then define $\bar{g}(x) = \pm x_1{}^2 \pm \cdots \pm x_n{}^2$ where the signs in the definition of $\bar{g}$ are the same as those in the representation of g above. Hence $\bar{g}\cdot H = f\cdot\eta^{-1} - f(p)$ and $\bar{g}\cdot\alpha = f - f(p)$ or $f(q) = f(p) + \bar{g}(\alpha(q))$. Let $\alpha_1, \ldots, \alpha_n$ be the coordinates with the chart α, then $f(q) = f(p) + (\pm\alpha_1{}^2(q) \pm \cdots \pm \alpha_n{}^2(q))$. Finally

$$(d^2f)_p = (d^2f\cdot\alpha^{-1})_{\alpha(p)} = (d^2\bar{g})_0 = \begin{pmatrix} \pm 2 & & 0 \\ & \ddots & \\ 0 & & \pm 2 \end{pmatrix}.$$

Since $(d^2f)_p$ has index k, so does $(d^2\bar{g})_0$. Thus by a simple reordering of the α_i's we may assume that $f(q) = g(\alpha(q))$. $\square$

Having analyzed the structure of a function in the nbhd of a nondegenerate critical point, we can now make a statement about the critical values.

Proposition 6.13. *Let X be a smooth manifold. The set of Morse functions all of whose critical values are distinct form a residual set in $C^\infty(X, \mathbf{R})$.*

Proof. Let $S = (S_1 \times S_1) \cap J_2{}^1(X, \mathbf{R}) \cap (\beta^2)^{-1}(\Delta\mathbf{R})$. We claim S is a submanifold of the multijet bundle $J_2{}^1(X, \mathbf{R})$. Let U be an open coordinate nbhd in X diffeomorphic to $\mathbf{R}^n$. In these local coordinates $J_2^1(U, \mathbf{R}) \cong (\mathbf{R}^n \times \mathbf{R}^n - \Delta\mathbf{R}^n) \times (\mathbf{R} \times \mathbf{R}) \times \text{Hom}(\mathbf{R}^n, \mathbf{R})^2$ and $S \cong (\mathbf{R}^n \times \mathbf{R}^n - \Delta\mathbf{R}^n) \times (\Delta\mathbf{R}) \times (0, 0)$ which is clearly a submanifold. Moreover codim $S = 2n + 1$ where $n = \dim X$.

Apply the Multijet Transversality Theorem (4.13) to conclude that the set of mappings $f: X \to \mathbf{R}$ for which $j_2^1 f \pitchfork S$ is residual. Transversality in these relative dimensions means that $j_2^1 f(X \times X - \Delta X) \cap S = \varnothing$. Thus if p and q are distinct critical points of f, then $(j^1 f(p), j^1 f(q)) \in S_1 \times S_1 \cap J_2^1(X, \mathbf{R})$. The fact that $j_2^1 f(p, q) \notin S$ means that $f(p) \neq f(q)$, i.e., f has distinct critical values. Thus the proposition is proved.

§7. The Tubular Neighborhood Theorem

***Definition** 7.1.* *Let X be a submanifold of the smooth manifold Y. A* tubular neighborhood *of X in Y is an open subset Z of Y together with a submersion $\pi: Z \to X$ such that*

(a) *$Z \xrightarrow{\pi} X$ is a vector bundle, and*

(b) *$X \subset Z$ is the zero section of this vector bundle.*

***Theorem** 7.2.* *Let X be a submanifold of Y, then there exists a tubular nbhd of X in Y.*

We prove some preliminary results first.

***Proposition** 7.3.* *Let Y and Y' be smooth manifolds with $X \subset Y$ and $X' \subset Y'$ submanifolds. Let $f: Y \to Y'$ be smooth and satisfy:*

(a) *$f|X: X \to X'$ is a diffeomorphism.*

(b) *$(df)_x: T_x Y \to T_{f(x)} Y'$ is an isomorphism for every x in X.*

Then there is an open nbhd V of X in Y such that $f(V)$ is open in Y' and $f|V$ is a diffeomorphism.

Proof. If X is a point, then Proposition 7.3 is just the Inverse Function Theorem. In any case, since X is second countable, there is a countable covering of X by open sets $U_1, U_2, \ldots$ in Y so that $f_i = f|\overline{U}_i$ is a diffeomorphism; i.e., f is a diffeomorphism on a nbhd of $\overline{U}_i$. Moreover, since X' is paracompact, there is a locally finite covering of X' by open sets $W_1, W_2, \ldots$ in Y' which is a refinement of the open covering $f_1(U_1), f_2(U_2), \ldots$ By replacing the U's with the sets $U_j = f^{-1}(W_j)$ where $W_j \subset f(U_i)$, we may assume that $W_i = f_i(U_i)$. Let $W = \bigcup_{i=1}^{\infty} W_i$ and

$$F = \{y \in W \mid \text{if } y \in \overline{W}_i \cap \overline{W}_j, \quad \text{then} \quad f_i^{-1}(y) = f_j^{-1}(y)\}$$

clearly contains X'. Moreover, we claim that F contains an open nbhd G of X' in Y'. For each x' in X', there is an open nbhd $G_{x'}$ which intersects only finitely many W_i's, say $W_{i_1}, \ldots, W_{i_s}$, since the covering $\{W_i\}$ is locally finite. By making G_x even smaller we may assume that x' is in $\overline{W}_{i_1} \cap \cdots \cap \overline{W}_{i_s}$. Now $f_{i_j}^{-1}$ is a local inverse for f near x'. Thus there is an open nbhd of x', H, such that $f_{i_1}{}^{-1}|H = \cdots = f_{i_s}{}^{-1}|H$ using the uniqueness of inverse functions and the fact that $f_{i_1}{}^{-1}(x') = \cdots = f_{i_s}{}^{-1}(x')$. Then $\tilde{G}_{x'} = H \cap G_{x'}$ is an open nbhd of x' in Y' and $\tilde{G}_{x'} \subset F$. Let $G = \bigcup_{x' \in X'} \tilde{G}_{x'}$.

Now define $g: G \to X$ by $g(y) = f_i^{-1}(y)$ if y is in W_i. g is well-defined since $G \subset F$ and smooth since locally $g = f_i^{-1}$ for some i. Also $f \cdot g = id_G$ so g is a diffeomorphism. Let $V = g(G)$. □

Lemma 7.4. *Let $E \xrightarrow{\pi} X$ be a vector bundle. Let V be an open nbhd of the zero section in E. Then there exists a diffeomorphism $h: E \to V$ (into) such that $\pi \cdot h = \pi$.*

Proof. Choose a metric t on E. The sets $B_x(a) = \{v \in E_x \mid t(x)(v, v) < a\}$ form a nbhd basis of 0 in E_x where a is in R^+. Since $V \cap E_x$ is an open nbhd of 0 in E_x, there is an $a > 0$ for which $B_x(a) \subset V \cap E_x$. Since t is smooth we may choose, by a partition of unity argument, a smooth function $\delta: X \to (0, 1)$ such that $B_x(\delta(x)) \subset V$ for all x in X. The mapping $h(v) = \delta(x)v/(1 + t(x)(v, v))^{1/2}$ where $x = \pi(v)$ is a diffeomorphism of E into V whose inverse is given by $w \mapsto w/(\delta(x)(1 - t(x)(w, w)^{1/2}))$ where $x = \pi(w)$. □

Combining the last two results we have:

Proposition 7.5. *Let $E \xrightarrow{\pi} X$ be a vector bundle and let U be an open nbhd of the zero section X_0 in E. Suppose that X is a submanifold of a smooth manifold Y and that $f: U \to Y$ is smooth and satisfies:*

(a) *$f|X_0: X_0 \to X$ is a diffeomorphism.*

(b) *$(df)_x: T_xE \to T_{f(x)}Y$ is an isomorphism for every x in X_0. Then there is a tubular nbhd of X in Y.*

Proof. By Proposition 7.3, there is an open nbhd V of X_0 such that $f|V: V \to Y$ is a diffeomorphism into. Let $h: E \to V$ be a diffeomorphism guaranteed by Lemma 7.4. Then $Z = f \cdot h(E)$ is a tubular nbhd of X in Y with projection mapping $\pi' = f \cdot h \cdot \pi \cdot (f \cdot h)^{-1}$. □

Lemma 7.6. *Let X be a n-submanifold of $\mathbf{R}^p$, then there is a tubular nbhd of X in $\mathbf{R}^p$.*

Proof. Equip $\mathbf{R}^p$ with the standard inner product. Let E_x be the $(p - n)$-plane of normal vectors to X at x. $E_x - x$ is a plane through the origin in $\mathbf{R}^p$ and has a vector space structure which we give to E_x. Let $E = \bigcup_{x \in X} E_x$. Then E is a vector bundle over X, since, in effect, E is just the complementary sub-bundle to TX in $T_X\mathbf{R}^p = X \times \mathbf{R}^p$. (See I, 5.12). The explicit construction of E gives a smooth mapping $f: E \to \mathbf{R}^p$ such that $f|X_0 = f|X = id_X$. To show the existence of a tubular nbhd we will apply Proposition 7.5. So we must show that $(df)_x: T_x E \to T_{f(x)}\mathbf{R}^p$ is an isomorphism for all x in E. Since $\dim T_xE = \dim T_x\mathbf{R}^p$ we need only show that $\operatorname{Ker}(df)_x = 0$. Let $v \in \operatorname{Ker}(df)_x$. Since $T_xE \cong T_xX \oplus T_xE_x$, $v = v_1 + v_2$ where $v_1 \in T_xX$ and $v_2 \in T_xE_x$. $(df)_xv = 0$ implies that $v_1 = 0$ since $f|X = id_X$. Represent v_2 by a curve c in E_x. $f \cdot c$ is then just the curve c in E_x when E_x is thought of as in $\mathbf{R}^p$. Thus $(df)_x(v_2) = 0$ implies that $v_2 = 0$. □

Proof of Theorem 7.2. Using the Whitney Embedding Theorem (Proposition 5.9), we may assume that $Y \subset \mathbf{R}^p$ for $p = 2 \dim Y + 1$. Thus X is a submanifold of $\mathbf{R}^p$ and by Lemma 7.6 there is a tubular nbhd of X in $\mathbf{R}^p$ which we call Z'. Let E_x be the set of vectors normal to X and tangent to Y at x for x in X and let $E = \bigcup_{x \in X} E_x$. E is a vector bundle over X since it is just the complementary subbundle to TX in $T_X Y$. As in Lemma 7.6, this explicit construction of E gives a smooth mapping $f: E \to \mathbf{R}^p$ such that $f|X_0 = f|X = id_X$. Let Z'' be a tubular nbhd of Y in $\mathbf{R}^p$ with projection map π''. Then $U = f^{-1}(Z' \cap Z'')$ is an open nbhd of X_0 in E. Consider $\pi'' \cdot f: U \to Y$ which is a smooth mapping. $\pi'' \cdot f|X = id_X$ since $X \subset Y$ and $\pi''|Y = id_Y$. We wish to apply Proposition 7.5 to obtain the desired result. In order to do so we need to know that $(d\pi'' \cdot f)_x : T_x E \to T_{f(x)} Y$ is an isomorphism for all x in X. Since $\dim E = \dim Y$ we need only show that $\operatorname{Ker} (d\pi'' \cdot f)_x = \{0\}$. Now $T_x E = T_x X \oplus T_x E_x$. Since $\pi'' \cdot f|X = id_X$ we have to show that if v is in $T_x E_x$ and $(d\pi'' \cdot f)_x(v) = 0$, then $v = 0$. Let c be a curve in E_x representing v, then $f \cdot c$ is a curve in $\mathbf{R}^p$ whose tangent at 0 is tangent to Y at x. Since $\pi''|Y = id_Y$, $\pi'' \cdot f \cdot c$ is a curve in Y whose tangent at 0 is the same as the tangent to $f \cdot c$ at 0. Thus $(d\pi'' \cdot f)_x(v) = 0$ implies that $v = 0$. ☐

Remark. If one traces through the proof of the Tubular Nbhd Theorem, one sees that the tubular nbhd Z is always isomorphic (as vector bundles) to a complementary subbundle of TX in $TY|X$. Such a complementary subbundle is always isomorphic to the quotient bundle N where $N_x = T_x Y / T_x X$ for each x in X. N is called the *normal bundle to X in Y* and N_x is called the *normal space to X in Y at x*. Thus a tubular nbhd is a realization of the normal bundle as an open nbhd of X in Y.

Chapter III

Stable Mappings

§1. Stable and Infinitesimally Stable Mappings

Definition 1.1.

(a) *Let f and f' be elements of $C^\infty(X, Y)$. Then f is* equivalent *to f' if there exist diffeomorphisms $g : X \to X$ and $h : Y \to Y$ such that the diagram*

$$\begin{array}{ccc} X & \xrightarrow{\;f\;} & Y \\ {\scriptstyle g}\downarrow & & \downarrow{\scriptstyle h} \\ X & \xrightarrow{\;f'\;} & Y \end{array}$$

commutes.

(b) *Let f be in $C^\infty(X, Y)$. Then f is* stable *if there is a nbhd W_f of f in $C^\infty(X, Y)$ such that each f' in W_f is equivalent to f.*

In other words f is stable if every nearby mapping f' is identical to f, after suitable changes of coordinates, both in the domain and the range of f'.

We now describe an alternative formulation of stability of mappings. Recall that a group G acts on a set A if there is a function $G \times A \to A$, written as $(g, a) \mapsto g \cdot a$, with the properties that $(gg') \cdot a = g \cdot (g' \cdot a)$ and $e \cdot a = a$ for every $g, g' \in G$ and $a \in A$ where e is the identity in G. The *orbit of* a in A is the set

$$G \cdot a \equiv \{b \in A \mid b = g \cdot a \quad \text{for some} \quad g \text{ in } G\}.$$

In the case at hand, let $G = \text{Diff}(X) \times \text{Diff}(Y)$ where $\text{Diff}(X)$ (resp. $\text{Diff}(Y)$) is the group of all diffeomorphisms on the manifold X (resp. Y) and let $A = C^\infty(X, Y)$. Then there is a natural action of G on A defined by $(g, h) \cdot f = h \cdot f \cdot g^{-1}$ (i.e., change of coordinates) where $g \in \text{Diff}(X)$, $h \in \text{Diff}(Y)$, and $f \in C^\infty(X, Y)$.

Lemma 1.2. *Let f be in $C^\infty(X, Y)$. Then f is stable iff the orbit of f in $C^\infty(X, Y)$ under the action of* $\text{Diff}(X) \times \text{Diff}(Y)$ *is an open subset.*

We recall the following:

Lemma 1.3. *Let X and Y be smooth manifolds with $g : X \to X$ a diffeomorphism. Then $g^* : C^\infty(X, Y) \to C^\infty(X, Y)$ given by $f \mapsto f \cdot g$ is continuous.*

Proof. Since g is proper, Note (2) after II,3.9 applies. □

Proof of Lemma 1.2. Let g be in Diff(X) and h be in Diff(Y). Let $\gamma_{(h,g)}: C^\infty(X, Y) \to C^\infty(X, Y)$ be induced by the action of Diff(X) $\times$ Diff(Y) on $C^\infty(X, Y)$. Thus $\gamma_{(g,h)} = (h)_*(g^{-1})^*$ and is continuous by the above Proposition and II,3.5. Moreover $\gamma_{(g,h)}$ is a homeomorphism since $\gamma_{(g^{-1},h^{-1})} \cdot \gamma_{(g,h)} = id_{C^\infty(X,Y)}$.

Now observe that f' is in the orbit of f iff f' is equivalent to f. Also the orbit of f is open iff there is an open nbhd of $C^\infty(X, Y)$ contained in the orbit of f (since any such nbhd can be translated by an element of Diff(X) $\times$ Diff(Y); i.e., $\gamma_{(g,h)}$, to an open nbhd around any other point in the orbit). These two facts taken together immediately yield the proof. □

This definition of stability proves difficult to apply in practice. However, using a criterion suggested by René Thom, John Mather has produced a theorem which provides a truly computable method for determining whether or not a mapping is stable. We now present that criterion.

Definition 1.4. *Let $f: X \to Y$ be smooth.*

(a) *Let $\pi_Y: TY \to Y$ be the canonical projection, and let $w: X \to TY$ be smooth. Then w is a* vector field along f *if the diagram*

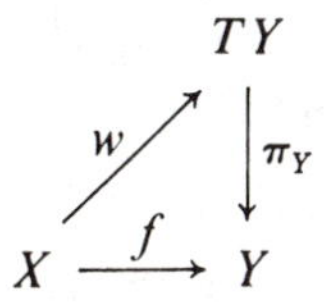

commutes. Let $C_f^\infty(X, TY)$ denote the set of vector fields along f.

(b) *f is* infinitesimally stable *if for every w, a vector field along f, there is a vector field s on X and a vector field t on Y such that*

$$(*) \qquad w = (df) \cdot s + t \cdot f$$

where $(df): TX \to TY$ is the Jacobian mapping of f.

Remark. Vector fields along f can be identified with sections of a certain vector bundle. Let E be a vector bundle over Y. Define $f^*E = \bigcup_{p \in X} E_{f(p)}$ (disjoint union) and let $\pi: f^*E \to X$ be the obvious projection. We claim that f^*E can be made into a vector bundle over X, the *pull-back bundle of E by f*, as follows. Let V be an open nbhd of Y such that $E|V$ is trivial, say $E|V \cong V \times \mathbf{R}^k$. Then make f^*E into a smooth manifold by demanding that $f^*E|f^{-1}(V) \cong f^{-1}(V) \times \mathbf{R}^k$. That the transition functions are smooth follows from the fact that E is a smooth vector bundle over Y. Now let s be a section of $f^*(TY)$. Then $s(x) \in f^*(TY)_x = T_{f(x)}Y$ so s may be thought of as a smooth mapping of $X \to TY$ such that $\pi_Y \cdot s = f$, (i.e., can be identified with a vector field along f). The converse is also clear. Thus we can identify $C_f^\infty(X, TY)$ with $C^\infty(f^*(TY))$.

Theorem 1.5. (*Mather*). *Let X be a compact manifold and $f: X \to Y$ be smooth. Then f is stable iff f is infinitesimally stable.*

Note. It is sufficient to assume that f is a proper mapping and to drop the assumption on X.

The proof of this theorem will be given in Chapter V. For the moment we will content ourselves with explaining what originally motivated the introduction of the concept of infinitesimal stability. This will require a sketchy development of the theory of Frechet manifolds.

Definition 1.6. *Let V be a topological vector space, i.e., a vector space with a topology in which addition and scalar multiplication are continuous. Let $|\cdot|: V \to \mathbf{R}$ be continuous and satisfy for all x and y in V:*

(a) $|x| \geq 0$

(b) $|x| = 0$ *iff* $x = 0$

(c) $|x + y| \leq |x| + |y|$

(d) $|x| = |-x|$ (not $|\lambda x| = |\lambda||x|$ *for arbitrary* $\lambda \in \mathbf{R}$).

We can now define a metric d on V by setting $d(x, y) = |x - y|$. If V is complete with respect to this metric, then it is a Frechet space.

Notes. (1) The norm on a Banach space satisfies axioms (a) through (d), so every Banach space is a Frechet space.

(2) Let V_1 and V_2 be Frechet spaces. Then $L(V_1, V_2)$, the set of all continuous linear mappings of V_1 into V_2, is a Frechet space if we define $|f| = \sup_{|x|=1} |f(x)|$. (This is well-defined by the linearity and the continuity of $|\cdot|$ and f.)

Definition 1.7. *Let V_1 and V_2 be Frechet spaces and U an open subset of V_1. Let $f: U \to V_2$ be continuous and p be a point in U. Then f is* differentiable at *p iff there is a linear mapping $A_p: V_1 \to V_2$ such that*

$$\operatorname*{Lim}_{|t| \to 0} \frac{|f(p + tv) - f(p) - tA_p(v)|}{|t|} = 0$$

for every v in V_1.

Note that the linear mapping A_p is unique if it exists. Thus we may define $(df)_p = A_p$ when f is differentiable at p.

f is *differentiable on U* if f is differentiable at p for every p in U. f is k-*times differentiable* if $(df): U \to L(V_1, V_2)$ defined by $(df)(p) = (df)_p$ is $(k - 1)$-times differentiable. f is *smooth* if f is k-times differentiable for every k.

We note that the chain rule holds for differentiation on Frechet spaces.

Definition 1.8. *Let X be a Hausdorff topological space. Then X is a* Frechet manifold *if*

(a) *there is a covering $\{U_\alpha\}_{\alpha \in I}$ of X by open sets.*

(b) *for each α in I there exists a homeomorphism $h_\alpha: U_\alpha \to V_\alpha$ with V_α an open subset of some Frechet space. (The h_α's are called* charts.)

(c) *for every $\alpha, \beta \in I$, $h_\alpha \cdot h_\beta^{-1}$ is smooth where defined.*

If V_α is contained in a Banach space, then X is called a Banach manifold.

As in the ordinary manifold case, it is possible to define the tangent space to a point p on a Frechet manifold X. Let $S_p(X)$ be the set of smooth curves $c: \mathbf{R} \to X$ such that $c(0) = p$. (Note that smooth mappings between Frechet manifolds are defined exactly as in the finite dimensional case.) Define c_1 *is tangent to* c_2 *at* p if for every chart h_α, $(dh_\alpha \cdot c_1)_0 = (dh_\alpha \cdot c_2)_0$. Since the chain rule is valid, "is tangent to" is a well-defined concept. Finally, let T_pX = set of equivalence classes of $S_p(X)$ under the equivalence relation "is tangent to." As in the finite dimensional case T_pX is a vector space.

Note that a smooth map $f: X \to Y$ between Frechet manifolds induces a well-defined linear mapping $(df)_p: T_pX \to T_{f(p)}Y$ just as in the familiar case of Chapter I.

The following proposition will present our basic examples of Frechet spaces.

Proposition 1.9. *Let X be a compact finite dimensional manifold. Then*
(a) *$C^\infty(X, \mathbf{R})$ is a Frechet space,*
(b) *if E is a vector bundle over X, then $C^\infty(E)$ is a Frechet space, and*
(c) *let $f: X \to Y$ be smooth, then $C_f^\infty(X, TY)$—the set of vector fields along f—is a Frechet space.*

It is understood that in each case the topology induced by the Frechet "norm" is the Whitney C^∞ topology.

Proof.
(a) Cover X be a finite number of open sets U_α where $\overline{U}_\alpha$ is contained in a coordinate patch. This is possible since X is compact. Let $g: X \to \mathbf{R}$ be smooth. Define

$$|g|_k^{U_\alpha} = \sup_{u \in \overline{U}_\alpha} \left\{ \sum_{|\beta| \leq k} \left| \frac{\partial^{|\beta|} g}{\partial x^\beta}(u) \right| \right\}.$$

Then define $|g|_k = \sum_\alpha |g|_k^{U_\alpha}$ and, finally, define

$$|g| = \sum_{k=0}^{\infty} \frac{1}{2^k} \frac{|g|_k}{1 + |g|_k}.$$

It is now an easy exercise to check that $C^\infty(X, \mathbf{R})$ is a Frechet space with $|\cdot|$ defined in this way.

(b) Choose the U_α's so that $E | U_\alpha$ is trivial. Then let the Frechet norm $|\cdot|_k^{U_\alpha}$ be the supremum over the Frechet norm of the coordinate functions. Then continue as in (a).

(c) Since $C_f^\infty(X, TY)$ can be identified with $C^\infty(f^*TY)$, the result follows from (b). (See the Remark after 1.4.) □

We note that one gets different metrics on $C^\infty(E)$ for different choices of U_α but the underlying topology is the same, and that this underlying topology is just the one induced on $C^\infty(E)$ from the Whitney C^∞ topology on $C^\infty(X, E)$.

Proposition 1.10. *Let X be a smooth manifold. Then* Diff(X) *is an open subset of $C^\infty(X, X)$ and hence a Frechet manifold. (For simplicity in the proof, we assume that X is compact.)*

Proof. By II, Proposition 5.8 there is a nbhd of f which consists of 1:1 immersions. Let $g: X \to X$ be a 1:1 immersion and X^0 a connected component of X. $g(X^0)$ is closed since X^0 is compact, while it is open since g is a submersion. So $g(X^0)$ is a connected component of X. We claim that $g: X^0 \to g(X^0)$ is a diffeomorphism; this is enough to prove the proposition. Now $(g|X^0)^{-1}$ exists since g is 1:1. Also, at each point of $g(X^0)$, $(g|X^0)^{-1}$ is smooth by the Inverse Function Theorem. So g is a diffeomorphism. □

The following theorem shows our basic reason for studying Frechet manifolds.

Theorem 1.11. *Let X and Y be smooth manifolds with X compact. Then* $C^\infty(X, Y)$ *is a Frechet manifold.*

Proof. It is easy to check that $C^\infty(X, Y)$ is a Hausdorff space. Let f be in $C^\infty(X, Y)$. We wish to produce an open nbhd of f, U_f, which is homeomorphic to an open subset V_f, of a Frechet space. We do this via tubular nbhds. Let $X_f = \text{graph}(f) \subset X \times Y$. X_f is a submanifold of $X \times Y$ so we may apply the tubular nbhd theorem (II, 7.2) to find a tubular nbhd Z of X_f with projection π. V_f will be an open nbhd of the zero section in $C^\infty(Z)$.

Since $\pi: Z \to X_f$ is smooth $\pi_*: C^\infty(X, Z) \to C^\infty(X, X_f)$ is continuous (II, Proposition 3.5), so that $(\pi_*)^{-1}(\text{Diff}(X, X_f))$ is an open subset of $C^\infty(X, Z)$. (Apply Proposition 1.10.) Since $C^\infty(X, Z)$ is an open subset of $C^\infty(X, X \times Y)$ $= C^\infty(X, J^0(X, Y))$ we see that $U_f = (j^0)^{-1}(\pi_*)^{-1}(\text{Diff}(X, X_f))$ is an open nbhd of f in $C^\infty(X, Y)$. To define V_f we let $\pi_X: X \times Y \to X$ be projection. The mapping $(\pi_X)_*: C^\infty(Z) \to C^\infty(X_f, X)$ given by $s \mapsto \pi_X \cdot s$ is continuous since the topology on $C^\infty(Z)$ is the topology induced from $C^\infty(X_f, Z)$. (Again apply II, Proposition 3.5.) Thus $V_f = (\pi_X)_*{}^{-1}(\text{Diff}(X_f, X))$ is an open nbhd of the 0-section in $C^\infty(Z)$.

Next we define $h_f: U_f \to V_f$ by $h_f(g)(x, f(x)) = j^0 g \cdot (\pi \cdot j^0 g)^{-1}(x, f(x))$. Note that $\pi \cdot j^0 g: X \to X_f$ is a diffeomorphism since g is in U_f so that $h_f(g): X_f \to Z$ is well-defined. Since $\pi \cdot h_f(g) = id_{X_f}$, $h_f(g)$ is actually a section of the vector bundle Z. Finally $\pi_X \cdot h_f(g) = (\pi \cdot j^0 g)^{-1}$ which is in $\text{Diff}(X_f, X)$ so that $h_f(g)$ is in V_f. To see that h_f is a bijection, define $k_f: V_f \to U_f$ by $k_f(s)(x) = \pi_Y \cdot s \cdot (\pi_X \cdot s)^{-1}(x)$ where $\pi_Y: X \times Y \to Y$ is projection. Since $\pi \cdot j^0 k_f(s) = (\pi_X \cdot s)^{-1}$ we see that $k_f(s)$ is in U_f. It is an easy exercise to show that $k_f = h_f^{-1}$.

Since our reason for introducing Frechet manifolds was just to motivate the criterion of infinitesimal stability, we will leave the details of showing that h_f is a homeomorphism and that $h_f \cdot h_g^{-1}$ is smooth (where defined) to the "interested" reader. □

Now suppose that the Implicit Function Theorem were true for smooth functions between Frechet manifolds. Then consider the mapping

$$\gamma_f: \text{Diff}(X) \times \text{Diff}(Y) \to C^\infty(X, Y)$$

given by $(h, g) \mapsto g \cdot f \cdot h^{-1}$. Im γ_f is the orbit of f in $C^\infty(X, Y)$ under the action of $\text{Diff}(X) \times \text{Diff}(Y)$, so a reasonable criterion for the stability of f

would be that γ_f be a submersion. (Using Lemma 1.2.) In fact we would need to know only that γ_f is a submersion at the identity $e = (id_X, id_Y)$, since then $\text{Im}\,\gamma_f$ would contain an open set and, as we saw in the proof of Lemma 1.2, this would imply that $\text{Im}\,\gamma_f$ is itself open. We claim that f is infinitesimally stable iff $(d\gamma_f)_e$ is onto. To see this we need to identify the spaces

$$T_e(\text{Diff}(X) \times \text{Diff}(Y)) \quad \text{and} \quad T_f C^\infty(X, Y).$$

Lemma 1.12. *Let $t \mapsto f_t$ and $t \mapsto g_t$ be smooth curves in $C^\infty(X, Y)$ with $f_0 = g_0$. Then f_t and g_t are tangent at $t = 0$ iff for each p in X, the curves in Y $t \mapsto f_t(p)$ and $t \mapsto g_t(p)$ are tangent at $t = 0$.*

Proof. Let $h = f_0 = g_0$ and $X_h = \text{graph}\,(h) \subset X \times Y$. Let Z be a tubular nbhd of X_h in $X \times Y$. As we saw in the Proof of Theorem 1.11, nearby functions to f can be identified with sections in $C^\infty(Z)$. So for t small we can think of $t \mapsto f_t$ as a smooth curve of sections in $C^\infty(Z)$. The Frechet derivative of $t \mapsto f_t$ at $t = 0$ is given by

$$(df)_p(1) = \left(\lim_{t\to 0} \frac{|f_t - f_0|}{|t|}\right)(p) = \lim_{t \to 0} \frac{|f_t(p) - f_0(p)|}{|t|}.$$

So $(df)_p(1) = (df_t/dt(p))|_{t=0}$. Writing the same for g concludes the proof of the Lemma. ☐

Proposition 1.13.
(a) $T_f C^\infty(X, Y) \cong C_f^\infty(X, TY)$
(b) $T_{id_X} \text{Diff}(X) \cong C^\infty(TX)$.

Proof. (b) follows immediately from (a) since

$$T_{id_X} \text{Diff}(X) = T_{id_X} C^\infty(X, X) \cong C^\infty_{id_X}(X, TY) = C^\infty(TX).$$

To prove (a) let w be in $T_f C^\infty(X, Y)$ and let $t \to f_t$ be a curve representing w. Define $w' : X \to TY$ by $w'(p) = (df_t/dt(p))|_{t=0}$. By the last Lemma, this definition is independent of the choice of curves f_t. Using the identification of functions f_t with sections of $C^\infty(Z)$ we see that $\pi_Y \cdot w' = f$, so w' is in $C_f^\infty(X, TY)$. The smoothness of w' is left as an exercise. Lemma 1.12 also shows that $w \mapsto w'$ is 1:1. To show that this map is onto, it is sufficient to work locally. Let w' be in $C_f^\infty(X, TY)$. For each p in X choose a curve $t \to f_t(p)$ which represents $w'(p)$ and do this so that $f_t(p)$ is jointly smooth in p and t. (This can be done by integrating the vector field. See I, §6.)

The curve $t \mapsto f_t$ generates a section w in $T_f C^\infty(X, Y)$. Linearity of this mapping of $C_f^\infty(X, TY) \to T_f C^\infty(X, Y)$ follows from Lemma 1.12 and the pointwise linearity of vectors in TY. ☐

Thus $(d\gamma_f)_e : T_e(\text{Diff}(X) \times \text{Diff}(Y)) \to T_f C^\infty(X, Y)$ can be identified with a mapping $\alpha_f \oplus \beta_f : C^\infty(TX) \oplus C^\infty(TY) \to C_f^\infty(X, TY)$.

Theorem 1.14. *In the notation above, $\alpha_f : C^\infty(TX) \to C_f^\infty(X, TY)$ is given by $s \mapsto -(df)\cdot s$ and $\beta_f : C^\infty(TY) \to C_f^\infty(X, TY)$ is given by $t \mapsto t \cdot f$. Thus infinitesimal stability reduces to the following criterion: for every vector field w along f, there exist vector fields s on X and t on Y such that $w = (df)\cdot s + t \cdot f$.*

Proof. Let s be in $C^\infty(TX)$. Let $\tilde{\alpha}: \mathrm{Diff}(X) \to C^\infty(X, Y)$ be given by $h \mapsto f \cdot h^{-1}$; so $(d\tilde{\alpha})_{id_X} = \alpha_f$. Let $t \mapsto h_t$ be a curve in $\mathrm{Diff}(X)$ based at id_X representing s. Then $\alpha_f(s) = (d\tilde{\alpha})_{id_X}(s)$ is represented by the curve $t \mapsto f \cdot h_t^{-1}$. Thus

$$\frac{d}{dt}(f \cdot h_t^{-1})|_{t=0} = (df) \cdot \frac{dh_t^{-1}}{dt}\bigg|_{t=0}$$

by the chain rule and the pointwise definition of dh_t^{-1}/dt given by Lemma 1.12. To compute $(dh_t^{-1}/dt)|_{t=0}$, we consider the mappings $\Delta : \mathbf{R} \to \mathbf{R}^2$ given by $t \mapsto (t, t)$ and $\Gamma : \mathbf{R}^2 \to X$ given by $\Gamma(u, v) = h_u h_v^{-1}(p)$ where p is some fixed point in X. Clearly $\Gamma \cdot \Delta(t) = p$ for all t. Using the chain rule we see that

$$0 = \frac{\partial \Gamma}{\partial u}(0, 0) + \frac{\partial \Gamma}{\partial v}(0, 0) = \frac{dh_t}{dt}(p)\bigg|_{t=0} + \frac{dh_t^{-1}}{dt}(p)\bigg|_{t=0}$$

since $h_0 = h_0^{-1} = id_X$. Thus $(dh_t^{-1}/dt)|_{t=0} = -s$ and $\alpha_f(s) = -(df) \cdot s$. The computation of β_f is similar and is left as an exercise. □

To summarize, Mather's criterion for stability of a mapping, namely, infinitesimal stability, arises naturally when $C^\infty(X, Y)$ is viewed as a Frechet manifold. Unfortunately, it is known that the Implicit Function Theorem is not true for Frechet manifolds, so the equivalence of stability and infinitesimal stability cannot be proved along these lines, although it can be proved.

For a counter-example to the Implicit Function Theorem on Frechet Manifolds see J. F. Marsden, Hamiltonian Mechanics, Infinite Dimensional Lie Groups, etc., Berkeley Notes December 1969, p. 50.

John Mather has given a correct proof of this fundamental theorem. We will give a slightly modified version of his proof in Chapter V.

Exercises

(1) Identify all infinitesimally stable mappings of $\mathbf{R} \to \mathbf{R}$. (Hint: Look at Example B of the next section for inspiration.)

(2) Show that the mapping of $\mathbf{R}^2 \to \mathbf{R}^2$ given by $(x, y) \mapsto (x, y^2)$ is infinitesimally stable.

§2. Examples

In this section we always assume that X is a compact, smooth manifold.

A. Submersions

Proposition 2.1. *Let f be a submersion of a manifold X to a manifold Y. Then f is infinitesimally stable (and hence stable).*

Proof. We will show that $C^\infty(TX) \to C_f^\infty(X, TY)$ given by $s \mapsto (df)\cdot s$ is onto, which trivially implies that f is infinitesimally stable. Since f is a submersion, $(df)_x : T_xX \to T_{f(x)}Y$ is onto for every x in X. Hence Ker $(df)_x$ has constant dimension and forms a subbundle of TX by *I*, Proposition 5.11. By I, Proposition 5.12 there exists a subbundle H complementary to Ker (df) in TX. Clearly $(df)_x : H_x \to T_{f(x)}Y$ is an isomorphism (onto) and thus induces an isomorphism on sections, i.e., $C^\infty(H) \to C^\infty(TY)$. □

Note. Proposition 2.1 gives no information about stable mappings in $C^\infty(X, \mathbf{R})$, since a submersion of X into R would be a map without critical points, but every differential function defined on a compact manifold achieves a maximum and hence has a critical point.

B. Morse Functions

Proposition 2.2. *Let f be in $C^\infty(X, \mathbf{R})$ where X is a compact manifold. Then f is stable iff f is a Morse function all of whose critical values are distinct (i.e., if p and q are distinct critical points of f in X, then $f(p) \neq f(q)$).*

Proof. Let f be stable. Then there is a nbhd W_f of f in $C^\infty(X, \mathbf{R})$ in which each function is equivalent to f. However, by II, Proposition 6.13, Morse functions g whose critical values are distinct form a dense subset of $C^\infty(X, \mathbf{R})$, so W_f must contain such a function g. Hence f is equivalent to g, but it is easy to see that any function equivalent to a Morse function with distinct critical values is itself such a function, so f is such a function.

To prove the converse we will use infinitesimal stability. Let $f\colon X \to \mathbf{R}$ be a Morse function all of whose critical values are distinct. Let $w\colon X \to T\mathbf{R} = \mathbf{R} \times \mathbf{R}$ be a vector field along f. Then $w(x) = (f(x), \bar{w}(x))$ for every x in X where $\bar{w}$ is in $C^\infty(X, \mathbf{R})$. Let s be a vector field on X. Then $df\cdot s(x) = (f(x), (df)_x(s(x))) = (f(x), s[f](x))$ where $s[f]$ is just the directional derivative of the function f in the direction s. Let t be a vector field on R. Then $t(r) = (r, \bar{t}(r))$ for all r in $\mathbf{R}$, where $\bar{t}$ is in $C^\infty(\mathbf{R}, \mathbf{R})$. So $t\cdot f(x) = (f(x), \bar{t}(f(x)))$ for all x in X. The condition of infinitesimal stability reduces in this case to the following: for every w in $C^\infty(X, \mathbf{R})$ there exists a vector field s in $C^\infty(TX)$ and a function t in $C^\infty(\mathbf{R}, \mathbf{R})$ so that

$$(*) \qquad w = s[f] + t\cdot f.$$

With the given assumptions on f we show how to solve (*). Since f is a Morse function and X is compact, there is only a finite number of critical points of f. Choose t so that $t\cdot f(x) = w(x)$ for every critical point x of f. This is possible since the critical values are distinct. For instance use the Lagrange Interpolation Formula. To solve (*), it is sufficient to solve

$$(**) \qquad w = s[f]$$

for any w in $C^\infty(X, \mathbf{R})$ where $w(x) = 0$ whenever x is a critical point of f. (Since the w in (**) can be taken to be $w - t\cdot f$ from (*).) Given such a w we construct the vector field s.

Around each point p in X, choose an open nbhd U_p as follows:

(a) If p is a regular point, choose U_p so small that $(df)_q \neq 0$ for every q in U_p. Choose a vector field s^p on U_p such that $(df)(s^p) \neq 0$ on U_p.

(b) If p is a critical point, then choose U_p to be a coordinate nbhd with coordinates $x_1, \ldots, x_n$ so that $f|U_p$ is given by $c + \varepsilon_1 x_1^2 + \cdots + \varepsilon_n x_n^2$ where $\varepsilon_1 = \cdots = \varepsilon_n = \pm 1$. (See II, Theorem 6.9.)

The collection $\{U_p\}_{p\in X}$ forms an open covering of X. Since X is compact, there exists a finite subcovering $U_1, \ldots, U_m$ corresponding to $p_1, \ldots, p_m$. Let $\rho_1, \ldots, \rho_m$ be a partition of unity subordinate to this covering. Choose vector fields s^i on X $(1 \leq i \leq m)$ as follows:

(a) if p_i is a regular point, then let

$$s_i(x) = \begin{cases} \dfrac{w(x)\rho_i(x)s^{p_i}(x)}{s^{p_i}[f](x)} & \text{on } U_{p_i} \\ 0 & \text{off } U_{p_i} \end{cases}$$

(b) if p_i is a critical point, then $w(p_i) = 0$ and $\rho_i w = \sum_{j=1}^n h_j x_j$ for selected smooth functions h_j in the coordinates on U_i given above. (See II, Lemma 6.10.) Moreover, the h_j's are compactly supported functions in U_i since $\rho_i w$ is.

Let

$$s_i = \sum_{j=1}^{n} \frac{\varepsilon_j h_j}{2} \frac{\partial}{\partial x_j} \quad \text{on} \quad U_i$$

and extend it to be zero off U_i.

Finally, define $s = s_1 + \cdots + s_m$. Then $s[f] = s_1[f] + \cdots + s_n[f]$.

In case p_i is a regular point,

$$s_i[f] = \begin{cases} \dfrac{w\rho_i s^{p_i}[f]}{s^{p_i}[f]} & \text{on } U_i \\ 0 & \text{off } U_i \end{cases} = \begin{cases} w\rho_i & \text{on } U_i \\ 0 & \text{off } U_i \end{cases} = \rho_i w$$

since ρ_i is zero off U_i.

In case p_i is a critical point,

$$\begin{aligned} s_i[f] &= \begin{cases} \displaystyle\sum_{j=1}^{m} \frac{\varepsilon_j h_j}{2} \frac{\partial}{\partial x_j} (c + \varepsilon_1 x_1^2 + \cdots + \varepsilon_n x_n^2) & \text{on } U_i \\ 0 & \text{off } U_i \end{cases} \\ &= \begin{cases} \displaystyle\sum_{j=1}^{m} \varepsilon_j^2 h_j x_j & \text{on } U_i \\ 0 & \text{off } U_i \end{cases} = \begin{cases} \displaystyle\sum_{j=1}^{m} h_j x_j & \text{on } U_i \\ 0 & \text{off } U_i \end{cases} \\ &= \rho_i w. \end{aligned}$$

Therefore $s[f] = \rho_1 w + \cdots + \rho_m w = w \sum_{j=1}^m \rho_i = w$. ☐

Notes. (1) By definition stable mappings in $C^\infty(X, Y)$ always form an open subset; Proposition 2.2 tells us that in the case of $C^\infty(X, \mathbf{R})$ the stable mappings also form a dense set. A natural question is "Are stable mappings dense in $C^\infty(X, Y)$ for an arbitrary manifold Y?" The answer is unfortunately "no." In Chapter VI we will give a counter-example. The general answer turns out to depend on the relative dimensions of X and Y of which more will be said later.

(2) The stable functions in $C^\infty(X, \mathbf{R})$ have a particularly nice form, since they are just the classical Morse functions. We see that such functions take on only a certain type of singularity (i.e., have only non-degenerate critical points). Moreover they are determined by this, and a certain condition on the set of critical points (i.e., have distinct critical values). In general, it is true that stable mappings take on only certain types of singularities; again, more will be presented on this point in the sixth chapter.

C. Immersions (1:1)

Proposition 2.3. *If X is compact and $f: X \to Y$ is a* 1:1 *immersion, then f is stable.*

Proof. We show that f is infinitesimally stable. With the given assumptions, $\operatorname{Im} f$ is a submanifold of Y. A vector field w along f can be identified with a vector field $\bar{w}$ on $\operatorname{Im} f$, since $f: X \to f(X)$ is a diffeomorphism. Let t be any smooth extension of $\bar{w}$ to all of Y. (To see that $\bar{w}$ has a smooth extension, construct it locally and use a partition of unity argument on $f(X)$.) Then $t \cdot f = \bar{w} \cdot f = w$. So $w = (df)(0) + t \cdot f$ and hence f is infinitesimally stable. □

Proposition 2.4. *Let X be compact and assume that* $\dim Y \geq 2 \cdot \dim X + 1$. *Then $f: X \to Y$ is stable iff f is a* 1:1 *immersion.*

Proof. We first note that any mapping equivalent to a 1:1 immersion is also a 1:1 immersion. If f is stable, then there is an open nbhd W_f of f in $C^\infty(X, Y)$ in which each mapping is equivalent to f. By the Whitney Embedding Theorem, there exists a 1:1 immersion in W_f. Hence f is a 1:1 immersion.

The converse is given by the last proposition. □

It is easy to see that not all immersions are stable.

(a) Consider $S^1 \to \mathbf{R}^2$ given by

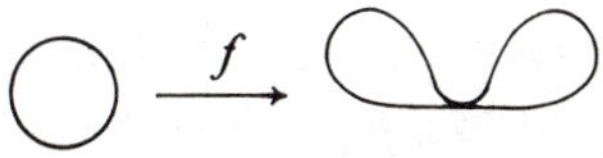

This can be perturbed slightly to $f': S^1 \to \mathbf{R}^2$ pictorially represented by

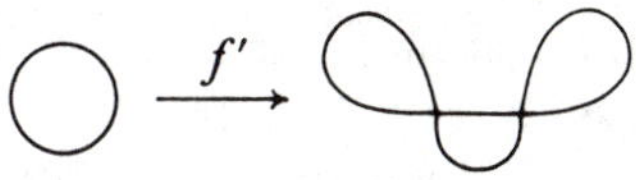

or

It is clear that f' and f'' are not equivalent to f since the number of self-intersections is an invariant of equivalence. So the first problem is that the self-intersections of f are not transversal.

(b) Consider $S^1 \to \mathbf{R}^2$ given by the trefoil

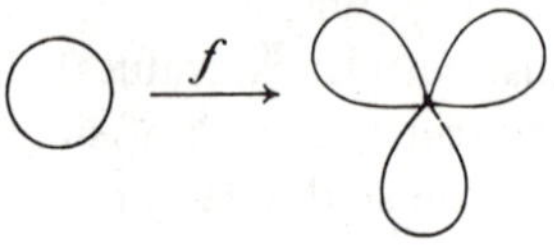

Perturb f slightly to $f' : S^1 \to \mathbf{R}^2$ given by

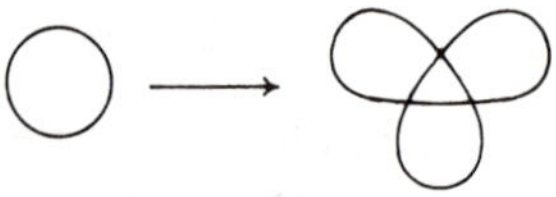

The number of self-intersections does not change (when counted with the proper multiplicity) but still f is not stable since the number of points in the image where there are crossings changes. Note, however, that each self-intersection of f is transversal.

So we find that even when $\dim X < \dim Y$ and $f: X \to Y$ has no singularities, there are still problems. It turns out that we need precise information on how the image of X under f sits inside Y.

§3. Immersions with Normal Crossings

Note that we are still assuming that X is compact.

Definition 3.1. *Let $f: X \to Y$ be smooth and $f^{(s)}: X^{(s)} \to Y^s$ the restriction of* $\underbrace{f \times \cdots \times f}_{s\text{-times}}: X^s \to Y^s$ *to $X^{(s)}$. (See* II, §4 *for the notation.) Let* $\Delta Y^s = \{(y, \ldots, y) \in Y^s \mid y \in Y\}$. *Then f is a* mapping with normal crossings *if for every $s > 1$, $f^{(s)} \pitchfork \Delta Y^s$.*

It is easy to see that the two examples at the end of the last section are not immersions with normal crossings, while the small perturbations are.

Proposition 3.2. *The set of mappings of X into Y with normal crossings is dense in $C^\infty(X, Y)$.*

Proof. Let $f: X \to Y$ be an immersion. Let $\beta^s : J_s^0(X, Y) \to Y^s$ be the multijet target mapping. Since β^s is a submersion $W^s = (\beta^s)^{-1}(\Delta Y^s)$ is a submanifold of $J_s^0(X, Y)$. It is easy to check that f is a mapping with normal crossings iff $j_s^0 f \pitchfork W^s$. So f is a mapping with normal crossings iff f satisfies a countable number of multijet transversality conditions. Applying the multijet transversality theorem and the fact that $C^\infty(X, Y)$ is a Baire space, we have the result. ∎

Corollary 3.3. *Immersions with normal crossings are dense in the set of all immersions.*

Proposition 3.4. *If $f: X \to Y$ is an immersion which is stable, then f is an immersion with normal crossings.*

Proof. Any mapping equivalent to an immersion with normal crossing is an immersion with normal crossings. □

We shall now proceed to prove the converse of this proposition, but first we need some preliminaries.

Definition 3.5. *Let V be a vector space and let $H_1, \ldots, H_r$ be subspaces of V. Then $H_1, \ldots, H_r$ are said to be* in general position *if for every sequence of integers $i_1, \ldots, i_s$ with $1 \le i_1 < \cdots < i_s \le r$.*

$$\operatorname{codim}(H_{i_1} \cap \cdots \cap H_{i_s}) = \operatorname{codim}(H_{i_1}) + \cdots + \operatorname{codim}(H_{i_s}).$$

Note. In the case $r = 2$, then H_1 and H_2 are in general position iff $H_1 + H_2 = V$. For

$$\begin{aligned}\dim(H_1 + H_2) &= \dim H_1 + \dim H_2 - \dim(H_1 \cap H_2)\\ &= n - (\operatorname{codim} H_1 + \operatorname{codim} H_2 - \operatorname{codim}(H_1 \cap H_2)).\end{aligned}$$

So $\dim(H_1 + H_2) = n$ iff $\operatorname{codim}(H_1 \cap H_2) = \operatorname{codim}(H_1) + \operatorname{codim}(H_2)$.

Lemma 3.6. *Let $f: X \to Y$ be an immersion with normal crossings. Choose q in Y. Let $f^{-1}(q) = \{p_1, \ldots, p_r\}$ all distinct points. ($f^{-1}(q)$ is finite since f is an immersion and X is compact.) Then the spaces $(df)_{p_1}(T_{p_1}X), \ldots, (df)_{p_r}(T_{p_r}X)$ are in general position as subspaces of T_qY.*

Proof. Choose a sequence of integers $i_1, \ldots, i_s$ such that $1 \le i_1 < \cdots < i_s \le r$. Let $H_j = (df)_{p_{i_j}}(T_{p_{i_j}}X)$ $(1 \le j \le s)$. Now $\bar{p} = (p_{i_1}, \ldots, p_{i_s})$ is a point in $X^{(s)}$ and $f(\bar{p}) = \bar{q} = (q, \ldots, q)$ is in ΔY^s. By the transversality condition of normal crossings of f, we have that

$$T_{\bar{q}}Y^s = (df)^{(s)}T_{\bar{p}}X^{(s)} + T_{\bar{q}}\Delta Y^s = H_1 \oplus \cdots \oplus H_s + T_{\bar{q}}\Delta Y^s.$$

Therefore

$$\begin{aligned}s \cdot \dim Y &= \dim T_{\bar{q}}Y^s\\ &= \dim(H_1 \oplus \cdots \oplus H_s) + \dim Y - \dim(H_1 \oplus \cdots \oplus H_s \cap T_{\bar{q}}\Delta Y^s)\end{aligned}$$

So $\operatorname{codim} H_1 + \cdots + \operatorname{codim} H_s = \operatorname{codim}(H_1 \oplus \cdots \oplus H_s \cap T_{\bar{q}}\Delta Y^s)$. But $H_1 \oplus \cdots \oplus H_s \cap T_{\bar{q}}\Delta Y^s \cong H_1 \cap \cdots \cap H_s$. □

Lemma 3.7. *Let $H_1, \ldots, H_r$ be subspaces of V in general position. Let $D = H_1 \cap \cdots \cap H_r$. Then there exist subspaces $F_1, \ldots, F_r$ of V satisfying*

(a) $V = D \oplus F_1 \oplus \cdots \oplus F_r$
(b) $H_i = D \oplus \sum_{j \ne i} F_j$
(c) $V = F_i \oplus H_i$.

Proof. Let $D_i = \bigcap_{j \neq i} H_j$. Choose a complementary subspace F_i to D in D_i $(1 \leq i \leq r)$. First note that $\dim F_i = \operatorname{codim} H_i$, since

$$\begin{aligned}\dim F_i &= \dim D_i - \dim D \\ &= \dim V - \operatorname{codim}\Big(\bigcap_{j \neq i} H_j\Big) - \dim V + \operatorname{codim}(H_1 \cap \cdots \cap H_r) \\ &= \operatorname{codim}(H_i)\end{aligned}$$

since the H_i's are in general position.

So

$$\begin{aligned}\dim D + \dim F_1 + \cdots + \dim F_r &= \dim D + \operatorname{codim} H_1 + \cdots + \operatorname{codim} H_r \\ &= \dim D + \operatorname{codim}(H_1 \cap \cdots \cap H_r) \\ &\qquad \text{(by general position)} \\ &= \dim V \qquad (\text{since } D = H_1 \cap \cdots \cap H_r).\end{aligned}$$

Thus to show (a) we need only show that the sum $D + F_1 + \cdots + F_r$ is direct. Suppose $d + f_2 + \cdots + f_r$ is in F_1 where f_j in F_j for $2 \leq j \leq r$. Note that each f_j is in $D_j - D$; so f_j is in H_i for $i \neq j$. Hence $d + f_2 + \cdots + f_r$ is in H_1. Now $F_1 \subset D_1 - D = H_2 \cap \cdots \cap H_r - H_1 \cap \cdots \cap H_r$. Hence $F_1 \cap H_1 = 0$ and thus $d + f_2 + \cdots + f_r = 0$.

Suppose $d + f_1 + \cdots + f_r = 0$; then $d + f_2 + \cdots + f_r \in F_1$ and by the above equals 0. By a simple induction argument we know that each $f_i = 0$ $(1 \leq i \leq r)$ and thus that $d = 0$.

To show (b) we just note that $D \subset H_i$ and that $F_j \subset H_i$ for $i \neq j$. So $H_i \supset D \oplus \sum_{j \neq i} F_j$. But $\operatorname{codim}(D \oplus \sum_{j \neq i} F_j) = \dim F_i = \operatorname{codim} H_i$. (c) follows from (a) and (b). □

We need two more preparatory lemmas, but first some definitions: let X_0 be a submanifold of X.

(a) An *X vector field s on X_0* is a section of $TX|X_0$.

(b) s is *tangent to X_0* if for every p in X_0 $s(p) \in T_pX_0 \subset T_pX$.

Lemma 3.8. *Let $H_1, \ldots, H_r$ be subspaces of $\mathbf{R}^n$ in general position. Regard each H_i as a submanifold of $\mathbf{R}^n$. Let t_j be an $\mathbf{R}^n$ vector field on H_j $(1 \leq j \leq r)$. Then there is a vector field t on $\mathbf{R}^n$ such that for every j, $t - t_j$ (restricted to H_j) is tangent to H_j.*

Proof. Choose $F_1, \ldots, F_r$ as in Lemma 3.7. Choose an inner product on $\mathbf{R}^n$ so that $D, F_1, \ldots, F_r$ are mutually orthogonal subspaces. Define $\pi_j : \mathbf{R}^n \to H_j$ to be orthogonal projection. View t_j as a map of H_j into $\mathbf{R}^n$. Let $\bar{t}_j$ be the vector field on $\mathbf{R}^n$ given by $\bar{t}_j = t_j \cdot \pi_j$.

Let $\zeta_j = \bar{t}_j - \pi_j \cdot \bar{t}_j$ $(1 \leq j \leq r)$. ζ_j is the normal component of $\bar{t}_j$ to H_j. Indeed, $\pi_j \zeta_j = \pi_j \bar{t}_j - \pi_j^2 \bar{t}_j = 0$. since $\pi_j^2 = \pi_j$. So $\operatorname{Im}(\zeta_j)$ is in F_j. Define $t = \zeta_1 + \cdots + \zeta_r$. We claim that $t|H_j - t_j$ is tangent to H_j. It is sufficient to show that $\pi_j(t - t_j) = t - t_j$ on H_j. Note that $\pi_j(\zeta_i) = \zeta_i$ for $i \neq j$ since $\operatorname{Im}(\zeta_i) \subset F_i \subset H_j$ and $\pi_j|H_j = id_{H_j}$. Hence on H_j,

$$\begin{aligned}\pi_j(t - t_j) &= t - \zeta_j - \pi_j t_j \\ &= t - \bar{t}_j + \pi_j \bar{t}_j - \pi_j t_j \\ &= t - t_j\end{aligned}$$

since $t_j = \bar{t}_j$ on H_j. □

Definition 3.9. *Let Y be a manifold with submanifolds $Y_1, \ldots, Y_r$. Suppose $q \in Y_1 \cap \cdots \cap Y_r$. Then $Y_1, \ldots, Y_r$ are in* general position *at q if $T_qY_1, \ldots, T_qY_r$ are in general position in T_qY.*

Lemma 3.10. *Assume that the submanifolds $Y_1, \ldots, Y_r$ of Y are in general position at a point q. Then there exists a nbhd W of q in Y, a chart $\phi: W \to \mathbf{R}^n$ and subspaces $H_1, \ldots, H_r$ such that $Y_i \cap W = \phi^{-1}(H_i)$ $(1 \leq i \leq r)$ where $n = \dim Y$.*

Note. This lemma says that in a nbhd of a point of general position, the submanifolds $Y_1, \ldots, Y_r$ can be simultaneously linearized.

Proof. Let $m_i = \operatorname{codim}(Y_i)$ in Y. There is a nbhd W^i of q and functions $f_{i,1}, \ldots, f_{i,m_i}$ such that

$$Y_i \cap W^i = \{p \in W^i \mid f_{i,1}(p) = \cdots = f_{i,m_i}(p) = 0\}$$

since Y_i is a submanifold of Y. Let $\overline{W} = \bigcap_{i=1}^r W^i$. By general position, we know that

$$\begin{aligned}\operatorname{codim}(T_qY_1 \cap \cdots \cap T_qY_r) &= \operatorname{codim}(T_qY_1) + \cdots + \operatorname{codim}(T_qY_r)\\ &= m_1 + \cdots + m_r \leq n.\end{aligned}$$

Let $l = n - m_1 - \cdots - m_r$. Consider the functions $\{f_{i,j}\}_{(1 \leq i \leq r)(1 \leq j \leq m_i)}$. The number of such functions is $m_1 + \cdots + m_r$. We claim that $(df_{ij})_q$ form a linearly independent set in T_q^*Y. Clearly the subspace of T_qY annihilated by all the $(df_{i,j})_q$ is just $T_qY_1 \cap \cdots \cap T_qY_r$ which has dimension $m_1 + \cdots + m_r$. Since there are exactly that many $(df_{i,j})_q$, they must be linearly independent.

Now choose functions $g_1, \ldots, g_l$ defined on $\overline{U}$ so that $(dg_1)_q, \ldots, (dg_l)_q$, $(df_{i,j})_q$ form a basis of T_q^*Y. Consider $\phi: \overline{W} \to \mathbf{R}^n$ given by

$$p \mapsto (f_{1,1}(p), \ldots, f_{r,m_r}(p), g_1(p), \ldots, g_l(p)).$$

By construction the Jacobian of this map is nonsingular at q. Choose W a nbhd of q, $W \subset \overline{W}$ on which ϕ is a diffeomorphism. In terms of this chart $Y_i \cap W$ is given by the linear equations $f_{i,j} = 0$ $(1 \leq j \leq m_i)$. □

Theorem 3.11. *Let $f: X \to Y$ be an immersion. Then f is stable iff f has only normal crossings.*

Proof. We assume that f has only normal crossings and prove that f is infinitesimally stable. The converse has already been shown in Proposition 3.4. Let $q \in Y$ and $\{p_1, \ldots, p_r\} = f^{-1}(q)$. We claim that there exists a nbhd W_q of q in Y and nbhds U_i of p_i in X $(1 \leq i \leq r)$ satisfying

(a) $U_i \cap U_j = \varnothing \qquad 1 \leq i < j \leq r.$

(b) $f|U_i$ is a 1:1 proper immersion.

(c) $f(U_i) \subset W_q$.

(d) $f^{-1}(W_q) = \bigcup_{i=1}^r U_i$.

(e) W_q can be chosen as small as desired.

It is easy to choose V_i satisfying (a) and (b). Also, there is a nbhd W_q of q such that $f^{-1}(W_q) \subset \bigcup_{i=1}^r V_i$. W_q can be chosen as small as wished, for if not, there is a sequence of points $x_1, x_2, \ldots$ in $X - \bigcup_{i=1}^r V_i$ such that $f(x_i)$ converges to q. Since $X - \bigcup_{i=1}^r V_i$ is compact we can assume that $\{x_j\}$ converges to some point p not in $f^{-1}(q)$. The continuity of f guarantees that $f(p) = q$, a contradiction. Let $U_i = V_i \cap f^{-1}(W_q)$. Now $Y_i = f(U_i)$ $(1 \leq i \leq r)$ are submanifolds of Y since $f|U_i$ is a 1:1 proper immersion. By Lemma 3.6 $Y_1, \ldots, Y_r$ are in general position at q since f is assumed to have only normal crossings. Choose W_q small enough so that it satisfies the conditions of Lemma 3.10, i.e., $Y_1, \ldots, Y_r$ are simultaneously linearized in W_q.

The collection $\{W_q\}_{q \in Y}$ form an open covering of Y. Hence $f^{-1}(W_q)$ is an open covering of X. By the compactness of X, there is a finite subcover of X, given by $f^{-1}(W_{q_1}), \ldots, f^{-1}(W_{q_m})$. Choose a partition of unity ρ_i $(1 \leq i \leq m)$ subordinate to this covering. Let ω be a vector field along f. Then $\rho_i\omega$ is also in $C_f^\infty(X, TY)$. Since $\omega = \sum_{i=1}^m \rho_i\omega$, it is sufficient to show that the criterion for infinitesimal stability holds for vector fields along f whose support lies in a given $f^{-1}(W_q)$. So let $\omega \in C_f^\infty(X, TY)$ with supp $\omega \subset f^{-1}(W_q)$ for some fixed q.

In W_q, we have the submanifolds $Y_i = f(U_i)$. Define t_i a vector field on Y_i by $t_i = \omega \cdot (f|U_i)^{-1}$. This is possible since $f|U_i: U_i \to Y_i$ is a diffeomorphism. Moreover each t_i is compactly supported. By the general position of $Y_1, \ldots, Y_r$ at q and Lemmas 3.8 and 3.10, there exists a vector field t on W_q (which is compactly supported since each t_i is compactly supported) such that $t|f(U_i) - t_i$ is tangent to Y_i. Extend t to a vector field on Y by setting $t \equiv 0$ off W_q. Consider $\omega' = \omega - t \cdot f$. ω' has the property that for every p in U_i, $\omega'(p) = \omega(p) - t(f(p))$ is tangent to $f(U_i)$ at $f(p)$. So there exists a unique vector field s_i on U_i such that $(df) \cdot s_i = \omega'$ since f is a 1:1 immersion from $U_i \to f(U_i)$. Moreover s_i is compactly supported so there exists a vector field $s = s_i$ on U_i and $s = 0$ off $f^{-1}(W_q)$. By construction $\omega = (df) \cdot s + t \cdot f$. □

Proposition 3.12. *Let* dim $Y = 2$ dim X, *X compact. Then* $f: X \to Y$ *is stable iff f is an immersion with normal crossings.*

Proof. By the Whitney Immersion Theorem (Theorem II, 5.6) the immersions of $X \to Y$ are open and dense, so stable mappings must be immersions. Using Corollary 3.3 any stable map is an immersion with normal crossings. The converse is given by Theorem 3.11. □

Having settled the question of stability for mappings without singularities (i.e., submersions and immersions) we will now focus our attention on mappings with singularities. As our study of Morse functions suggested, in order to understand singularities it is useful to describe the behavior of a function in a nbhd of a given singularity by a fixed "normal form." In the next section of this chapter we will investigate the stability of a class of mappings similar to Morse functions and in doing so again demonstrate the usefulness of normal forms.

Exercise. Interpret what it means for an immersion of $X \to Y$ to have normal crossings when $\dim Y = 2 \dim X$ in terms of the number of self-crossings.

In general show that if $f: X \to Y$ is a stable immersion then the number of points in $f^{-1}(q)$—for any q in Y—is bounded by $\dim Y/(\dim Y - \dim X)$.

4. Submersions with Folds

Let X and Y be smooth manifolds with $\dim X \geq \dim Y$. Let $k = \dim X - \dim Y$. Let $f: X \to Y$ satisfy $j^1 f \pitchfork S_1$ where S_1 is the submanifold of $J^1(X, Y)$ of jets of corank 1 as defined in II, §5. Applying II, Theorem 4.4, we see that $S_1(f) = (j^1 f)^{-1}(S_1)$ is a submanifold of X with $\operatorname{codim} S_1(f) = \operatorname{codim} S_1 = k + 1$ (II, Theorem 5.4). Note that at a point x in $S_1(f)$, $\dim \operatorname{Ker} (df)_x = k + 1$; that is, the tangent space to $S_1(f)$ and the kernel of $(df)_x$ have complementary dimensions.

Definition 4.1. *Let $f: X \to Y$ satisfy $j^1 f \pitchfork S_1$. Then x in $S_1(f)$ is a* fold point *if $T_x S_1(f) + \operatorname{Ker} (df)_x = T_x X$.*

Definition 4.2. (a) *A mapping $f: X \to Y$ is a* submersion with folds *if the only singularities of f are fold points. (Note that a submersion with folds, f, satisfies $j^1 f \pitchfork S_1$.)*

(b) *Let $f: X \to Y$ be a submersion with folds. Then $S_1(f)$ is called the* fold locus *of f.*

Example. In the case $Y = \mathbf{R}$, the set of submersions with folds is precisely the set of Morse functions on X.

The first obvious fact to observe about submersions with folds is:

Lemma 4.3. *Let $f: X \to Y$ be a submersion with folds, then f restricted to its fold locus is an immersion.*

The main theorem for this section describes a simple criterion to determine when a submersion with folds is stable.

Theorem 4.4. *Let $f: X \to Y$ be a submersion with folds. Then f is stable iff $f|S_1(f)$ is an immersion with normal crossings.*

Notes. (1) The criterion that f restricted to its fold locus is an immersion with normal crossings in the case that f is a Morse function is just the criterion that f has distinct critical values.

(2) We shall actually prove Theorem 4.4 with "stable" replaced by "infinitesimally stable" and appeal to the as yet unproved Theorem 1.5 of Mather to obtain the desired result.

Example. The following is an example of the necessity of this criterion

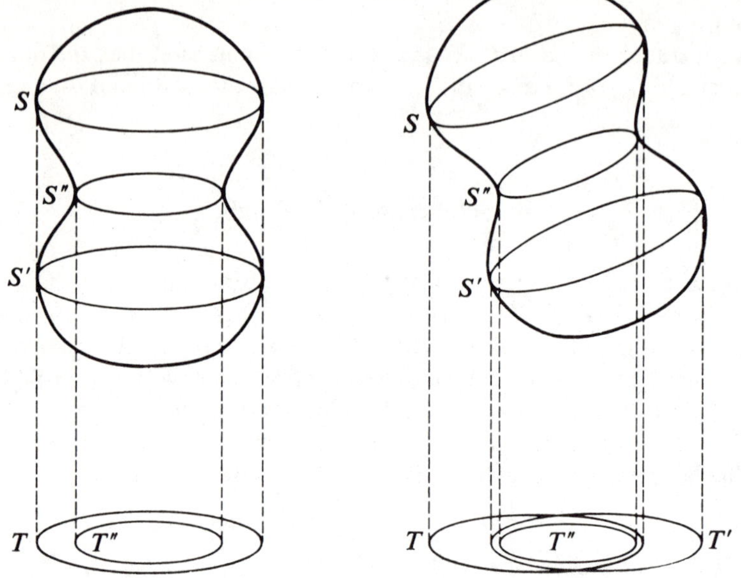

Figure 1

X is a sphere dented at the equator, Y is $\mathbf{R}^2$, and $f: X \to Y$ is the projection of $\mathbf{R}^3$ onto $\mathbf{R}^2$, restricted to X. The singular set is a curve of fold points consisting of the three circles S, S', and S'' while the image forms two circles T and T''. f is obviously not stable. For example, we can perturb the picture above so that the image of S and S' intersect transversely at isolated points.

Proof of Theorem 4.4. Necessity. We know that $f|S_1(f)$ is an immersion and—with the assumption that f is infinitesimally stable—we show that $f|S_1(f)$ has normal crossings. Applying Theorem 3.11 it is enough to show that $f|S_1(f)$ is infinitesimally stable.

Let τ be a vector field along $g = f|S_1(f)$. Extend τ to a vector field $\bar{\tau}$ along f. (Check that this can be done locally and use a partition of unity argument.) Since f is infinitesimally stable there exist vector fields $\bar{\zeta}$ on X and η on Y so that $\bar{\tau} = (df)(\bar{\zeta}) + \eta \cdot f$. On $S_1(f)$, we have $\tau = (dg)(\bar{\zeta}|S_1(f)) + \eta \cdot g$. Now $\bar{\zeta}|S_1(f) = \zeta + \zeta'$ where ζ is a vector field on $S_1(f)$ and ζ' is in $\operatorname{Ker}(dg)$ since by the definition of fold points $TX|S_1(f) = TS_1(f) \oplus \operatorname{Ker}(dg)$. Thus $\tau = (dg)(\zeta) + \eta \cdot g$ and $g = f|S_1(f)$ is infinitesimally stable. □

Before proving the sufficiency part of Theorem 4.4 we shall need a normal form theorem for the local behavior of a submersion with folds near a fold point similar to II, Theorem 6.9 for Morse functions.

Theorem 4.5. *Let $f: X \to Y$ be a submersion with folds and let p be in $S_1(f)$. Then there exist coordinates $x_1, \ldots, x_n$ centered at p and $y_1, \ldots, y_m$ centered at $f(p)$ so that in these coordinates f is given by*

$$(x_1, \ldots, x_n) \mapsto (x_1, \ldots, x_{m-1}, x_m{}^2 \pm \cdots \pm x_n{}^2).$$

Remark. From this local normal form we see the reason for the nomenclature "fold point." Take the particularly simple example of 2-manifolds. In this case the normal form is given by $(x_1, x_2) \mapsto (x_1, x_2^2)$. This map is depicted in Figure 2. We first map the (x_1, x_2)-plane onto the parabolic cylinder $x_3 = x_2^2$ in (x_1, x_2, x_3) space by the map $(x_1, x_2) \mapsto (x_1, x_2, x_2^2)$ and then follow this by the projection onto the (x_1, x_3) plane.

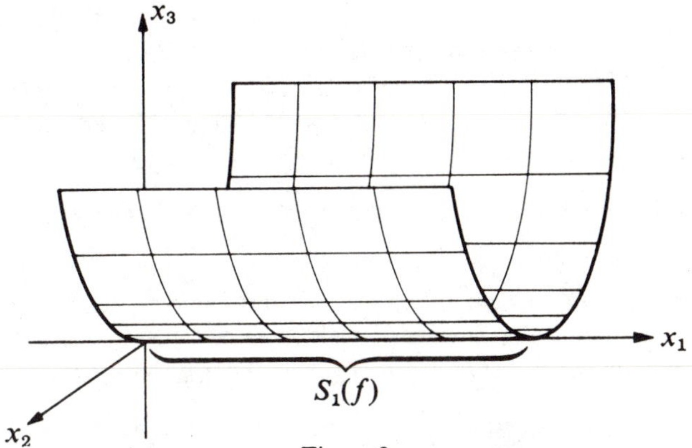

Figure 2

Proof of Theorem 4.5. First choose coordinates $y_1, \ldots, y_m$ centered at $f(p)$ so that the image of $S_1(f)$ under f is described by the equation $y_m = 0$. Since $f|S_1(f)$ is an immersion the image is locally an $m - 1$-dimensional manifold so this choice of y's is possible. (See I, Theorem 2.10.) Since $f: S_1(f) \to \{y_m = 0\}$ is a diffeomorphism locally near p, we can choose coordinates $x_1, \ldots, x_n$ near p so that $x_i = y_i \cdot f$ for $1 \leq i \leq m - 1$ and $S_1(f)$ is defined near p by the equations $x_m = \cdots = x_n = 0$. In those coordinates f has the form $(x_1, \ldots, x_m) \mapsto (x_1, \ldots, x_{m-1}, f_m(x))$. Of course $S_1(f)$ is also described by the equations $\partial f_m/\partial x_i = 0$ $(m \leq i \leq n)$. So these partial derivatives vanish when $x_m = \cdots = x_n = 0$. Now f_m itself vanishes when $x_m = \cdots = x_n = 0$ since the equation $y_m = 0$ describes the image of $S_1(f)$. These two facts imply that

$$f_m(x) = \sum_{i,j \geq m} h_{ij}(x) x_i x_j$$

where $h_{ij}(x)$ are smooth functions. (This is similar to II, Lemma 6.10). Suppose that $\{h_{ij}(0)\}$ is a nonsingular $(n - m + 1) \times (n - m + 1)$ matrix. The rest of the proof is similar to the proof of II, Theorem 6.9 about Morse functions. In particular, using arguments as in II, Lemma 6.11 show that by a change of coordinates in $(x_m, \ldots, x_n)$ we can assume that f_m has the form $x_m^2 \pm \cdots \pm x_n^2$. □

We prove the supposition in the following lemma and in doing so emphasize that point in the proof where the transversality hypothesis is used.

Lemma 4.6. *The matrix $h_{ij}(0)$ is nonsingular.*

Proof. Suppose it were singular. Then we could make a linear change of coordinates in the variables $x_m, \ldots, x_n$ so that this matrix is in diagonal form with entries ± 1 and 0 along the diagonal and at least one of the diagonal entries is zero. Thus we could assume that f has the same 2-jet at 0 as the map

$$(*) \qquad (x_1, \ldots, x_n) \mapsto (x_1, \ldots, x_{m-1}, x_m^2 \pm \cdots \pm x_r^2)$$

where $r < n$. Now the condition that $j^1 f \pitchfork S_1$ at p is a condition on $j^1 f(p)$ and $(dj^1 f)_p$, i.e., on the 2-jet of f at p. Thus if the transversality condition is satisfied by $j^1 f$ at p it is satisfied by any other map with the same 2-jet at p as f. For the mapping (*) the set $S_1(*)$ is given by the equations $x_m = \cdots = x_r = 0$ and thus has codimension $r - m + 1 < n - m + 1$. If the transversality condition were satisfied it would have codimension $n - m + 1$; so we have a contradiction and the matrix is nonsingular. □

We now isolate the main step in the proof of the sufficiency part of Theorem 4.4 with the following:

Lemma 4.7. *Let p be a fold point of f and let τ be a vector field along f defined on some nbhd U of p such that $\tau|(S_1(f) \cap U) = 0$. Then there exists a vector field ζ defined on a nbhd $V \subset U$ of p such that $\tau = (df)(\zeta)$ on V.*

Proof. Choose coordinates $x_1, \ldots, x_n$ centered at p and coordinates $y_1, \ldots, y_m$ centered at $f(p)$ satisfying Theorem 4.5. In these coordinates τ is just an m-tuple of smooth functions $\tau = (\tau_1, \ldots, \tau_m)$ (that is $\tau(x) = \sum_{i=1}^m \tau_i(x)(\partial/\partial y_i)$ and a vector field ζ is given by an n-tuple of functions $(\zeta_1, \ldots, \zeta_n)$. Given the normal form of f in these coordinates solving the equation $\tau = (df)(\zeta)$ is equivalent to solving the system

$$\tau_i = \zeta_i \qquad 1 \le i \le m - 1$$

and

$$\tau_m = x_m \zeta_m + \cdots + x_n \zeta_n.$$

The first equations are trivially solvable and the last equation is solvable providing that $\tau_m = 0$ on the points $x_m = \cdots = x_n = 0$. But these equations describe the fold locus $S_1(f) \cap U$ and by assumption $\tau_m \equiv 0$ on this set. □

Proof of Theorem 4.4. Sufficiency. We assume that f is a submersion with folds and that $f|S_1(f)$ is an immersion with normal crossings. We will show that f is infinitesimally stable. To do this let τ be a vector field along f; we must find vector fields ζ in X and η on Y so that $\tau = (df)(\zeta) + \eta \cdot f$. Since $g = f|S_1(f)$ is an immersion with normal crossings, g is infinitesimally stable so there exist vector fields $\bar{\zeta}$ on $S_1(f)$ and η on Y so that $\tau|S_1(f) = (dg)(\bar{\zeta}) + \eta \cdot g$. Extend $\bar{\zeta}$ to a vector field ζ on X and consider a new $\tau = \tau - (df)(\zeta) - \eta \cdot f$. This new τ has the property that $\tau|S_1(f) = 0$.

Applying Lemma 4.7 we can assume that around each point p in $S_1(f)$ there is a nbhd V of p and a vector field ζ on V so that $\tau = (df)(\zeta)$ on V. At points p not in $S_1(f)$ there exists a nbhd of p such that $f|V$ is a submersion. On these nbhds the equation $\tau = (df)(\zeta)$ is clearly solvable. Using a partition of unity argument we obtain a global solution by patching. □

Chapter IV

The Malgrange Preparation Theorem

§1. The Weierstrass Preparation Theorem

In this chapter we will prove a technical theorem about smooth functions which will be used to prove Mather's Theorem about stable mappings and to establish the existence of normal forms for singularities of certain stable mappings. In order to make the theorem palatable, we first state and prove the corresponding but less complicated result for analytic functions of several *complex* variables.

Theorem 1.1. (*Weierstrass Preparation Theorem*). *Let F be a complex-valued holomorphic function defined on a nbhd of* 0 *in* $\mathbf{C} \times \mathbf{C}^n$ *satisfying:*

(a) $F(w, 0) = w^k g(w)$ *where* $(w, 0) \in \mathbf{C} \times \mathbf{C}^n$ *and g is a holomorphic function of one variable in some nbhd of* 0 *in* $\mathbf{C}$, *and*

(b) $g(0) \neq 0$.

Then there exists a complex-valued holomorphic function q defined on a nbhd of 0 *in* $\mathbf{C} \times \mathbf{C}^n$ *and complex-valued holomorphic functions* $\lambda_0, \ldots, \lambda_{k-1}$ *defined on a nbhd of* 0 *in* $\mathbf{C}^n$ *such that*

(i) $(qF)(w, z) = w^k + \sum_{i=0}^{k-1} \lambda_i(z)w^i$ *for all* (w, z) *in some nbhd of* 0 *in* $\mathbf{C} \times \mathbf{C}^n$, *and*

(ii) $q(0) \neq 0$.

Remark. The reader may well ask what such a theorem is good for. Before we proceed we point out one trivial consequence. Given a nonzero holomorphic function F of $n + 1$ complex variables, we may assume (by a linear change of coordinates) that $F = F(w, z)$ is in the form above. Then the Weierstrass Preparation Theorem states that the zero set of F equals the zero set of the function

$$w^k + \sum_{i=0}^{k-1} \lambda_i(z)w^i$$

which is just a "branched covering surface" over the z hyperplane.

We will actually prove a more general result.

Theorem 1.2. (*Weierstrass Division Theorem*). *Let F, g, and k be as above and let G be any complex-valued holomorphic function defined on a nbhd of* 0 *in* $\mathbf{C} \times \mathbf{C}^n$. *Then there exist complex-valued holomorphic functions q and r defined on a nbhd of* 0 *in* $\mathbf{C} \times \mathbf{C}^n$ *such that*

(i) $G = qF + r$, *and*

(ii) $r(w, z) = \sum_{i=0}^{k-1} r_i(z)w^i$ *for all* (w, z) *in some nbhd of* 0 *in* $\mathbf{C} \times \mathbf{C}^n$ *where each r_i is a holomorphic function defined on a nbhd of* 0 *in* $\mathbf{C}^n$.

(iii) *q and r are unique (on some nbhd of* 0).

Proof that Theorem 1.2 ⇒ Theorem 1.1. Let $G(w, z) = w^k$ and apply Theorem 1.2. Setting λ_i equal to r_i proves (i). To show that $q(0) \neq 0$, consider

$$\begin{aligned} w^k &= q(w, 0)F(w, 0) + r(w, 0) \\ &= q(w, 0)w^k g(w) + \sum_{i=0}^{k-1} r_i(0)w^i. \end{aligned}$$

Since both sides are analytic functions of w, we may use power series techniques to conclude that $q(0)g(0) = 1$ and thus $q(0) \neq 0$. □

Proof of Theorem 1.2 (iii). Uniqueness. Suppose

$$G = qF + r = q_1F + r_1.$$

Then $(q - q_1)F = r_1 - r$. Fix z in $\mathbf{C}^n$; then $r_1 - r$ is a polynomial of degree $\leq k - 1$ in w and has at most $k - 1$ roots (including multiplicity). We shall show that there is a nbhd of 0 in $\mathbf{C}^n$ such that for every z in this nbhd $(q - q_1)F$ has at least k zeroes when viewed as a function of w. Thus we can conclude that $r_1 = r$ and $q = q_1$ (since $F \not\equiv 0$ on a nbhd of 0).

It is clearly enough to show that $F(\cdot, z)$ has k zeroes. Let $\bar{F}(w) = F(w, 0)$. Since the zeroes of a nonzero complex analytic function of one variable are isolated, and $\bar{F}(0) = 0$, there is a constant $\delta > 0$ for which $\bar{F}(w) \neq 0$ whenever $0 < |w| \leq \delta$. Let $\varepsilon = \inf_{|w|=\delta} |\bar{F}(w)|$. Since F is continuous there is a constant $\sigma > 0$ for which $|F(w, z) - \bar{F}(w)| < \varepsilon$ whenever $|z_j| < \sigma$ for $j = 1, \ldots, n$ where $z = (z_1, \ldots, z_n)$ and $|w| = \delta$. Choose such a z and let $h(w) = F(w, z)$. Since

$$|h(w) - \bar{F}(w)| < \varepsilon \leq |\bar{F}(w)| \quad \text{when} \quad |w| = \delta$$

we can apply Rouché's Theorem [see Ahlfors, Complex Analysis, p. 152] and conclude that h and $\bar{F}$ have the same number of zeroes in the disk $|w| < \delta$. Since $\bar{F}(w) = w^k g(w)$ on a nbhd of 0, we know that $\bar{F}$ has k-zeroes (counting multiplicity) in this disk. □

Definition 1.3. *Let $P_k: \mathbf{C} \times \mathbf{C}^n \times \mathbf{C}^k \to \mathbf{C}$ be the polynomial $P_k(w, z, \lambda) = w^k + \sum_{i=0}^{k-1} \lambda_i w^i$ where $\lambda = (\lambda_0, \ldots, \lambda_{k-1})$.*

The heart of the proof of the Division Theorem lies in proving the theorem for the polynomials P_k.

Theorem 1.4. (*Polynomial Division Theorem*). *Let $G(w, z)$ be holomorphic on a nbhd of 0 in $\mathbf{C} \times \mathbf{C}^n$. Then there exist holomorphic functions $q(w, z, \lambda)$ and $r(w, z, \lambda)$ defined on a nbhd of 0 in $\mathbf{C} \times \mathbf{C}^n \times \mathbf{C}^k$ satisfying:*

(i) $G(w, z) = q(w, z, \lambda)P_k(w, \lambda) + r(w, z, \lambda)$, *and*

(ii) $r(w, z, \lambda) = \sum_{i=0}^{k-1} r_i(z, \lambda)w^i$ *where each r_i is a holomorphic function defined on a nbhd of 0 in $\mathbf{C}^n \times \mathbf{C}^k$.*

Proof that Theorem 1.4 ⇒ Theorem 1.2. Let F and G be as in the hypotheses of the Division Theorem. Using Theorem 1.4 choose holomorphic function q_F, r_F, q_G, and r_G satisfying

(*) $$F = q_F P_k + r_F \quad \text{and} \quad G = q_G P_k + r_G.$$

Also $r_F(w, z, \lambda) = \sum_{i=0}^{k-1} r_i^F(z, \lambda)w^i$. First we note that $r_i^F(0) = 0$ and $q_F(0) \neq 0$; for

$$\begin{aligned} w^k g(w) &= F(w, 0) = q_F(w, 0)P_k(w, 0) + r_F(w, 0) \\ &= q_F(w, 0)w^k + \sum_{i=0}^{k-1} r_i^F(0)w^i. \end{aligned}$$

Now apply a simple power series argument to both sides, recalling that $g(0) \neq 0$.

Next let $f_i(\lambda) = r_i^F(0, \lambda)$. We claim that $\det((\partial f_i/\partial\lambda_j)(0)) \neq 0$. For let $z = 0$, then

$$\begin{aligned} w^k g(w) &= F(w, 0) = q_F P_k + r_F \\ &= q_F(w, 0, \lambda)\left(w^k + \sum_{i=0}^{k-1} \lambda_i w^i\right) + \sum_{i=0}^{k-1} f_i(\lambda)w^i. \end{aligned}$$

Differentiate both sides with respect to λ_j and evaluate at $\lambda = 0$ to obtain

$$0 = \frac{\partial q_F}{\partial\lambda_j}(w, 0)w^k + q_F(w, 0)w^j + \sum_{i=0}^{k-1} \frac{\partial f_i}{\partial\lambda_j}(0)w^i.$$

Matching coefficient on w^i for $i < j$ we see that $(\partial f_i/\partial\lambda_j)(0) = 0$ and also $(\partial f_j/\partial\lambda_j)(0) = -q_F(0)$. So the matrix $((\partial f_i/\partial\lambda_j)(0))$ is lower triangular and has determinant equal to $(-1)^k q_F(0) \neq 0$ (shown above).

We now apply the Implicit Function Theorem (for holomorphic functions) [10, p. 17] to insure the existence of holomorphic functions $\theta_i(z)$ $(0 \leq i \leq k-1)$ satisfying

(a) $r_j^F(z, \theta(z)) \equiv 0$ where $\theta(z) = (\theta_0(z), \ldots, \theta_{k-1}(z))$,

and

(b) $\theta(0) = 0$ (since $r_j^F(0) = 0$).

Define $\bar{q}(w, z) = q_F(w, z, \theta(z))$, and $P(w, z) = P_k(w, \theta(z))$. Then

$$\begin{aligned} F(w, z) &= q_F(w, z, \theta(z))P(w, \theta(z)) + r_F(w, z, \theta(z)) \\ &= \bar{q}(w, z)P(w, z). \end{aligned}$$

Moreover $\bar{q}(0) = q_F(0) \neq 0$, so $P(w, z) = F(w, z)/\bar{q}(w, z)$ on a nbhd of 0 in $\mathbf{C} \times \mathbf{C}^n$.

From the second equation in (*) we obtain

$$\begin{aligned} G(w, z) &= q_G(w, z, \theta(z))P_k(w, \theta(z)) + r_G(w, z, \theta(z)) \\ &= q(w, z)F(w, z) + r(w, z) \end{aligned}$$

where

$$q(w, z) = \frac{q_G(w, z, \theta(z))}{\bar{q}(w, z)}$$

and

$$r(w, z) = r_G(w, z, \theta(z)) = \sum_{i=0}^{k-1} r_i^G(z, \theta(z))w^i.$$

Finally let $r_i(z) = r_i^G(z, \theta(z))$. $\square$

Proof of Theorem 1.4. Given a holomorphic function $G(w, z)$ we must produce q and r so that

$$G(w, z) = q(w, z, \lambda)P_k(w, \lambda) + r(w, z, \lambda)$$

where r is of the form $\sum_{i=0}^{k-1} r_i(z, \lambda)w^i$. We recall the following form of the Cauchy Integral Formula:

$$G(w, z) = \frac{1}{2\pi i}\int_\gamma \frac{G(\eta, z)}{(\eta - w)}\,d\eta$$

where γ is a simple closed curve in the complex plane with w in the interior of γ. Now note that for appropriately defined holomorphic functions $s_i(w, \lambda)$,

$$P_k(\eta, \lambda) - P_k(w, \lambda) = (\eta - w)\sum_{i=0}^{k-1} s_i(\eta, \lambda)w^i,$$

or

$$(*) \qquad \frac{P_k(\eta, \lambda)}{\eta - w} = \frac{P_k(w, \lambda)}{\eta - w} + \sum_{i=0}^{k-1} s_i(\eta, \lambda)w^i.$$

Thus

$$G(w, z) = \frac{1}{2\pi i}\int_\gamma \frac{G(\eta, z)}{(\eta - w)}\frac{P_k(\eta, \lambda)}{P_k(\eta, \lambda)}\,d\eta$$

(Using (*))

$$= \left(\frac{1}{2\pi i}\int_\gamma \frac{G(\eta, z)}{P_k(\eta, \lambda)(\eta - w)}\,d\eta\right)P_k(w, \lambda) + \sum_{i=0}^{k-1}\left(\frac{1}{2\pi i}\int_\gamma \frac{G(\eta, z)}{P_k(\eta, \lambda)}s_i(\eta, \lambda)\,d\eta\right)w^i$$

if the appropriate integrals are in fact well-defined. Thus we should like to set

$$q(w, z, \lambda) = \frac{1}{2\pi i}\int_\gamma \frac{G(\eta, z)}{P_k(\eta, \lambda)(\eta - w)}\,d\eta$$

and

$$r_i(z, \lambda) = \frac{1}{2\pi i}\int_\gamma \frac{G(\eta, z)}{P_k(\eta, \lambda)}s_i(\eta, \lambda)\,d\eta.$$

But these integrals give well-defined functions as long as the zeroes of $P_k(\eta, \lambda)$ do not occur on the curve γ for λ near 0 in $\mathbf{C}^k$. Such a γ is easily chosen. ∎

2. The Malgrange Preparation Theorem

The proof given in §1 of the Weierstrass Preparation Theorem can be adapted to a corresponding theorem about real smooth functions, the difficulties in the adaptation appearing in the Polynomial Division Theorem (1.4). Our proof follows Nirenberg [41].

Theorem 2.1. (*Mather Division Theorem*). *Let F be a smooth real-valued function defined on a nbhd of* 0 *in* $\mathbf{R} \times \mathbf{R}^n$ *such that* $F(t, 0) = g(t)t^k$ *where* $g(0) \neq 0$ *and g is smooth on some nbhd of* 0 *in* $\mathbf{R}$. *Then given any smooth real-valued function G defined on a nbhd of* 0 *in* $\mathbf{R} \times \mathbf{R}^n$, *there exist smooth functions q and r such that*

(i) $G = qF + r$ *on a nbhd of* 0 *in* $\mathbf{R} \times \mathbf{R}^n$, *and*

(ii) $r(t, x) = \sum_{i=0}^{k-1} r_i(x)t^i$ *for* $(t, x) \in \mathbf{R} \times \mathbf{R}^n$ *near* 0.

Notes. (1) The Malgrange Preparation Theorem which states that there exists a smooth q with $q(0) \neq 0$ such that $(qF)(t, x) = t^k + \sum_{i=0}^{k-1} \lambda_i(x)t$ follows from 2.1 in precisely the same way that Theorem 1.1 follows from Theorem 1.2.

(2) In the complex analytic theorem q and r are unique; this is not necessarily true in the real C^∞ case. As an example, let $F(t, x) = t^2 - x$ and $G(t, x) \equiv 0$. Then $q_1 = 0 = r_1$ and

$$r_2(t, x) = \begin{cases} e^{-1/x^2} & x \leq 0 \\ 0 & x \geq 0 \end{cases} \quad \text{and} \quad q_2 = -\frac{r_2}{F}$$

are two pairs of q and r which satisfy the conclusions of the Division Theorem. This is not surprising when one realizes that the proof of the uniqueness part of the Weierstrass Division Theorem used methods that depended crucially on complex variable theory (of one variable). It is possible to state a division theorem for formal power series algebras and in this setting uniqueness also holds. [See Zariski and Samuel; Commutative Algebra, Vol. II, p. 139].

(3) For the case when $k = 1$, however, q and r are unique; in fact, the Mather Division Theorem follows from the Implicit Function Theorem. By the Implicit Function Theorem (I, 2.4) there exists a unique real-valued smooth function $\psi(t, x)$ such that $F(\psi(t, x), x) = t$ and $\psi(0) = 0$. Suppose that $G = qF + r$, then

$$G(\psi(t, x), x) = q(\psi(t, x), x)t + r(x).$$

Setting $t = 0$, we see that $r(x) = G(\psi(0, x), x)$ is uniquely determined. Now suppose that $G = q_1F + r$ also, then $(q - q_1)F \equiv 0$ and, in particular, $H(\psi(t, x), x)t \equiv 0$ where $H = q - q_1$. Now $(t, x) \mapsto (\psi(t, x), x)$ maps a nbhd of 0 in $\mathbf{R} \times \mathbf{R}^n$ onto a nbhd of 0 in $\mathbf{R} \times \mathbf{R}^n$ so that $H \equiv 0$ on a nbhd of 0 and q is also uniquely defined.

It is now easy to see how to prove the Malgrange Theorem in this special case. Choose ψ as above and define $r(x) = G(\psi(0, x), x)$ and $q = (G - r)/F$. We leave the verification that q is a smooth function as an exercise.

(4) For the case of one variable t ($n = 0$) the Malgrange Theorem is trivial and q and r are uniquely defined. This is left as an exercise—use Taylor expansions of order k.

(5) The proof of the Division Theorem given by Mather [26] yields a somewhat more general result; namely, the choice of q and r can be made to depend linearly and continuously (in the Whitney C^∞ topology) on F and G.

This proof can be used to modify and extend Nirenberg's proof [32] to obtain these extra results. For our purposes the Division Theorem as stated is sufficient.

(6) Mather also proves a global division theorem not just a local one [26].

As in the complex analytic case, the crucial theorem is the following:

Theorem 2.2 (*Polynomial Division Theorem*). *Let $G(t, x)$ be a complex-valued function defined and smooth on a nbhd of 0 in $\mathbf{R} \times \mathbf{R}^n$. Then there exist smooth, complex-valued functions $q(t, x, \lambda)$ and $r(t, x, \lambda)$ defined on a nbhd of 0 in $\mathbf{R} \times \mathbf{R}^n \times \mathbf{R}^k$ satisfying:*

(i) $G(t, x) = q(t, x, \lambda)P_k(t, \lambda) + r(t, x, \lambda)$, and

(ii) $r(t, x, \lambda) = \sum_{i=0}^{k-1} r_i(x, \lambda)t^i$ where each r_i is a smooth function defined on a nbhd of 0 in $\mathbf{R}^n \times \mathbf{R}^k$.

Moreover if G is real-valued then q and r may be chosen to be real-valued.

Note. The "moreover" part is obtained by equating the real parts of both sides of equation (i) since P_k is real-valued. Also each r_i is easily seen to be real-valued.

Proof that Theorem 2.2 ⇒ Theorem 2.1. This proof is word for word the same as the proof that Theorem 1.4 ⇒ Theorem 1.2 with the single exception that smooth is substituted for holomorphic throughout. □

We shall use the same idea to prove 2.2 as we used to prove 1.4 but first we need an analogue of the Cauchy Integral Formula. This is provided by Green's Theorem from Advanced Calculus.

Let $z = x + iy$ be a complex coordinate on $\mathbf{R}^2$. Then we can solve for x and y in terms of z and $\bar{z}$ $(= x - iy)$; namely, $x = \frac{1}{2}(z + \bar{z})$ and $y = (1/2i)(z - \bar{z})$. Let $f: \mathbf{C} \to \mathbf{R}$ and define $\partial f/\partial \bar{z}$ so that the chain rule holds; that is,

$$\frac{\partial f}{\partial \bar{z}} = \frac{\partial f}{\partial x}\frac{\partial x}{\partial \bar{z}} + \frac{\partial f}{\partial y}\frac{\partial y}{\partial \bar{z}} = \frac{1}{2}\left(\frac{\partial f}{\partial x} + i\frac{\partial f}{\partial y}\right).$$

Now suppose that $F: \mathbf{C} \to \mathbf{C}$ is given by $f + ig$ where $f, g: \mathbf{C} \to \mathbf{R}$. Then

$$(*) \qquad \frac{\partial F}{\partial \bar{z}} = \frac{\partial f}{\partial \bar{z}} + i\frac{\partial g}{\partial \bar{z}} = \frac{1}{2}\left(\left(\frac{\partial f}{\partial x} - \frac{\partial g}{\partial y}\right) + i\left(\frac{\partial f}{\partial y} + \frac{\partial g}{\partial x}\right)\right).$$

Thus $dF/\partial \bar{z} \equiv 0$ iff F satisfies the Cauchy-Riemann Equations iff F is holomorphic. It is easy to check that the standard rules of differentiation work when differentiating with respect to $\bar{z}$. We also make the convention that $dz \wedge d\bar{z} = -2i\, dx \wedge dy$.

Lemma 2.3. *Let $F: \mathbf{C} \to \mathbf{C}$ be a smooth function (when viewed as a mapping of $\mathbf{R}^2 \to \mathbf{R}^2$). Let γ be a simple closed curve in $\mathbf{C}$ whose interior is D. Then for w in D*

$$F(w) = \frac{1}{2\pi i}\int_\gamma \frac{F(z)}{z - w}\, dz + \frac{1}{2\pi i}\iint_D \frac{\partial F}{\partial \bar{z}}(z)\frac{dz \wedge d\bar{z}}{z - w}$$

Note. If F is holomorphic in D, then this formula reduces to the Cauchy Integral Formula.

Proof. Let w be in D and choose ε less than the distance from w to γ. Let $D_\varepsilon = D -$ (disk of radius ε about w) and $\gamma_\varepsilon =$ boundary D_ε. Now recall Green's Theorem for $\mathbf{R}^2$. Let $M, N: D_\varepsilon \to \mathbf{R}$ be smooth and continuous on γ_ε, then

$$\int_{\gamma_\varepsilon} M\,dx + N\,dy = \iint_{D_\varepsilon} \left(\frac{\partial N}{\partial x} - \frac{\partial M}{\partial y}\right) dx \wedge dy.$$

Note that the formula still holds if M and N are complex-valued since we integrate the real and imaginary parts separately. Apply Green's Theorem and (*) above to $F = f + ig$ and obtain

$$(**) \qquad \int_{\gamma_\varepsilon} F\,dz = \int_{\gamma_\varepsilon} (f + ig)(dx + i\,dy) = 2i \iint_{D_\varepsilon} \frac{\partial F}{\partial \bar{z}}\, dx \wedge dy$$

$$= -\iint_{D_\varepsilon} \frac{\partial F}{\partial \bar{z}}\, dz \wedge d\bar{z}$$

Finally, apply (**) to $F(z)/(z - w)$. Since $1/(z - w)$ is holomorphic on D_ε,

$$\frac{\partial}{\partial \bar{z}}\left(\frac{F(z)}{z - w}\right) = \frac{(\partial F/\partial \bar{z})(z)}{z - w}.$$

Thus

$$(***) \quad -\iint_{D_\varepsilon} \frac{\partial F}{\partial \bar{z}}(z)\frac{dz \wedge d\bar{z}}{z - w} = \int_{\gamma_\varepsilon} \frac{F(z)}{z - w}\,dz = \int_{\gamma} \frac{F(z)}{z - w}\,dz - \int_{S_\varepsilon} \frac{F(z)}{z - w}\,dz$$

where S_ε is the circle of radius ε about w. Using polar coordinates centered at w, one obtains

$$\int_{S_\varepsilon} \frac{F(z)}{z - w}\,dz = \int_0^{2\pi} F(w + \varepsilon e^{i\theta}) i\,d\theta.$$

So letting $\varepsilon \to 0$, we see that the RHS of (***) goes to

$$\int_{\gamma} \frac{F(z)}{z - w}\,dz - 2\pi i F(w)$$

while the LHS converges to

$$-\iint_{D} \frac{\partial F}{\partial \bar{z}}(z)\frac{dz \wedge d\bar{z}}{z - w}.$$

(Note this last limit exists since $\partial F/\partial \bar{z}$ is bounded on D and $1/(z - w)$ is integrable over D.) Thus taking (***) to the limit proves the lemma. □

Proof of Theorem 2.2. Let $G(t, x)$ be a smooth complex-valued function defined on a nbhd of 0 in $\mathbf{R} \times \mathbf{R}^n$; we need to show that for appropriate choices of q and r, $G = qP_k + r$. Let $\tilde{G}(z, x, \lambda)$ be a smooth function defined on a nbhd of 0 in $\mathbf{C} \times \mathbf{R}^n \times \mathbf{C}^k$ so that G is an extension of G, that is,

$\tilde{G}(t, x, \lambda) = G(t, x)$ for all real t. Then $\tilde{G} = qP_k + r$ (on $\mathbf{C} \times \mathbf{R}^n \times \mathbf{C}^k$) where

$$q(w, x, \lambda) = \frac{1}{2\pi i}\int_\gamma \frac{\tilde{G}(\eta, x, \lambda)}{P_k(\eta, \lambda)} \frac{d\eta}{(\eta - w)} + \frac{1}{2\pi i}\iint_D \frac{(\partial\tilde{G}/\partial\bar{z})(\eta, x, \lambda)}{P_k(\eta, \lambda)} \frac{d\eta \wedge d\bar{\eta}}{(\eta - w)}$$

and

$$r_i(x, \lambda) = \frac{1}{2\pi i}\int_\gamma \frac{\tilde{G}(\eta, x, \lambda)}{P_k(\eta, \lambda)} s_i(\eta, \lambda)\, d\eta + \frac{1}{2\pi i}\iint_D \frac{(\partial\tilde{G}/\partial\bar{z})(\eta, x, \lambda)}{P_k(\eta, \lambda)} s_i(\eta, \lambda)\, d\eta \wedge d\bar{\eta}$$

if these integrals are well-defined and yield smooth functions. (This follows from Lemma 2.3 as the calculations in Theorem 1.4 follow from the Cauchy Integral Formula.) The first integral in the definition of both q and r is well-defined and smooth for the same reasons as for the corresponding integral in Theorem 1.4. □

The problem lies with the second integrals since D may contain zeroes of P_k. But if it is possible to choose a smooth extension $\tilde{G}$ of G so that $\partial\tilde{G}/\partial\bar{z}$ vanishes on the zeroes of P_k and for real z, then we will have q and r well-defined. Yet this is not enough, since q and r must be smooth functions. Since the integrands are bounded we can differentiate under the integral sign and then quickly see that an appropriate condition for insuring that q and r are smooth is the existence of a smooth extension $\tilde{G}$ of G such that $\partial\tilde{G}/\partial\bar{z}$ vanishes to infinite order on the zeroes of P_k and for z real. Thus the last detail—in fact, the crucial detail—is to show the existence of such an extension. This we shall now do.

Proposition 2.4. (*Nirenberg Extension Lemma*). *Let $G(t, x)$ be a smooth complex-valued function defined on a nbhd of* 0 *in* $\mathbf{R} \times \mathbf{R}^n$. *Then there exists a smooth complex-valued function $\tilde{G}(z, x, \lambda)$ defined on a nbhd of* 0 *in* $\mathbf{C} \times \mathbf{R}^n \times \mathbf{C}^k$ *satisfying*

(1) $\tilde{G}(t, x, \lambda) = G(t, x)$ *for all real* t,
(2) $\partial\tilde{G}/\partial\bar{z}$ *vanishes to infinite order on* $\{\operatorname{Im} z = 0\}$, *and*
(3) $\partial\tilde{G}/\partial\bar{z}$ *vanishes to infinite order on* $\{P_k(z, \lambda) = 0\}$.

As a first step in proving the Nirenberg Extension Lemma we recall a more elementary extension lemma due to Emile Borel.

Lemma 2.5. *Let $f_0(x), f_1(x), \ldots$ be a sequence of smooth functions defined on a given nbhd of* 0 *in* $\mathbf{R}^n$. *Then there is a smooth function $F(t, x)$ defined on a nbhd of* 0 *in* $\mathbf{R} \times \mathbf{R}^n$ *such that* $(\partial^l F/\partial t^l)(x, 0) = f_l(x)$ *for all* l.

Proof. Let $\rho : \mathbf{R} \to \mathbf{R}$ be a smooth function such that

$$\rho(t) = \begin{cases} 1 & |t| \le \frac{1}{2} \\ 0 & |t| \ge 1. \end{cases}$$

Set

$$(*) \qquad F(t, x) = \sum_{l=0}^{\infty} \frac{t^l}{l!} \rho(\mu_l t) f_l(x)$$

where the μ_l's are an increasing sequence of real numbers such that $\operatorname{Lim}_{l\to\infty}\mu_l = \infty$. We will choose the μ_l's so that $F(t, x)$ is smooth on a nbhd of 0 in $\mathbf{R} \times \mathbf{R}^n$. First observe that the RHS of (*) is well-defined for all t (since for any given t only finitely many terms are nonzero) and is a smooth function of t when $t \neq 0$ (since for $t \neq 0$ only finitely many terms are not identically zero on a nbhd of t). Next choose a compact nbhd K of 0 in $\mathbf{R}^n$ contained in the common domain of the f_l's and let $M_l = \sup_{x\in K} |f_l(x)|$. Now differentiate the terms on the RHS of (*) s times with respect to t; the resulting series is dominated in K by

$$C_s \sum_{l=s}^{\infty} \frac{|t|^{l-s}}{(l-s)!}\,\rho(\mu_l t)\mu_l{}^s M_l$$

and its first s derivatives. To see this note that $\rho(\mu_l t) = 0$ for $|t| > 1/\mu_l$ and that $\rho^{[q]}(t) \leq K_q \rho(t)$ for some constant K_q. This series is itself dominated by

$$C_s \sum_{p=0}^{\infty} \frac{M_{p+s}}{\mu_{p+s}^{p-s}\, p!},$$

which will converge for all s if the μ_l's tend toward infinity rapidly enough. This shows that if the RHS of (*) is differentiated with respect to t termwise (s-times) the resulting series converges uniformly on $\mathbf{R} \times K$. The corresponding result obtained by differentiating with respect to the x-variables is clear. Suppose that you want to show that the series for $(\partial^s/\partial t^s)(\partial^{|\alpha|}/\partial x^\alpha)F$ converges uniformly. Let

$$M_l{}^\alpha = \sup_{x\in K} \left| \frac{\partial^{|\alpha|} f_l}{\partial x^\alpha}(x) \right|$$

and proceed as before. Finally to do this for all possible mixed partials simultaneously we use the diagonal trick of I, Proposition 4.8. That is, let

$$M_l = \sup_{\substack{x\in K\\ |\alpha| \leq l}} \left| \frac{\partial^{|\alpha|} f_l}{\partial x^\alpha}(x) \right|.$$

Proceed as above choosing the μ_l's and note that now the series for each partial derivative converges uniformly on K and that this F is smooth on a nbhd of 0. □

Note. One can use Lemma 2.5 to show that for any power series about 0 in $\mathbf{R}^n$ there exists a smooth C^∞ real-valued functions whose Taylor series expansion at 0 is this given power series.

We will need another elementary extension lemma which is, in fact, a special case of the Whitney Extension Theorem. [59]

Lemma 2.6. *Let V and W be subspaces of $\mathbf{R}^n$ such that $V + W = \mathbf{R}^n$. Let g and h be smooth functions defined on a nbhd of 0 in $\mathbf{R}^n$. Assume that for all multi-indices α*

$$\frac{\partial^{|\alpha|} g}{\partial x^\alpha}(x) = \frac{\partial^{|\alpha|} h}{\partial x^\alpha}(x) \qquad \textit{for all } x \textit{ in } V \cap W.$$

Then there exists a smooth function F defined on a nbhd of 0 in $\mathbf{R}^n$ such that for all α

$$\frac{\partial^{|\alpha|}F}{\partial x^\alpha}(x) = \begin{cases} \dfrac{\partial^{|\alpha|}g}{\partial x^\alpha}(x) & \text{if } x \text{ is in } V \\[2ex] \dfrac{\partial^{|\alpha|}h}{\partial x^\alpha}(x) & \text{if } x \text{ is in } W \end{cases}$$

Proof. We observe first that it is sufficient to prove the lemma for $h \equiv 0$. For let F_1 be the extension for $g - h$ and 0, then $F = F_1 + h$ is the required extension for g and h. So we do assume that $h \equiv 0$.

Next choose coordinates $y_1, \ldots, y_n$ on $\mathbf{R}^n$ so that V is defined by the equations $y_1 = \cdots = y_j = 0$ and W is defined by the equations $y_{j+1} = \cdots = y_k = 0$. This is possible since $V + W = \mathbf{R}^n$. Then set

$$F(y) = \sum_{\substack{|\alpha|=0 \\ \alpha=(a_1,\ldots,a_j,0,\ldots,0)}}^{\infty} \frac{y^\alpha}{\alpha!}\frac{\partial^{|\alpha|}g}{\partial y^\alpha}(0,\ldots,0,y_{j+1},\ldots,y_n)\rho\left(\mu_{|\alpha|}\sum_{i=1}^{j} y_i^2\right)$$

where ρ is the same smooth functions which appeared in the proof of the last lemma and the sequence $\{\mu_l\}_{l=0}^\infty$ is increasing to infinity. As in the last lemma the μ_l's can be chosen to increase rapidly enough to infinity to insure that F is a smooth function on a nbhd of 0 in $\mathbf{R}^n$.

We need only check that F has the desired properties. If $y = (y_1, \ldots, y_n)$ is in W, then $y_{j+1} = \cdots = y_k = 0$ and every term of $(\partial^{|\beta|}F/\partial y^\beta)(y)$ contains a factor of the form $(\partial^{|\gamma|}g/\partial y^\gamma)(0, \ldots, 0, y_{k+1}, \ldots, y_n)$. Since $(0, \ldots, 0, y_{k+1}, \ldots, y_n)$ is in $V \cap W$ that factor equals zero by assumption. Thus $(\partial^{|\beta|}F/\partial y^\beta)(y) = 0$. On the other hand, if y is in V, then $y_1 = \cdots = y_j = 0$ and

$$\frac{\partial^{|\gamma|}}{\partial y^\gamma}\rho\left(\mu_{|\alpha|}\sum_{i=1}^{j} y_i^2\right)\Bigg|_{y_1=\cdots=y_j=0} = \begin{cases} 1 & \gamma = 0 \\ 0 & \gamma \neq 0 \end{cases}.$$

Thus

$$\frac{\partial^{|\beta|}F}{\partial y^\beta}(y) = \sum_{|\alpha|=0}^{\infty} \frac{\partial^{|\beta|}}{\partial y^\beta}\left(\frac{y^\alpha}{\alpha!}\frac{\partial^{|\alpha|}g}{\partial y^\alpha}(y)\right)\Bigg|_{y_1=\cdots=y_j=0}.$$

Let $\beta = (b_1, \ldots, b_n)$. It is easy to see that if $b_i \neq a_i$ for some $i \leq j$, then the given term in the series is 0. In fact, the only nonzero term is $(\partial^{|\beta|}g/\partial y^\beta)(y)$. So F is the desired extension. □

Finally we need to solve formally an "initial value" problem for certain partial differential equations.

Lemma 2.7. *Let $f(x)$ be a smooth complex-valued function defined on a nbhd of 0 in $\mathbf{R}^n$ and let X be a vector field on $\mathbf{R}^n$ with complex coefficients. Then there exists a smooth complex-valued function F defined on a nbhd of 0 in $\mathbf{R} \times \mathbf{R}^n$ such that*

(a) *$F(0, x) = f(x)$ for all x in $\mathbf{R}^n$, and*
(b) *$\partial F/\partial t$ agrees to infinite order with XF at all points $(0, x)$ in $\mathbf{R} \times \mathbf{R}^n$.*

Proof. An obvious candidate for such a solution is the formal expression

$$\bar{F}(t, x) = e^{tX}f \equiv \sum_{k=0}^{\infty} \frac{t^k}{k!} X^k f.$$

In fact, by differentiating the LHS of this equation term by term and evaluating at $t = 0$ we see that (b) holds. Clearly (a) holds. The only problem is that $\bar{F}$ need not be smooth. Now by the Borel Theorem (Lemma 2.5) we may choose a smooth function F of the form

$$F = \sum_{k=0}^{\infty} \frac{t^k}{k!} \rho(\mu_k t) X^k f$$

having the same power series expansion as $\bar{F}$ at $t = 0$. This F will solve our "initial value" problem. □

Proof of Proposition 2.4. The proof will be done by induction on k. When $k = 0$, $P_k(z) \equiv 1$, so we need only show that there exists a smooth function $\tilde{G}(z, x)$ such that $\tilde{G}(t, x) = G(t, x)$ for real t and $(\partial\tilde{G}/\partial\bar{z})(t, x)$ vanishes to infinite order for real t. Let $z = s + it$. Then

$$i\frac{\partial}{\partial\bar{z}} = \frac{1}{2}\left(-\frac{\partial}{\partial t} + i\frac{\partial}{\partial s}\right)$$

and the existence of such a $\tilde{G}$ follows from the last lemma by taking $X = (i/2)(\partial/\partial s)$.

We now assume that the case $k - 1$ has been proved and attempt to prove the proposition for k. In particular, we will show that there exist smooth functions $E(z, x, \lambda)$ and $F(z, x, \lambda)$ satisfying

(i) E and F agree to infinite order on the set $\{P_k(z, \lambda) = 0\}$

(ii) F is an extension of G

(iii) $\partial F/\partial\bar{z}$ vanishes to infinite order on $\{\operatorname{Im} z = 0\}$

(iv) Let $M = F \mid \{P_k(z, \lambda) = 0\}$. Then $\partial M/\partial\bar{z}$ vanishes to infinite order on $\{(\partial P_k/\partial z)(z, \lambda) = 0\}$ and

(v) $\partial E/\partial\bar{z}$ vanishes to infinite order on $\{P_k(z, \lambda) = 0\}$.

First we show that the existence of E and F is sufficient to prove the proposition. Set $u = P(z, \lambda) \equiv P_k(z, \lambda)$, and let $\lambda' = (\lambda_1, \ldots, \lambda_{k-1})$. Then consider the change of coordinates $(z, \lambda_0, \lambda') \mapsto (z, u, \lambda')$ on $\mathbf{C} \times \mathbf{C} \times \mathbf{C}^{k-1}$. Recalling that $P(z, \lambda) = z^k + \sum_{i=0}^{k-1} \lambda_i z^i$ we see that $\partial u/\partial\lambda_0 \equiv 1$ so that this is a legitimate change of coordinates. In these new coordinates the hypersurface $\{P_k(z, \lambda) = 0\}$ is given by the simple equation $u = 0$. The coup de grâce is then administered by Lemma 2.6. By this lemma there exists a function $\tilde{G}$ which agrees to infinite order with E on $u = 0$ and to infinite order with F on $\operatorname{Im} z = 0$. (Note that $u = 0$ and $\operatorname{Im} z = 0$ are subspaces of $\mathbf{R}^{2k+2}$ which intersect transversely. Along with (i) this guarantees that Lemma 2.6 is applicable.) Properties (ii), (iii), and (v) then imply that $\tilde{G}$ is the desired extension of G.

Now to show the existence of E and F. First we assume the existence of F and construct E. Consider again the coordinates (z, u, λ') and notice that in these coordinates the vector field $\partial/\partial z$ has the form $\partial/\partial z + (\partial P/\partial z)(\partial/\partial u)$. Similarly, $\partial/\partial \bar{z}$ becomes $\partial/\partial \bar{z} + (\overline{\partial P/\partial z})(\partial/\partial \bar{u})$. So our problem in these coordinates is to find $E(z, x, u, \lambda')$ such that

(a) $E = F$ to infinite order on $\{u = 0\}$, and

(b) $(\partial/\partial \bar{z} + (\overline{\partial P/\partial z})(\partial/\partial \bar{u})E) = 0$ to infinite order on the set $\{u = 0\}$.

Let $X = -(\overline{\partial P/\partial z})^{-1}(\partial/\partial \bar{z})$. (We will deal with the problem of the zeroes of $\partial P/\partial z$ in a moment.) Then this problem can be reformulated in a form analogous to Lemma 2.7, i.e. find a smooth function E satisfying (a) and

(b′) $\partial E/\partial \bar{z} = XE$ to infinite order on $\{u = 0\}$.

This admits the same sort of formal solution as Lemma 2.7; namely,

$$(*) \qquad E = \sum_{l=0}^{\infty} \frac{(\bar{u})^l}{l!} \rho(\mu_l|\bar{u}|^2)X^l M(z, x, \lambda)$$

where ρ is the same bump function used before. Now since $\partial M/\partial \bar{z} = 0$ to infinite order on the set $\{(\partial P/\partial z)(z, \lambda') = 0\}$ (assumption (iv)) we see that $X^l M$ is a smooth function in z, x, and λ' for all l. Hence we can choose μ_l's which increase to infinity rapidly enough to guarantee that the RHS of (*) is a smooth function of z, x, u, and λ. Thus the zeroes of $\partial P/\partial z$ cause no problem and this E is then the desired function.

Thus to complete the proof we need to construct a smooth function F satisfying (in the z, x, u, λ' coordinates)

(ii) $F(t, x, u, \lambda') = G(t, x)$ for all real t,

(iii) $\partial F/\partial \bar{z} = XF$ to infinite order on $\{\text{Im } z = 0\}$, and

(iv) Let $M = F \mid \{u = 0\}$. Then $\partial M/\partial \bar{z} = 0$ to infinite order on $\{\partial P_k/\partial z = 0\}$.

Consider the hyperplane $u = 0$ and the change of coordinates $\lambda' = (\lambda_1, \ldots, \lambda_{k-1}) \mapsto (\lambda_1/1, \ldots, \lambda_{k-1}/k - 1) = \lambda''$. These conditions translate to finding a smooth function $M(z, x, \lambda'')$ (which will be F restricted to $\{u = 0\}$) satisfying

(1) $M(t, x, \lambda'') = G(t, x)$ for all real t,

(2) $\partial M/\partial \bar{z}$ vanishes to infinite order on $\{\text{Im } z = 0\}$, and

(3) $\partial M/\partial \bar{z}$ vanishes to infinite order on $\{P_{k-1}(z, \lambda'') = 0\}$.

By our induction hypothesis such a smooth function M of the variables z, x, and λ'' exists and we can view M as a smooth function of z, x, and λ'.

Finally we define F using M and (*); that is,

$$F(z, x, u, \lambda) = \sum_{l=0}^{\infty} \frac{(\bar{u})^l}{l!} \rho(\mu_l|\bar{u}|^2)X^l M(z, x, \lambda').$$

Again since $\partial M/\partial \bar{z}$ vanishes to infinite order when $(\partial P/\partial z)(z, \lambda') = 0 = P_{k-1}(z, \lambda'')$, $X^l M$ is smooth in z, x, and λ'. Thus the μ_l's can be chosen so that F is a smooth function satisfying (ii) and (iii). Also on $\{u = 0\}$ $F = M$ so $\partial F/\partial \bar{z}$ vanishes to infinite order and F satisfies (iv). $\square$

Thus we have proved the Nirenberg Extension Lemma and the Malgrange Preparation Theorem.

Exercise. Let $f: \mathbf{R} \to \mathbf{R}$ be a smooth even function. Show that there exists a smooth function $g: \mathbf{R} \to \mathbf{R}$ such that $f(x) = g(x^2)$. Hint: Use the trick used to prove Lemma 2.5.

§3. The Generalized Malgrange Preparation Theorem

Our purpose is to generalize the Malgrange Preparation Theorem to a statement about certain local rings. We will only discuss rings which are commutative and have a multiplicative identity.

Definition 3.1. *Let X be a smooth manifold and let p be a point in X.*

(a) *Two smooth real-valued functions f and g, defined on nbhds of p, are* equivalent near p *iff $f = g$ on some nbhd of p.*

(b) *Let $f: U \to \mathbf{R}$ be a smooth function where U is some nbhd of p. Then* $[f]_p \equiv$ germ of f at p *is the equivalence class of f in the equivalence relation defined in* (a). *Let $C_p^\infty(X)$ be the set of all germs of smooth, real-valued functions defined on a nbhd of p.*

(c) *A* local ring *is a ring with a unique maximal ideal.*

Lemma 3.2. *$C_p^\infty(X)$ is a local ring if the ring operations are given by $[f]_p + [g]_p = [f + g]_p$ and $[f]_p \cdot [g]_p = [fg]_p$ where $f + g$ and fg are assumed to be defined on $\operatorname{dom} f \cap \operatorname{dom} g$ which is a nbhd of p. Let $\mathcal{M}_p(X) = \{[f]_p \in C_p^\infty(X) \mid f(p) = 0\}$. Then $\mathcal{M}_p(X)$ is the unique maximal ideal.*

Proof. It is easy to convince oneself that these operations are well-defined and that $C_p^\infty(X)$ is a commutative ring with multiplicative identity. It is also a trivial exercise to see that $\mathcal{M}_p(X)$ is an ideal in $C_p^\infty(X)$. As for unique maximality, let $\mathcal{M}$ be any other ideal in $C_p^\infty(X)$. Suppose $[f]_p \in \mathcal{M} - \mathcal{M}_p(X)$. Then $[1/f]_p$ is defined since $f(0) \neq 0$ and therefore $[1/f]_p \cdot [f]_p = [1]_p \in \mathcal{M}$. So $\mathcal{M} = C_p^\infty(X)$. Thus $\mathcal{M}_p(X)$ is the unique maximal ideal in $C_p^\infty(X)$. $\square$

Note. We will sometimes omit the brackets when discussing germs, and use the same symbols as for functions. The context should allay any possible confusion.

Lemma 3.3. *Let $\mathcal{M}_p^2(X)$ be the ideal generated by germs of the form fg where $f, g \in \mathcal{M}_p(X)$. Then $\mathcal{M}_p(X)/\mathcal{M}_p^2(X)$ is a vector space (over* **R***) canonically isomorphic with T_p^*X. This isomorphism is induced by the mapping $\psi: \mathcal{M}_p(X) \to T_p^*X$ given by $[f]_p \mapsto (df)(p)$.*

Proof. The facts that $\mathcal{M}_p(X)/\mathcal{M}_p^2(X)$ is a vector space and that ψ is well-defined and linear are easy to verify. Equally easy to see is that ψ is onto. For let $x_1, \ldots, x_n$ be local coordinates on X based at p, and v be a cotangent vector in T_p^*X, so that $v = (v_1 dx_1 + \cdots + v_n dx_n)|_{x=p}$. If we let $f(x) = \sum_{i=1}^n v_i x_i$ which is defined on a nbhd of p, then $(df)(p) = v$. Finally we show that $\operatorname{Ker} \psi = \mathcal{M}_p^2(X)$. It is an easy calculation to show that $\mathcal{M}_p^2(X) \subset \operatorname{Ker} \psi$.

So let $[f]_p \in \operatorname{Ker} \psi$. Since $f(0) = 0$, $f(x) = \sum_{i=1}^n x_i f_i(x)$ where $f_i(0) = (\partial f/\partial x_i)(0)$ by II, Lemma 6.10. Since $(df)(p) = 0$, $f_i(0) = 0$. Thus $[f]_p$ is in $\mathscr{M}_p{}^2(X)$. □

Lemma 3.4. *Let $\phi: X \to Y$ be a smooth mapping with $q = \phi(p)$. Then ϕ induces a ring homomorphism $\phi^*: C_q^\infty(Y) \to C_p^\infty(X)$ given by $[f]_q \mapsto [f\cdot\phi]_p$. Moreover ϕ is locally (near p) a diffeomorphism iff ϕ^* is an isomorphism.*

Proof. It is easy to see that ϕ^* is well-defined. To show that ϕ^* is a ring homomorphism and that if ϕ is a local diffeomorphism then ϕ^* is an isomorphism is also easy. (Note that $(\phi^*)^{-1} = (\phi^{-1})^*$.) So we assume that ϕ^* is an isomorphism. Clearly ϕ^* induces an isomorphism of $\mathscr{M}_q(Y)/\mathscr{M}_q{}^2(Y) \to \mathscr{M}_p(X)/\mathscr{M}_p{}^2(X)$ so that by Lemma 3.3 dim Y = dim X. Choose local coordinates $x_1, \ldots, x_n$ on X based at p corresponding to the chart η. Let $[x_i]_p = \phi^*[h_i]_q$ for some smooth functions h_i. Define

$$H: (\operatorname{dom} h_1 \cap \cdots \cap \operatorname{dom} h_n) \to \mathbf{R}^n$$

by $H(y) = (h_1(y), \ldots, h_n(y))$. H is smooth and $\eta = H\cdot\phi$ on a small nbhd of p. Applying the chain rule, we have $(d\eta)_p = (dH)_q(d\phi)_p$. Since $(d\eta)_p$ is invertible $(d\phi)_p$ is 1:1. Apply the Inverse Function Theorem to see that ϕ is a local diffeomorphism. □

Let $\mathscr{R}$ be a ring (commutative with identity) and A an abelian group (with the group operation denoted by +). Then we recall that A is an $\mathscr{R}$-*module* if there is a mapping of $\mathscr{R}$ into the set of homomorphisms on A. We denote the action of r in $\mathscr{R}$ on a in A by ra and demand that the following relations hold for all r_1, r_2 in $\mathscr{R}$ and a in A: $(r_1 + r_2)a = r_1a + r_2a$, $(r_1r_2)a = r_1(r_2a)$, $r(a_1 + a_2) = ra_1 + ra_2$, and $1\cdot a = a$. Note that if $\mathscr{R}$ is a field, A is simply a vector space over $\mathscr{R}$. Recall also that an $\mathscr{R}$-module A is *finitely generated over* $\mathscr{R}$ if there is a finite number of elements $a_1, \ldots, a_n$ in A such that each element a in A can be written as a linear combination $a = r_1a_1 + \cdots + r_na_n$ for some r_i's in $\mathscr{R}$. (*Warning*: In an arbitrary module the linear combination need not be unique, even if the generating set $\{a_1, \ldots, a_n\}$ is minimal.)

We assume in what follows that the reader is familiar with such elementary notions as submodules, quotient modules, module homomorphism, etc.

We will need two lemmas about $\mathscr{R}$-modules.

Lemma 3.4. (*Nakayama*). *Let $\mathscr{R}$ be a commutative local ring with identity and let $\mathscr{M}$ be the maximal ideal in $\mathscr{R}$. Let A be an $\mathscr{R}$-module. Assume that*

(i) *A is finitely generated, and*
(ii) *$A = \mathscr{M}A$ (= the set of* sums *of elements of the form ra with $r \in \mathscr{M}$ and $a \in A$.)*

Then $A = \{0\}$.

Proof. Let $e_1, \ldots, e_n$ be a finite set of generators for A over $\mathscr{R}$. We will show that each $e_k = 0$. First we may write $e_k = m_1a_1 + \cdots + m_sa_s$ where each m_i is in $\mathcal{M}$ since $A = \mathcal{M}A$. But since $e_1, \ldots, e_n$ are a set of generators we may also write $a_i = \sum_{j=1}^n r_{ij}e_j$. Thus $e_k = \sum_{j=1}^n s_{kj}e_j$ where $s_{kj} = \sum_{i=1}^s m_i r_{ij}$ is in $\mathcal{M}$. Using the Kronecker delta we have that $\sum_{j=1}^n (\delta_{kj} - s_{kj})e_j = 0$ for each k. This is a system of n linear equations in n unknowns $e_1, \ldots, e_n$. Now note that if the matrix $(\delta_{ij} - s_{ij})$ is invertible then the system of equations has only the trivial solution $e_1 = \cdots = e_n = 0$. Now a matrix D (over a commutative ring with unit) is invertible iff $\det(D)$ is invertible in the ring. (To review the theory of determinants of matrices over a commutative ring with identity, see Chapter V of Hoffman and Kunze, *Linear Algebra*.) Now by using the standard expansion of $\det(\delta_{ij} - s_{ij})$ by permutations it is easy to see that $\det(\delta_{ij} - s_{ij}) = 1 + s$ where s is in $\mathcal{M}$. Also, in a local ring, the maximal ideal is precisely the set of noninvertible elements: Suppose t is in $\mathcal{M}$; then t is not invertible. For if t were invertible then $1 = tt^{-1}$ would be in $\mathcal{M}$. Conversely if t is not in $\mathcal{M}$, then t is invertible, for the ideal generated by t is not contained in $\mathcal{M}$ so it must be all of $\mathcal{R}$. Thus there is t' in $\mathcal{R}$ such that $tt' = 1$. From this we may conclude that $1 + s$ (which is not in $\mathcal{M}$) is invertible and, therefore, $e_1 = \cdots = e_n = 0$. □

Remark. A more sophisticated way of formulating the last half of the above argument is that the quotient ring $\mathcal{R}/\mathcal{M}$ is a *field*.

Corollary 3.5. *Let A be a finitely generated $\mathcal{R}$-module. Then $A/\mathcal{M}A$ is a finite dimensional vector space over the field $\mathcal{R}/\mathcal{M}$. Let $\phi: A \to A/\mathcal{M}A$ be the natural projection and $v_1, \ldots, v_n$ a basis for this vector space. Choose $e_1, \ldots, e_n$ in A so that $\phi(e_i) = v_i$. Then $e_1, \ldots, e_n$ form a set of generators of A over $\mathcal{R}$.*

Proof. Since the action of $\mathcal{R}$ on A clearly induces an action of $\mathcal{R}/\mathcal{M}$ on $A/\mathcal{M}A$ we see that $A/\mathcal{M}A$ is a module over the field $\mathcal{R}/\mathcal{M}$; i.e. a vector space. To show that $\dim_{\mathcal{R}/\mathcal{M}} A/\mathcal{M}A < \infty$, let $a_1, \ldots, a_k$ be a set of generators of A over $\mathcal{R}$ and let v be in $A/\mathcal{M}A$. Choose a in A so that $\phi(a) = v$ and choose r_i in $\mathcal{R}$ so that $a = r_1a_1 + \cdots + r_ka_k$. Then $v = [r_1]\phi(a_1) + \cdots + [r_k]\phi(a_k)$ where $[r_i]$ denotes the equivalence class of r_i in $\mathcal{R}/\mathcal{M}$. Thus $\phi(a_1), \ldots, \phi(a_k)$ form a set of generators of $A/\mathcal{M}A$.

Conversely let $v_1, \ldots, v_n$ be a basis for $A/\mathcal{M}A$ with $e_1, \ldots, e_n$ chosen as in the statement of the corollary. Let B be the submodule of A generated by $e_1, \ldots, e_n$ and let C be the quotient module A/B. Since A is finitely generated over $\mathcal{R}$, C is finitely generated over $\mathcal{R}$. Now $A = B + \mathcal{M}A$. For if a is in A, then $\phi(a) = [r_1]v_1 + \cdots + [r_n]v_n$. So $a = r_1e_1 + \cdots + r_ne_n + s$ where s is in $\mathcal{M}A$. Thus $C = A/B = (B + \mathcal{M}A)/B = \mathcal{M}(A/B) = \mathcal{M}C$. Use arguments about cosets to check these equalities. Finally apply Nakayama's Lemma to show that $C = 0$ and thus $A = B$. □

Returning to our local ring of interest, suppose that A is a $C_p^\infty(X)$ module and that $\phi: X \to Y$ with $q = \phi(p)$ is a smooth mapping. The induced ring homomorphism ϕ^* allows us to view A as a $C_q^\infty(Y)$ module. More specifically if a is in A and $[f]_q$ is in $C_q^\infty(Y)$, then we define $[f]_q a \equiv \phi^*[f]_q a$. We now state the local ring generalization of the Malgrange Preparation Theorem.

Theorem 3.6. (Generalized Malgrange Preparation Theorem). *Let X and Y be smooth manifolds and $\phi : X \to Y$ be a smooth mapping with $\phi(p) = q$. Let A be a finitely generated $C_p^\infty(X)$-module. Then A is a finitely generated $C_q^\infty(Y)$-module iff $A/\mathscr{M}_q(Y)A$ is a finite dimensional vector space over* $\mathbf{R}$.

Notes. Consider the mapping of $C_p^\infty(X) \to \mathbf{R}$ given by $[f]_p \mapsto f(p)$ to see that

(1) $\mathbf{R} \cong C_p^\infty(X)/\mathscr{M}_p(X) \cong C_q^\infty(Y)/\mathscr{M}_q(Y)$ so that $A/\mathscr{M}_q(Y)A$ is a real vector space.

(2) As we noted in Corollary 3.5, the fact that A is finitely generated over $C_q^\infty(Y)$ automatically implies that $A/\mathscr{M}_q(Y)A$ is a finite dimensional vector space.

(3) We will now prove that Theorem 3.6 is, in fact, a generalization of the Malgrange Preparation Theorem (2.1). Let $X = \mathbf{R} \times \mathbf{R}^n$, $Y = \mathbf{R}^n$, and $\pi : \mathbf{R} \times \mathbf{R}^n \to \mathbf{R}^n$ be given by $\pi(t, x) = x$. Also let $p = 0 = q$. Let $[F]_0$ be in $C_0^\infty(\mathbf{R} \times \mathbf{R}^n)$ and assume that $F(t, 0) = t^k g(t)$ where $g(0) \neq 0$. Let $[G]_0$ be a germ of some other smooth function in $C_0^\infty(\mathbf{R} \times \mathbf{R}^n)$. We must find $[q]_0$ and $[r]_0$ satisfying the appropriate conditions for the Malgrange Theorem.

Let $A = C_0^\infty(\mathbf{R} \times \mathbf{R}^n)/(F)$ where (F) is the ideal in $C_0^\infty(\mathbf{R} \times \mathbf{R}^n)$ generated by $[F]_0$. A is clearly a $C_0^\infty(\mathbf{R} \times \mathbf{R}^n)$-module and is finitely generated. (In fact, it is generated by the image of $[1]_0$ in A.) The vector space $A/\mathscr{M}_0(\mathbf{R}^n)A = C_0^\infty(\mathbf{R} \times \mathbf{R}^n)/(F, x_1, \ldots, x_n)$ since $\mathscr{M}_0(\mathbf{R}^n) = (x_1, \ldots, x_n)$ where $(\ ,\ldots,\ )$ indicates the ideal generated by germs of the indicated smooth functions. (Consider the mapping of $A \to C_0^\infty(\mathbf{R} \times \mathbf{R}^n)/(F, x_1, \ldots, x_n)$ given by $[g]_0 + (F) \mapsto [g]_0 + (F, x_1, \ldots, x_n)$ to obtain the above identification.)

We will show that $A/\mathscr{M}_0(\mathbf{R}^n)A$ is a finite dimensional vector space. First we claim that $(F, x_1, \ldots, x_n) = (t^k, x_1, \ldots, x_n)$. Let $h : \mathbf{R} \to \mathbf{R}$ be given by $h(s) = F(t, sx)$ where (t, x) is fixed in $\mathbf{R} \times \mathbf{R}^n$. Then

$$F(t, x) - F(t, 0) = h(1) - h(0) = \int_0^1 \frac{dh}{dt}(s)\, ds$$

$$= \int_0^1 \sum_{i=1}^n x_i \frac{\partial F}{\partial x_i}(t, sx)\, ds = \sum_{i=1}^n x_i g_i(t, x)$$

where $g_i(t, x) = \int_0^1 \partial F/\partial x_i\,(t, sx)\, ds$. Thus $F(t, x) = t^k g(t) + r$ where r is in $(x_1, \ldots, x_n)$. Since $g(0) \neq 0$, $[g]_0$ is invertible and we obtain $(F, x_1, \ldots, x_n) = (t^k, x_1, \ldots, x_n)$. Thus a basis for the vector space $C_0^\infty(\mathbf{R} \times \mathbf{R}^n)/(F, x_1, \ldots, x_n)$ is given by the images of $[1]_0, [t]_0, \ldots, [t^{k-1}]_0$. Applying the Generalized Malgrange Preparation Theorem, we deduce that A is a finitely generated $C_0^\infty(\mathbf{R}^n)$ module. By Corollary 3.5 the images of $[1]_0, [t]_0, \ldots, [t^{k-1}]_0$ generate A as a $C_0^\infty(\mathbf{R}^n)$-module. Thus

$$[G]_0 = [r_0]_0[1]_0 + [r_1]_0[t]_0 + \cdots + [r_{k-1}]_0[t^{k-1}]_0 + V$$

where V is in (F). Let $V = [qF]_0$, and then, on a nbhd of 0 in $\mathbf{R} \times \mathbf{R}^n$,

$$G(t, x) = q(t, x)F(t, x) + r_0(x) + r_1(x)t + \cdots + r_{k-1}(x)t^{k-1}. \qquad \square$$

The proof of 3.6 will be given in two special cases, immersions and submersions, and then done in general.

Lemma 3.7. *Let $\pi: X \to Y$ be a submersion with* $\dim X = n = \dim Y + 1$ *and* $q = \pi(p)$. *Let A be a finitely generated $C_p^\infty(X)$-module. If $V = A/\mathscr{M}_q(Y)A$ is a finite dimensional vector space over* $\mathbf{R}$ *then A is a finitely generated $C_q^\infty(Y)$-module (via π^*).*

Proof. Since this is a local result, we may assume, with a proper choice of charts, that $X = \mathbf{R}^n$, $Y = \mathbf{R}^{n-1}$, $p = 0 = q$, and $\pi: \mathbf{R}^n \to \mathbf{R}^{n-1}$ is given by $(x_1, \ldots, x_n) \mapsto (x_2, \ldots, x_n)$. Let $\psi: A \to V$ be the canonical projection and choose $e_1, \ldots, e_p$ in A so that $\{\psi(e_1), \ldots, \psi(e_p)\}$ is a basis for V.

Step I: $e_1, \ldots, e_p$ generate A as a $C_0{}^\infty(\mathbf{R}^n)$-module. To see this note that $\mathscr{M}_0(\mathbf{R}^{n-1}) \subset \mathscr{M}_0(\mathbf{R}^n)$ so that there is a natural surjection $\eta: A/\mathscr{M}_0(\mathbf{R}^{n-1})A \to A/\mathscr{M}_0(R^n)A$. Thus $\eta\cdot\psi(e_1), \ldots, \eta\cdot\psi(e_p)$ is a set of generators for $A/\mathscr{M}_0(\mathbf{R}^n)A$ and Step I follows from Corollary 3.5.

Step II: All elements of A have the form $\sum_{i=1}^p (c_i e_i + f_i e_i)$ where c_i is a scalar in $\mathbf{R}$ and f_i is in $\mathscr{M}_0(\mathbf{R}^{n-1})C_0^\infty(\mathbf{R}^n)$. Since $\psi(e_1), \ldots, \psi(e_p)$ is a basis for V, we see that if a is in A, then $a = \sum_{i=1}^p c_i e_i + \bar{e}$ where $\bar{e}$ is in $\mathscr{M}_0(\mathbf{R}^{n-1})A$. Thus $\bar{e} = \sum_{j=1}^m g_j a_j$ where g_j is in $\mathscr{M}_0(\mathbf{R}^{n-1})$ and a_j is in A. But by Step I $a_j = \sum_{i=1}^p h_i e_i$ where h_i is in $C_0^\infty(\mathbf{R}^n)$. Thus $\bar{e} = \sum_{i=1}^p (\sum_{j=1}^m g_i h_j) e_j$ and by letting $f_i = \sum_{j=1}^m g_i h_j$. Step II is proved.

Now we prove the Lemma. By Step II, $x_1 e_i = \sum_{j=1}^p (c_{ij} + f_{ij}) e_j$ with c_{ij} and f_{ij} in the appropriate places. Using the Kronecker delta we obtain the n-linear equations in n-unknowns $e_1, \ldots, e_p$

$$(*) \qquad \sum_{j=1}^{p} (x_1 \delta_{ij} - c_{ij} - f_{ij}) e_j = 0.$$

Let $P(x_1, \ldots, x_n)$ be the determinant of the matrix $(x_1\delta_{ij} - c_{ij} - f_{ij})$. By Cramer's Rule $Pe_i = 0$ for each i. Now note that $f_{ij}(x_1, 0, \ldots, 0) = 0$ since f_{ij} is in $\mathscr{M}_0(\mathbf{R}^{n-1})C_0^\infty(\mathbf{R}^n)$. Therefore, $P(x_1, 0, \ldots, 0) = \det(x_1\delta_{ij} - c_{ij})$ which is a polynomial in x_1 of degree $\leq p$. Hence there exists $k \leq p$ such that $P(x_1, 0, \ldots, 0) = x_1{}^k g(x_1)$ and $g(0) \neq 0$. By Step II, if a is in A, then $a = \sum_{i=1}^p (c_i e_i + f_i e_i)$. Apply the Malgrange Preparation Theorem to f_i and P to obtain

$$f_i = Q_i P + \sum_{j=0}^{k-1} R_{ij}(x_2, \ldots, x_n) x_1{}^j.$$

Since $Pe_i = 0$, we have that $f_i e_i = \sum_{j=0}^{k-1} R_{ij} x_1{}^j e_i$ and

$$a = \sum_{i=1}^{p} \left(c_i e_i + \sum_{j=1}^{k-1} R_{ij} x_1{}^j e_i \right).$$

Thus A is generated by the pk elements $e_1, \ldots, e_p, x_1 e_1, \ldots, x_1 e_p, \ldots, x_1{}^k e_p$ as a module over $C_0{}^\infty(\mathbf{R}^{n-1})$ since R_{ij} is in $C_0^\infty(\mathbf{R}^{n-1})$. ☐

Lemma 3.8. *Let $\phi: X \to Y$ be an immersion with $q = \phi(p)$. Let A be a finitely generated $C_p^\infty(X)$-module. Then A is a finitely generated $C_q^\infty(Y)$-module.*

Proof. $\phi^* : C_q^\infty(Y) \to C_p^\infty(X)$ is onto, since there is a nbhd U of p in X such that $\phi|U$ is a 1:1 immersion and $\phi(U)$ is a submanifold of Y. Clearly, using the definition of submanifold, every smooth function on $\phi(U)$ can be extended to be a smooth function on a nbhd of $\phi(U)$ in Y. (Since we are working locally we can assume that $\phi(U)$ is small enough to be made a k-plane in $\mathbf{R}^m$ by one chart on Y.) The surjectivity of ϕ^* implies the stated result. □

Proof of Theorem 3.6. Define $\tilde{\phi}: X \to X \times Y$ by $\tilde{\phi}(x) = (x, \phi(x))$. Using charts, we may assume that $X = \mathbf{R}^n$ and that $p = 0$. Let $\pi_i : \mathbf{R}^i \times Y \to \mathbf{R}^{i-1} \times Y$ be given by $(x_1, \ldots, x_i, y) \mapsto (x_2, \ldots, x_i, y)$. Then $\phi = \pi_1 \cdot \cdots \cdot \pi_n \cdot \tilde{\phi}$ (locally). Since $\tilde{\phi}$ is an immersion Lemma 3.8 applies and A is a finitely generated $C_{(0,q)}^\infty(\mathbf{R}^n \times Y)$-module. Now assume that $A/\mathscr{M}_q(Y)A$ is a finite dimensional vector space. Since $\mathscr{M}_{(0,q)}(\mathbf{R}^{n-1} \times Y)A \supset \mathscr{M}_q(Y)A$, there is a natural surjection of $A/\mathscr{M}_q(Y)A \to A/\mathscr{M}_{(0,q)}(\mathbf{R}^{n-1} \times Y)A$ so that the latter space is finite dimensional. Thus the hypotheses of Lemma 3.7 are satisfied for π_n and we may conclude that A is a finitely generated $C_{(0,q)}^\infty(\mathbf{R}^{n-1} \times Y)$-module. A simple induction argument implies that A is a finitely generated $C_q^\infty(Y)$-module. □

Examples

(A) Let $f: \mathbf{R} \to \mathbf{R}$ be a smooth even function; then there exists a smooth function $g: \mathbf{R} \to \mathbf{R}$ satisfying $f(x) = g(x^2)$. This is easy to prove if f is assumed real or complex analytic near 0 (using Taylor series) but is not quite so obvious in the stated case. (This fact was first proved by Hassler Whitney—*Duke Journal of Mathematics*, volume 10, 1943.)

Proof. Let $p = 0$ in the domain and $q = 0$ in the range. Let $A = C_p^\infty(\mathbf{R})$ which is clearly a finitely generated module over $C_p^\infty(\mathbf{R})$. Let $\phi(x) = x^2$. Then via ϕ, $C_p^\infty(\mathbf{R})$ becomes a $C_q^\infty(\mathbf{R})$-module, the module action being given by $(ba)(x) = b(x^2)a(x)$ where a is in $C_p^\infty(\mathbf{R})$ and b is in $C_q^\infty(\mathbf{R})$. Observe that $\mathscr{M}_q(\mathbf{R})C_p^\infty(\mathbf{R}) = (x^2)$ and that the images of 1 and x span the vector space $C_p^\infty(\mathbf{R})/\mathscr{M}_q(\mathbf{R})C_p^\infty(\mathbf{R})$. Apply Theorem 3.6 and Corollary 3.5 to see that $f(x) = g(x^2) + xh(x^2)$. Since f is even, $h(x^2) \equiv 0$ and $f(x) = g(x^2)$ on a nbhd of 0. It is easy to see how to make this equality a global one. □

(B) Let $g_1, \ldots, g_n$ be the n *elementary symmetric polynomials* in n-variables; that is,

$$g_1(x_1, \ldots, x_n) = x_1 + \cdots + x_n$$
$$g_2(x_1, \ldots, x_n) = x_1x_2 + \cdots + x_1x_n + \cdots + x_{n-1}x_n$$
$$g_n(x_1, \ldots, x_n) = x_1 \cdots x_n.$$

Let $f(x_1, \ldots, x_n)$ be any smooth symmetric function; that is if σ is any permutation on n-letters, then $f(x_1, \ldots, x_n) = f(x_{\sigma(1)}, \ldots, x_{\sigma(n)})$. Then there exists a smooth function $h: \mathbf{R}^n \to \mathbf{R}$ satisfying

$$f(x) = h(g_1(x), \ldots, g_n(x)).$$

This global result was proved originally by Glaeser [9]. We shall only prove the corresponding local result. Define $g: \mathbf{R}^n \to \mathbf{R}^n$ by $g(x) =$

$(g_1(x), \ldots, g_n(x))$. Using the same convention concerning p and q in (A) we see that via g $C_p^\infty(\mathbf{R}^n)$ is a $C_q^\infty(\mathbf{R}^n)$-module. Let B be the set of multi-indices $\beta = (\beta_1, \ldots, \beta_{n-1}, 0)$ where $\beta_i < n$. Then the set of nomomials $\{x^\beta \mid \beta \in B\}$ is a generating set for the vector space $C_p^\infty(\mathbf{R}^n)/\mathscr{M}_q(\mathbf{R}^n)C_p^\infty(\mathbf{R}^n)$. (To see this note that

$$(x - x_1)\cdots(x - x_n) = x^n + g_1(x_1, \ldots, x_n)x^{n-1} + \cdots + g_n(x_1, \ldots, x_n).$$

Substituting x_i into this polynomial we have that x_i^n is in the submodule $\mathscr{M}_q(\mathbf{R}^n)C_p^\infty(\mathbf{R}^n)$. Also $x_n = -x_1 - \cdots - x_{n-1}$ modulo $\mathscr{M}_q(\mathbf{R}^n)C_p^\infty(\mathbf{R}^n)$.) Applying Theorem 3.6 and Corollary 3.5 we see that

$$f(x) = h(g(x)) + \sum_{\beta \in B} h_\beta(g(x))x^\beta.$$

Since f is symmetric and x_n is not a factor of any of the monomials we see that each $h_\alpha(g(x)) \equiv 0$ and that $f(x) = h(g(x))$. □

In all of our applications of the Malgrange Theorem we shall be dealing with modules of smooth functions. The most obvious problem in dealing with such functions (as distinct from analytic functions) is that the Taylor series about a point does not necessarily converge to the given function. Thus, in order to show that a module is finitely generated it would be nice to show that the prospective generators need only generate the module in question up to some finite order, thus eliminating the problem of what happens to the smooth functions "at the tail". We now show that this is, in fact, the case.

Define inductively a sequence of ideals $\mathscr{M}_p{}^k(X)$ in $C_p^\infty(X)$ by letting $\mathscr{M}_p{}^1(X)$ be $\mathscr{M}_p(X)$, and $\mathscr{M}_p{}^k(X)$ be the vector space generated by germs of the form fg where f is in $\mathscr{M}_p(X)$ and g is in $\mathscr{M}_p{}^{k-1}(X)$.

Lemma 3.9. *$\mathscr{M}_0{}^k(\mathbf{R}^n)$ consists precisely of germs of smooth functions f whose Taylor series at 0 begin with terms of degree k, i.e., $\partial^\alpha f/\partial x^\alpha(0) = 0$ for $|\alpha| \le k - 1$. Thus $C_0^\infty(\mathbf{R}^n)/\mathscr{M}_0{}^k(\mathbf{R}^n)$ can be identified with the vector space of polynomials in n variables of degree $\le k - 1$.*

The proof of this Lemma is a simple induction argument based on II, Lemma 6.10 and is left to the reader.

Theorem 3.10. *Let A be a finitely generated $C_p^\infty(X)$-module. Let $\phi: X \to Y$ be smooth with $q = \phi(p)$ and let $e_1, \ldots, e_k$ be elements of A. Then $e_1, \ldots, e_k$ generate A as a $C_q^\infty(Y)$-module iff $\eta(e_1), \ldots, \eta(e_k)$ generate $A/\mathscr{M}_p{}^{k+1}(X)A$ as a $C_q^\infty(Y)$-module where $\eta: A \to A/\mathscr{M}_p{}^{k+1}(X)A$ is the obvious projection.*

Proof. The forward implication is obvious, so assume that $\eta(e_1), \ldots, \eta(e_k)$ generate $A/\mathscr{M}_p{}^{k+1}(X)A$ as a $C_q^\infty(Y)$-module. Let

$$B = A/(\mathscr{M}_p{}^{k+1}(X)A + \mathscr{M}_q(Y)A).$$

Note that $\mathscr{M}_q(Y)$ acts trivially on B; thus we may consider B as a module over $C_q^\infty(Y)/\mathscr{M}_q(Y) = \mathbf{R}$, i.e., a real vector space. Since the images of $e_1, \ldots, e_k$ generate B, $\dim_R B \le k$. Consider the sequence of vector spaces $B \supset$

$\mathcal{M}_p^1(X)B \supset \cdots \supset \mathcal{M}_p^{k+1}(X)B = 0$. There are $k + 2$ vector spaces in this decreasing sequence. Applying the "pigeon-hold principle" we see that there must be $i \leq k$ such that $\mathcal{M}_p^i(X)B = \mathcal{M}_p^{i+1}(X)B$. Thus $\mathcal{M}_p^i(X)A + \mathcal{M}_q(Y)A = \mathcal{M}_p^{i+1}(X)A + \mathcal{M}_q(Y)A$, since $\mathcal{M}_p^{k+1}(X) \subset \mathcal{M}_p^{i+1}(X) \subset \mathcal{M}_p^i(X)$. Finally consider the $C_p^\infty(X)$-module $C = A/\mathcal{M}_q(Y)A$ and note that $\mathcal{M}_p^i(X)C = \mathcal{M}_p^{i+1}(X)C = \mathcal{M}_p(X)\mathcal{M}_p^i(X)C$. Thus we may apply Nakayama's Lemma and deduce that $\mathcal{M}_p^i(X)C = 0$. (Note that as a $C_p^\infty(X)$-module $\mathcal{M}_p^i(X)C$ is finitely generated. This follows since $\mathcal{M}_p^i(X)$ is a finitely generated ideal in $C_p^\infty(X)$ and C is a finitely generated module.) But $C = A/\mathcal{M}_q(Y)A$, so $\mathcal{M}_p^i(X)A \subset \mathcal{M}_q(Y)A$ for some $i \leq k$. Since the images of $e_1, \ldots, e_k$ generate $A/\mathcal{M}_p^{k+1}(X)A$, $\eta(e_1), \ldots, \eta(e_k)$ must generate $A/\mathcal{M}_q(Y)A$ as a $C_q^\infty(Y)$-module and thus as a $C_q^\infty(Y)/\mathcal{M}_q(Y) = \mathbf{R}$-module. Said differently, $\dim_R A/\mathcal{M}_q(Y)A \leq k$. We can now apply the generalized Malgrange Preparation Theorem to conclude that A is a finitely generated $C_q^\infty(Y)$-module and Corollary 3.5 to conclude that $e_1, \ldots, e_k$ generate A. ☐

The usefulness of Theorem 3.10 is illustrated by the following:

Corollary 3.11. *If the projections of $e_1, \ldots, e_k$ form a spanning set of vectors in the vector space $A/(\mathcal{M}_p^{k+1}(X)A + \mathcal{M}_q(Y)A)$, then $e_1, \ldots, e_k$ form a set of generators for A as a $C_q^\infty(Y)$-module.*

Chapter V

Various Equivalent Notions of Stability

§1. Another Formulation of Infinitesimal Stability

In this section we have three objectives: to show that

(1) Infinitesimal stability is locally a condition of finite order; i.e., if the equations can be solved locally to order dim Y then they can be solved for smooth data.

(2) Infinitesimal stability is globally equivalent to a multijet version of local infinitesimal stability.

(3) Infinitesimally stable mappings form an open set.

We won't be able to achieve our last objective just yet; but, at least, we shall be able to give a sufficient condition for the existence of a neighborhood of infinitesimally stable mappings around a given infinitesimally stable mapping.

Let X and Y be smooth manifolds with p in X and q in Y. Denote by $C^\infty(X, Y)_{p,q}$ the germs at p of mappings of $X \to Y$ which also map p to q. Recall that a germ at p is an equivalence class of mappings where two mappings are equivalent if they agree on a neighborhood of p. (We shall use the symbol $[f]_p$ to indicate the germ of $f: X \to Y$ at p—at least at those times when pedagogy overwhelms natural instincts.)

Let E be a vector bundle over X. Denote by $C^\infty(E)_p$ the germs of smooth sections of E at p. Since sections are mappings of $X \to E$ this makes sense according to the above prescription. In particular, we can speak of germs of vector fields along f as germs of sections of $f^*(TY)$ using the identification of $C_f^\infty(X, TY)$ with $C^\infty(f^*(TY))$ discussed in III, §1 after Definition 1.4.

Definition 1.1. *Let $f: X \to Y$, let p be in X, and let $q = f(p)$ in Y.*

(a) *the germ $[f]_p$ is* infinitesimally stable *if for every germ of a vector field along f, $[\tau]_p$, there exist germs of vector fields $[\zeta]_p$ in $C^\infty(TX)_p$ and $[\eta]_q$ in $C^\infty(TY)_q$ so that*

$$(*) \qquad [\tau]_p = [(df)(\zeta)]_p + [f^*\eta]_p.$$

(b) *f is* locally infinitesimally stable *at p if $[f]_p$ is infinitesimally stable.*

It is clear that if $f: X \to Y$ is infinitesimally stable, then f is locally infinitesimally stable.

Choose coordinates $x_1, \ldots, x_n$ on X based at p and coordinate $y_1, \ldots, y_m$ on Y based at q. We will compute equation (*) in these coordinates. If τ is a vector field along f, then we can write $\tau(x) = \sum_{i=1}^m \tau_i(x)(\partial/\partial y_i)$. So equation (*) becomes

$$(**) \qquad \tau_i = \sum_{j=1}^n \frac{\partial f_i}{\partial x_j} \zeta_j + \eta_i(f_1, \ldots, f_m) \qquad 1 \le i \le n$$

where

$$\zeta = \sum_{j=1}^{n} \zeta_j \frac{\partial}{\partial x_j}, \qquad \eta = \sum_{i=1}^{m} \eta_i \frac{\partial}{\partial y_i},$$

and $f_1, \ldots, f_m$ are the coordinate functions of f.

Equations (**) can be solved to order k if for each set of germs $\tau_1, \ldots, \tau_m$ in $C_0^\infty(\mathbf{R}^n)$, there exists germs $\zeta_1, \ldots, \zeta_n$ in $C_0^\infty(\mathbf{R}^n)$ and germs $\eta_1, \ldots, \eta_m$ in $C_0^\infty(\mathbf{R}^m)$ so that

$$\tau_i = \sum_{j=1}^{n} \frac{\partial f_i}{\partial x_j} \zeta_j + \eta_i(f_1, \ldots, f_m) + O(|x|^{k+1}).$$

Theorem 1.2. *Let $f: X \to Y$ with p in X, $q = f(p)$ in Y, and $m =$* dim *Y. Then $[f]_p$ is infinitesimally stable iff equations (**) can be solved to order m.*

Proof. We shall use the Generalized Malgrange Preparation Theorem. First note that $C^\infty(f^*TY)_p = \bigoplus_{i=1}^m C_p^\infty(X)$ since $\tau(x) = \sum_{i=1}^m \tau_i(x)(\partial/\partial y_i)$ (locally). Thus $C^\infty(f^*TY)_p$ is a finitely generated $C_p^\infty(X)$ module. Let $A = \{(df)(\zeta) \mid \zeta \in C^\infty(TX)_p\}$. A is a submodule of $C^\infty(f^*TY)_p$ so that $M_f{}^p = C^\infty(f^*(TY))_p/A$ is a finitely generated $C_p^\infty(X)$-module. Via f^* we can view $M_f{}^p$ as a $C_q^\infty(Y)$-module. Let e_i be the projection of $f^*(\partial/\partial y_i)$ in $M_f{}^p$. We first observe that $[f]_p$ is infinitesimally stable iff $e_1, \ldots, e_m$ generate $M_f{}^p$ as a $C_q^\infty(Y)$-module. Recall that $\mathcal{M}_p{}^{m+1}(X)$ consists of germs of functions at p whose Taylor series start with terms of order $m + 1$ or greater (IV, Lemma 3.9). Now apply IV, Theorem 3.10 to see that $[f]_p$ is infinitesimally stable iff the module $M_f{}^p/\mathcal{M}_p{}^{m+1}(X)M_f{}^p$ is generated over $C_q^\infty(Y)$ by the projections of $e_1, \ldots, e_m$. This last statement is easily seen to be equivalent to solving equations (**) to order m, for if $[\tau]_p$ is in $C^\infty(f^*TY)_p$, then $\tau = \sum_{i=1}^m (\eta_i \cdot f)e_i + (df)(\zeta) + g$ where ζ and η are defined as usual and g is in $\mathcal{M}_p{}^{m+1}(X)C^\infty(TY)_q$; i.e., g is $O(|x|^{m+1})$. $\square$

Theorem 1.2 makes it clear that whether or not f is infinitesimally stable at p is determined by $j^{m+1}f(p)$. We shall formalize this notion. Let E be a vector bundle over X and let $J^k(E) = \{\sigma \in J^k(X, E) \mid \sigma$ is represented by a section of $E\} =$ *k-jet bundle of sections of E*. Let $\pi : E \to X$ be the projection. Then $\pi_* : J^k(X, E) \to J^k(X, X)$ is a submersion. Let I be the submanifold of $J^k(X, X)$ given by $\{\sigma \in I \mid \sigma$ is represented by $id_X\}$. Then $J^k(E) = (\pi_*)^{-1}(I)$ and is thus a submanifold of $J^k(X, E)$. The source map $\alpha : J^k(X, E) \to X$ restricts to a map of $J^k(E) \to X$. It is not hard to see that this is a fiber map. Let $J^k(E)_p =$ fiber of $J^k(E)$ at p. This has a natural vector space structure. In fact, given two elements of $J^k(E)_p$ we can find sections that represent them. Add these sections and take the k-jet of the sum. We let the reader check that this operation is well-defined; i.e. independent of the choice of sections and that this gives $J^k(E)$ a vector bundle structure over X. *Hint:* Do this first for the trivial bundle whose sections are just maps of $X \to \mathbf{R}^n$.

The following is just a restatement of Theorem 1.2.

Corollary 1.3. *Let $f: X \to Y$ with p in X, $q = f(p)$ in Y, and $m = \dim Y$. Then f is locally infinitesimally stable at p iff*

$$J^m(f^*TY)_p = (df)_p J^m(TX)_p + f^* J^m(TY)_q$$

where $(df)_p$ and f^* are the obvious mappings into $J^m(f^*TY)_p$ induced by the action of (df) and f^* on vector fields.

We wish to obtain conditions analogous to those in Corollary 1.3 which will be equivalent to infinitesimal stability. We have local conditions but these are not sufficient; for what happens at the self intersections of a function is not taken into account. In particular, the choice of η in $J^m(TY)_q$ might be forced in two different ways at two different pre-image points. We shall extend our results to take care of these cases.

Let $f: X \to Y$ be smooth and let q be in Y. Let $S = \{p_1, \ldots, p_k\} \subset f^{-1}(q)$. Define $C_S^\infty(X) = \bigoplus_{i=1}^k C_{p_i}^\infty(X)$ and note that $C_S^\infty(X)$ is a ring where the operations are done coordinatewise. Since f induces a ring homomorphism $f^*: C_q^\infty(Y) \to C_{p_i}^\infty(X)$ for each i, it induces a ring homomorphism of $C_q^\infty(Y) \to C_S^\infty(X)$ which we also denote by f^*. So if A is a $C_S^\infty(X)$-module, then via f^* A becomes a $C_q^\infty(Y)$-module. We reformulate the Generalized Malgrange Preparation Theorem so that it is applicable to these modules.

Lemma 1.4. *Let A_i $(1 \le i \le k)$ be a finitely generated $C_{p_i}^\infty(X)$ module. Then $A = A_1 \oplus \cdots \oplus A_k$ is a finitely generated $C_S^\infty(X)$ module (where the action of $C_{p_i}^\infty(X)$ on A_j for $i \neq j$ is zero). Let $e_1, \ldots, e_m$ be in A so that the projections of the e_i's in $A/\mathcal{M}_q(Y)A$ span this vector space. Then $e_1, \ldots, e_m$ generate A as a $C_q^\infty(Y)$ module.*

Proof. Since $A/\mathcal{M}_q(Y)A = (A_1/\mathcal{M}_q(Y)A_1) \oplus \cdots \oplus (A_k/\mathcal{M}_q(Y)A_k)$ we can apply the Malgrange Theorem (IV, 3.6) coordinatewise. ☐

Remark. Let $\mathcal{M}_S(X) = \mathcal{M}_{p_1}(X) \oplus \cdots \oplus \mathcal{M}_{p_k}(X)$. Then it is enough to know that the projections of $e_1, \ldots, e_m$ span the vector space

$$A/(\mathcal{M}_q(Y)A + \mathcal{M}_S{}^{m+1}(X)A).$$

Just apply III, Corollary 3.11 in the above Lemma instead of III, Theorem 3.6.

Let E be a vector bundle over X and let $J^m(E)_S = \bigoplus_{i=1}^k J^m(E)_{p_i}$. Now $f: X \to Y$ induces mappings $f^*: J^m(TY)_q \to J^m(f^*TY)_{p_i}$ and thus a mapping $f^*: J^m(TY)_q \to J^m(f^*TY)_S$ given by $f^*[\eta]_q = ([\eta \cdot f]_{p_1}, \ldots, [\eta \cdot f]_{p_k})$. Also f induces $(df): J^m(TX)_{p_i} \to J^m(f^*TY)_{p_i}$ and thus induces a mapping $(df): J^m(TX)_S \to J^m(f^*TY)_S$.

Let $f: X \to Y$ be smooth and $S = \{P_1, \ldots, P_k\} \subset f^{-1}(q)$. Then f is *simultaneously locally infinitesimally stable at* $p_1, \ldots, p_k$ if given germs of vector fields along f $[\tau_1]_{p_1}, \ldots, [\tau_k]_{p_k}$, there exist germs of vector fields $[\zeta_1]_{p_1}, \ldots, [\zeta_k]_{p_k}$ and $[\eta]_q$ such that

$$[(df)(\zeta_i)]_{p_i} + [\eta \cdot f]_{p_i} = [\tau_i]_{p_i}$$

for all i.

Note. The "simultaneously" is to emphasize that one vector field germ of Y is being chosen along with the k vector field germs on X to solve the equations of infinitesimal stability for k germs of vector fields along f.

Proposition 1.5. *Let $f: X \to Y$ be smooth and $S = \{p_1, \ldots, p_k\} \subset f^{-1}(q)$. Then f is simultaneously locally infinitesimally stable at $p_1, \ldots, p_k$ iff $J^m(f^*TY)_S = (df)J^m(TX)_S + f^*J^m(TY)_q$.*

Proof. For S consisting of a single point this result is given by Corollary 1.3. The proof for general S is exactly as in the single point case except that we substitute Lemma 1.4 and the subsequent remark for the Generalized Malgrange Preparation Theorem. In particular choose coordinates on X at $p_1, \ldots, p_k$ (with disjoint domains) and coordinates on Y at q and write down the equations generalizing (**) to solve the local infinitesimal stability condition simultaneously to order m at $p_1, \ldots, p_k$. Let $M_f^S = \bigoplus_{i=1}^k M_f^{p_i}$. Continue as before—except for the substitution of Lemma 1.4 and its subsequent remark. □

Theorem 1.6. *Let $f: X \to Y$ be smooth. Then f is infinitesimally stable iff* (†) *for every q in Y and every finite subset S of $f^{-1}(q)$ with no more than $(m + 1)$ points*

$$J^m(f^*TY)_S = (df)(J^m(TX)_S) + f^*(J^m(TY)_q).$$

The necessity part of this theorem is obvious. Before proving the sufficiency we need some preparatory lemmas.

Lemma 1.7. *Let $H_1, \ldots, H_k$ be subspaces of a finite dimensional vector space V. Then $H_1, \ldots, H_k$ are in general position* (see III, Definition 3.5) *iff* (*) *given $v_1, \ldots, v_k$ in V, there exists h_i in H_i and z in V such that $v_i = z + h_i$ for all i.*

Proof. Let $\pi_i: V \to V/H_i$ be the natural projection and let

$$\pi: V \to V/H_1 \oplus \cdots \oplus V/H_k$$

be given by $\pi(v) = (\pi_1(v), \ldots, \pi_k(v))$. Clearly $\operatorname{Ker} \pi = H_1 \cap \cdots \cap H_k$ so that the sequence

$$0 \to H_1 \cap \cdots \cap H_k \to V \xrightarrow{\pi} V/H_1 \oplus \cdots \oplus V/H_k$$

is exact. Now π is onto iff $\operatorname{codim}(H_1 \cap \cdots \cap H_k) = \sum_{i=1}^k \dim V/H_i$ iff $H_1, \ldots, H_k$ are in general position. But clearly π is onto iff given $v_1, \ldots, v_k$ in V, there exists z in V such that $\pi_i(z) = v_i$, i.e. there exists h_i in H_i so that $v_i = z + h_i$. So π is onto iff condition (*) holds. □

Lemma 1.8. *Let $f: X \to Y$ satisfy* (†) *and let $S = \{p_1, \ldots, p_k\} \subset f^{-1}(q)$. Let $H_i = (df)_{p_i}(T_{p_i}X)$ for $1 \leq i \leq k$. Then $H_1, \ldots, H_k$ are in general position as subspaces of T_qY.*

Proof. Let $z_1, \ldots, z_k$ be in T_qY. By Lemma 1.7 we must show that there exist h_i in H_i and y in T_qY so that $z_i = h_i + y$ for all i. Choose a vector field τ along f so that $\tau(p_i) = z_i$. By (†) choose vector fields ζ on X and η on Y so that $\tau = (df)(\zeta) + f^*\eta$ on a neighborhood of S. (Use Proposition 1.5.) Let $h_i = (df)_{p_i}(\zeta_i)$ and $y = \eta_q$. □

For our current purposes we shall call p in X a *critical point* of $f: X \to Y$ if $(df)_p: T_pX \to T_{f(p)}Y$ is not onto. Thus a critical point is either a singularity in case dim $Y \geq$ dim X or an arbitrary point if dim $X <$ dim Y.

Lemma 1.9. *Let $f: X \to Y$ satisfy (†), let $m =$ dim Y, and let q be in Y. Then the number of critical points in $f^{-1}(q)$ is $\leq m$.*

Proof. We shall argue by contradiction. Suppose $S = \{p_1, \ldots, p_{m+1}\}$ consists of distinct critical points of f in $f^{-1}(q)$. Let $H_i = (df)_{p_i}(T_{p_i}X)$. The last lemma states that $H_1, \ldots, H_{m+1}$ are in general position as subspaces of T_qY. Thus $m \geq \operatorname{codim}(H_1 \cap \cdots \cap H_{m+1}) = \sum_{i=1}^{m+1} \operatorname{codim} H_i \geq m + 1$. The last inequality holds since if p is a critical point $\operatorname{codim}(df)_p(T_pX) \geq 1$. □

Proof of Theorem 1.6. Sufficiency. We assume that f satisfies (†). Let Σ be the critical point set of f in X and let $\Sigma_q = \Sigma \cap f^{-1}(q)$. By the last lemma Σ_q is a finite set with $\leq m$ points. Let τ be a vector field along f. To prove that f is infinitesimally stable we need to show that there exist vector fields ζ on X and η on Y so that $\tau = (df)(\zeta) + f^*\eta$. We first show that this equation can be solved on a neighborhood of Σ.

We claim that there exist open sets $U_1, \ldots, U_N$ in X, $V_1, \ldots, V_N$ in Y, and $W_1, \ldots, W_N$ in Y; and vector fields ζ_i on U_i and η_i on V_i satisfying

(a) $f(\Sigma) \subset \bigcup_{i=1}^N W_i$
(b) $f(U_i) \subset V_i$
(c) $\tau = (df)(\zeta_i) + f^*\eta$ on U_i
(d) $f^{-1}(\overline{W}_i) \cap \Sigma \subset U_i$, and
(e) $\overline{W}_i \subset V_i$.

One need only construct U, V, W, ζ, and η for a given q in $f(\Sigma)$ (i.e., W must be a neighborhood of q) satisfying (b)–(e). Since $f(\Sigma)$ is compact, the necessary N will exist. By Proposition 1.5 we may choose open neighborhoods U of Σ_q in X and V of q in Y, and vector fields ζ on U and η on V so that (b) and (c) are satisfied. Next we choose W satisfying (d). Since $\Sigma_q = f^{-1}(q) \cap \Sigma \subset U_i$, there is a small open neighborhood W so that $f^{-1}(W) \cap \Sigma \subset U_i$. If not there would exist a sequence of critical points $x_1, x_2, \ldots$ with $\operatorname{Lim}_{i\to\infty} f(x_i) = q$ and $x_i \notin U_i$ for all i. Since X is compact we may asume that $x_i \to x$. By continuity $f(x) = q$ and x is a critical point. Thus x is in Σ_q. A contradiction since x is in $X - U$ and $(X - U) \cap \Sigma_q = \varnothing$. By shrinking W we may assume that $f^{-1}(\overline{W}) \cap \Sigma \subset U$ and that $\overline{W}_i \subset V$.

Next choose a partition of unity $\rho_1, \ldots, \rho_N$ on $W = \bigcup_{i=1}^N W_i$ with supp $\rho_i \subset W_i$. Choose an open neighborhood Z of Σ such that $f^{-1}(\overline{W}_i) \cap Z \subset U_i$. (This is possible. Otherwise there exists a sequence $x_1, x_2, \ldots$ converging to x in Σ with x_i in $f^{-1}(\overline{W}_i) \cap (X - U)$. Since both $f^{-1}(\overline{W}_i)$ and $X - U$ are closed x is in $f^{-1}(\overline{W}_i)$ and x is not in U. Contradiction.) Choose a smooth function $\rho: X \to \mathbf{R}$ such that supp $\rho \subset Z$ and $\rho \equiv 1$ on a neighborhood of Σ. Let $\zeta = \sum_{i=1}^N \rho f^*(\rho_i)\zeta_i$. This is well-defined on X since supp $\rho f^*(\rho_i) \subset$

$f^{-1}(\overline{W}_i) \cap Z \subset U_i$. Next let $\eta = \sum_{i=1}^N \rho_i\eta_i$. Then calculate that $\tau = (df)(\zeta) + f^*\eta$ on a neighborhood of Σ.

Thus we may assume that $\tau \equiv 0$ on a neighborhood U of Σ in X. If $\dim X < \dim Y$, then we are finished for $\Sigma = X$. In case $\dim Y \geq \dim X$, then f is a submersion on $X - \Sigma$. Thus there exists a vector field ζ on $X - \Sigma$ so that $(df)(\zeta) = \tau$. Use III, Proposition 2.1 which states that all submersions are infinitesimally stable. (Note the fact that X is compact was not used in that proof.) Choose a smooth function $\rho : X \to \mathbf{R}$ which is zero on Σ and 1 off U. Then $\rho\zeta$ is globally defined on X and $(df)(\rho\zeta) = \tau$ since $\operatorname{supp} \tau \subset X - U$. □

We now attack the last of our three objectives.

Proposition 1.10. *Let $f: X \to Y$ be infinitesimally stable. Then there exists a neighborhood W of f such that every g in W is locally infinitesimally stable.*

Proof. Let p be in X and let $q = f(p)$. Since f is infinitesimally stable $J^m(f^*TY)_p = (df)J^m(TX)_p + f^*J^m(TY)_q$. Now consider the mapping $\tilde{f}: J^m(TX)_p \oplus J^m(TY)_q \to J^m(f^*TY)_p$ given by $(df) + f^*$. This is just a linear mapping between vector spaces. In particular, if we choose chart neighborhoods U of p and V of q such that $f(\overline{U}) \subset V$, then in these local coordinates $\tilde{f}$ is a linear mapping of $B^m_{n,n} \oplus B^m_{m,m} \to B^m_{n,m}$ where $B^k_{n,m}$ is the vector space of polynomials of degree $\leq k$ from $\mathbf{R}^n \to \mathbf{R}^m$. Now clearly $\tilde{f}$ depends continuously on p and f (in fact on $j^{m+1}f$). Thus there is an open neighborhood U_p of p and an open neighborhood W_p of f such that if g is in W_p then $g(\overline{U}) \subset V$ (a C^0 condition) and if p' is in U_p, then $\tilde{g}$ is onto at p'. Thus g is infinitesimally stable at p' using Corollary 1.3. Since X is compact there is a finite covering of X by U_p's. The intersection of the corresponding W_p's is an open neighborhood of f with the desired properties. □

We would like to prove a corresponding theorem for (global) infinitesimal stability; unfortunately there are some difficulties. We have the global result given in Theorem 1.6 to use instead of Corollary 1.3 but the proof above will not work for $X^{(s)}$ is not compact (even though X is compact). To avoid these difficulties we need to make a definition. First some notation: Let $p = (p_1, \ldots, p_r)$ be in the generalized diagonal $X^r - X^{(r)}$. Let $q_1, \ldots, q_s$ be the distinct points among the p_i's each occuring with multiplicity m_i, $i = 1, \ldots, s$.

Definition 1.11. (a) *An infinitesimally stable mapping $f: X \to Y$ satisfies* condition $\mathcal{O}$ *at p if there exist neighborhoods U_j of q_j and a neighborhood W_p of f such that if g is in W_p and if $S \subset (\bigcup U_j) \cap g^{-1}(q)$ (with $\# S \cap U_j = m_j$), then*

$$(\dagger) \qquad J^m(g^*TY)_S = (dg)J^m(TX)_S + g^*J^m(TY)_q.$$

(b) *f satisfies* condition $\mathcal{O}$ *if it satisfies $\mathcal{O}$ at p for all p with $r \leq m + 1$.*

The object of this definition is to finesse—for the moment—the question of what happens in $X^{(s)}$ near the generalized diagonal. This is because of the following:

Lemma 1.12. *If $f: X \to Y$ is an infinitesimally stable mapping satisfying condition $\mathcal{O}$ then there exists an open neighborhood W of f which consists of infinitesimally stable mappings.*

Proof. We will show that for every $r \leq m + 1$ there exists a nbhd W_r such that if g is in W_r and S is as in Definition 1.11, then g satisfies (†). (Then $W = \bigcap_{r=1}^{m+1} W_r$ is the desired nbhd by Theorem 1.6.)

Note that W_1 is given by Proposition 1.10. Let $p = (p_1, \ldots, p_r)$ be in X^r. We claim there exist nbhds U_i of p_i and W_p of f such that if a set $S \subset (\bigcup_{i=1}^{r} U_i) \cap g^{-1}(q)$ consists of r points with $\# S \cap U_i = m_i$, then g satisfies (†) on S. For p in the generalized diagonal of X^r, condition $\mathcal{O}$ gives the desired nbhd. For p not in the generalized diagonal one may choose U_i of p_i and $\tilde{W}_i$ of f by Proposition 1.10 such that the U_i's are disjoint and $W_p = \tilde{W}_1 \cap \cdots \cap \tilde{W}_r$. As the nbhds $U_{p_1} \times \cdots \times U_{p_r}$ cover X^r and X^r is compact, we may choose a finite subcover indexed by the points $p^1, \ldots, p^N$ in X^r. The desired W_r is just the intersection of the W_{p^i}'s. □

Exercises

(1) Consider the mapping $f: \mathbf{R}^2 \to \mathbf{R}^2$ defined by $(x, y) \mapsto (x, xy - y^3)$. Try showing that f is infinitesimally stable using only the definition of infinitesimal stability. In doing so you should get to a functional equation which is rather difficult to solve on a nbhd of the origin, i.e., for every pair of smooth functions $\tau_1, \tau_2 : \mathbf{R}^2 \to \mathbf{R}$ there exist smooth functions ξ_1, ξ_2, η_1, and $\eta_2 : \mathbf{R}^2 \to \mathbf{R}$ such that

$$(*) \qquad \begin{bmatrix} \tau_1 = \xi_1 + \eta_1 \cdot f \\ \tau_2 = y\xi_1 + (x - 3y^2)\xi_2 + \eta_2 \cdot f. \end{bmatrix}$$

Use Theorem 1.2 to show that $[f]_0$ is infinitesimally stable by solving equations (*) to order 2. Try showing that f is infinitesimally stable by applying Theorem 1.6. (Since f is proper the theorem is still valid even though $X = \mathbf{R}^2$ is not compact.)

(2) Let $f: \mathbf{R}^2 \to \mathbf{R}^3$ be given by $f(x, y) = (x, xy, y^2)$. Show that f is locally infinitesimally stable at 0.

(3) It is possible to reduce even further the calculations needed to compute local infinitesimal stability using the following observation due to Arnold. Let $f: \mathbf{R}^n \to \mathbf{R}^m$ and assume that $f(0) = 0$. Say that f satisfies condition (Z) if for every germ $\phi \in C_0^\infty(\mathbf{R}^n)$ there exists an $n \times m$ matrix H of germs in $C_0^\infty(\mathbf{R}^n)$ and an $m \times m$ matrix K of germs in $C_0^\infty(\mathbf{R}^m)$ such that

$$\phi \mathrm{I}_m = (df)H + K \cdot f.$$

where I_m is the $m \times m$ identity matrix.

(a) Show that condition (Z) holds (to order m) at 0 iff equations (**) hold (to order m) at 0.

Hint: For (**) $\Rightarrow$ condition (Z). Let $\tau^l = (0, \ldots, \phi, \ldots, 0)$ where ϕ is in the lth position for $1 \leq l \leq m$. Using (**) obtain $\tau^l = (df)(\xi^l) + \eta^l \circ f$. Let $H = \{\xi_i^l\}$ and $K = \{\eta_i^l\}$ where $\xi^l = (\xi_1^l, \ldots, \xi_n^l)$ and $\eta^l = (\eta_1^l, \ldots, \eta_m^l)$. For condition ($Z$) $\Rightarrow$ (**), let $\tau = (\tau_1, \ldots, \tau_m)$ and solve $\tau_l = (df)H + K \circ f$. Let $\xi_i = \sum_{j=1}^m \eta_{ij}$ and $\eta_i = \sum_{j=1}^m k_{ij}$ where $H = (h_{ij})$ and $K = (k_{ij})$. Then let $\xi = (\xi_1, \ldots, \xi_n)$ and $\eta = (\eta_1, \ldots, \eta_m)$.

(b) Show that if condition (Z) can be solved for ϕ_1 and ϕ_2 in $C_0^\infty(\mathbf{R}^n)$, then condition (Z) holds for $\phi = \phi_1 \phi_2$.

Hint: Choose H_1, K_1 for ϕ_1 and H_2, K_2 for ϕ_2. Let $H = \phi_1 H_2 + H_1 K_2$ and let $K = K_1 K_2$.

(c) ***Proposition 1.13.*** Let $f: X \to Y$ be smooth and let p be in X. Choose coordinates $x_1, \ldots, x_n$ on X at p. Then f is locally infinitesimally stable at p iff equations (**) are solvable to order m for $\tau_l^k = (0, \ldots, x_k, \ldots, 0)$ where the x_k appears in the lth position for $1 \leq l \leq m$ and $1 \leq k \leq n$.

(4) Let $f: \mathbf{R}^n \to \mathbf{R}^n$ be given by $(x_1, \ldots, x_n) \mapsto (x_1, \ldots, x_{n-1}, x_1 x_n + x_2 x_n^2 + \cdots + x_{n-1} x_n^{n-1} + x_n^{n+1})$. Show that f is locally infinitesimally stable at 0.

§2. Stability Under Deformations

Our intention in this chapter is to prove (under the assumption that the domain is a compact manifold) that stability is equivalent to infinitesimal stability. To accomplish this task it seems necessary to introduce several other notions of stability—all of which are, in fact, equivalent to the ones just mentioned. The most natural of these is the concept of stability under deformations introduced by Thom and Levine.

Definition 2.1. *Let $f: X \to Y$ be smooth and let $I_\varepsilon = (-\varepsilon, \varepsilon)$. Then*

(a) *let $F: X \times I_\varepsilon \to Y \times I_\varepsilon$ be smooth. F is a* deformation *of f if*

(i) *for each s in $(-\varepsilon, \varepsilon)$, $F: X \times \{s\} \to Y \times \{s\}$. Denote by F_s the mapping of $X \mapsto Y$ defined by $F(x, s) = (F_s(x), s)$.*

(ii) $F_0 = f$.

(b) *Let $F: X \times I_\varepsilon \to Y \times I_\varepsilon$ be a deformation of f. Then F is* trivial *if there exist diffeomorphisms $G: X \times I_\delta \to X \times I_\delta$ and $H: Y \times I_\delta \to Y \times I_\delta$ (where $0 < \delta \leq \varepsilon$) such that G and H are deformations of id_X and id_Y, respectively, and such that the diagram*

$$\begin{array}{ccc} X \times I_\delta & \xrightarrow{F} & Y \times I_\delta \\ {\scriptstyle G}\downarrow & & \downarrow{\scriptstyle H} \\ X \times I_\delta & \xrightarrow[f \times id_{I_\delta}]{} & Y \times I_\delta \end{array}$$

commutes.

(c) f *is* stable under deformations (*or* homotopically stable) *if every deformation of f is trivial.*

Remarks. (1) In a trivial deformation F of f each F_t is equivalent to f. Also, if f is stable then for a given deformation F each F_t (for t small enough) is equivalent to f. Unfortunately, this is not enough to show that F is trivial, since the conjugating maps need not vary smoothly.

(2) By viewing a deformation of f as a mapping of $I_\varepsilon \mapsto C^\infty(X, Y)$ we can equate deformations of f with curves in $C^\infty(X, Y)$ based at f. Recall that in Chapter III, §1 we motivated the definition of infinitesimal stability in terms of Frechet manifolds. In particular, we showed that f is infinitesimally stable iff $(d\gamma_f)_{id}$ is onto where $\gamma_f : \text{Diff}(X) \times \text{Diff}(Y) \to C^\infty(X, Y)$ is defined by $\gamma_f(g, h) = h \cdot f \cdot g^{-1}$. Certainly $(d\gamma_f)_{id}$ is onto if for every curve $t \mapsto F_t$ in $C^\infty(X, Y)$ where $F_0 = f$, there is a curve $t \mapsto (G_t, H_t)$ in $\text{Diff}(X) \times \text{Diff}(Y)$ so that $\gamma_f(G_t, H_t) = F_t$ for all small t; i.e., for small t the diagram

$$\begin{array}{ccc} X & \xrightarrow{f} & Y \\ {\scriptstyle G_t}\big\downarrow & & \big\downarrow{\scriptstyle H_t} \\ X & \xrightarrow[F_t]{} & Y \end{array}$$

commutes. But this is just the condition that the deformation F be trivial. So it should come as no surprise that we will show that homotopic stability implies infinitesimal stability. In fact, they are equivalent notions and this also will be shown later.

(3) As for the relationship between stability under deformations and stability, the only fact which can be immediately proved is the following:

Lemma 2.2. *Let $f: X \to Y$ be stable under deformations. Suppose there exists an open nbhd W of f in $C^\infty(X, Y)$ such that each g in W is stable under deformations, then f is stable.*

Proof. By shrinking, if necessary, we can assume that W is "arc-wise connected"; that is, for each g in W, there is a deformation $F: X \times [-1, 1] \to Y \times [-1, 1]$ of f so that $F_1 = g$ and F_t is in W for all t in $[-1, 1]$. (In III, Theorem 1.12 we identified functions in a nbhd U of f with sections of a tubular nbhd of graph (f) in $X \times Y$. For functions in U the deformation is obvious. Thus by "shrinking" we mean replacing W by $W \cap U$.) Now let g be in W and choose such an F. Consider the equivalence relation on $[-1, 1]$ defined as follows: $s \sim t$ if F_s is equivalent to F_t as mappings of $X \to Y$. The assumption that each mapping in W is homotopically stable along with Remark (1) implies that each equivalence class is open. Since $[-1, 1]$ is connected there is only one equivalence class and thus g is equivalent to f. ☐

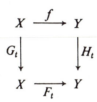

So if we know that infinitesimally stable maps form an open set and that infinitesimal stability is equivalent to homotopic stability, then we would know that infinitesimally stable mappings are stable. Recall that in the last section we showed (in Lemma 1.12) that if every infinitesimally stable mapping satisfies condition $\mathcal{O}$ then the set of infinitesimally stable mappings is open. We now generalize the concept of homotopic stability and show that a mapping which is both infinitesimally stable and satisfies this generalized homotopic stability criterion also satisfies condition $\mathcal{O}$.

***Definition* 2.3.** *Let $f: X \to Y$ be smooth and let U be a nbhd of 0 in $\mathbf{R}^k$.*

(a) *Let $F: X \times U \to Y \times U$ be smooth. F is a k-*deformation *of f if*

(i) *for each v in U, $F: X \times \{v\} \to Y \times \{v\}$. Denote by F_v the mapping of $X \to Y$ defined by $F(x, v) = (F_v(x), v)$.*

(ii) $F_0 = f$.

(b) *Let $F: X \times U \to Y \times U$ be a k-deformation of f. Then F is* trivial *if there exist diffeomorphisms $G: X \times V \to X \times V$ and $H: Y \times V \to Y \times V$ where V is an open nbhd of 0 contained in U such that G and H are deformations of id_X and id_Y respectively, and such that the diagram*

$$\begin{array}{ccc} X \times V & \xrightarrow{\quad F \quad} & Y \times V \\ {\scriptstyle G}\big\downarrow & & \big\downarrow{\scriptstyle H} \\ X \times V & \xrightarrow[f \times id_V]{\qquad} & Y \times V \end{array}$$

commutes.

(c) f *is* stable under k-deformations *if every k-deformation of f is trivial.*

Remarks. (1) Stability under 1-deformations = homotopic stability.

(2) If f is stable under k-deformations, then f is stable under l deformations for $l \leq k$. In particular, f is homotopically stable.

Before giving our discussion of condition $\mathcal{O}$ we make some preparatory lemmas. If $g: \mathbf{R}^n \to \mathbf{R}$ is smooth and K is a compact subset of $\mathbf{R}^n$, then define

$$\|g\|_s^K = \max_{\substack{x \in K \\ 0 \leq |\alpha| \leq s}} \left| \frac{\partial^{|\alpha|} g}{\partial x^\alpha}(x) \right|.$$

***Lemma* 2.4.** *Let p be in a convex compact subset K of $\mathbf{R}^n$ and let $g: \mathbf{R}^n \to \mathbf{R}$ be smooth. Let*

$$g(x) = \sum_{0 \leq |\alpha| \leq r} a_\alpha (x - p)^\alpha + \sum_{|\beta| = r+1} g_\beta(x)(x - p)^\beta$$

be the Taylor expansion with remainder term of order $r + 1$. Then if $\|g\|_s^K < \varepsilon$, then $\|g_\beta\|_{s-r-1}^K < \varepsilon$ for $r < s$ and $|\beta| = r + 1$.

Proof. Without loss of generality we may assume that $p = 0$. We proceed by induction on r. For $r = 0$, $g(x) = g(0) + \sum_{i=1}^n x_i g_i(x)$ where $g_i(x) = \int_0^1 (\partial g/\partial x_i)(tx)\, dt$. (See II, Lemma 6.10.) Thus

$$\left|\frac{\partial^{|\alpha|}g_i}{\partial x^\alpha}(x)\right| \le \int_0^1 \left|\frac{\partial^{|\alpha|}}{\partial x^\alpha}\frac{\partial g}{\partial x_i}(tx)\right| dt < \varepsilon \quad \text{for} \quad |\alpha| \le s - 1$$

since $|\alpha| + 1 \le s$ and $\|g\|_s^K < \varepsilon$. (Note that tx is in K since K is convex.) So $\|g_i\|_{s-1}^K < \varepsilon$.

For the general case just note that if

$$g(x) = \sum_{0 \le |\alpha| \le r-1} a_\alpha x^\alpha + \sum_{|\beta| = r} h_\beta(x)x^\beta,$$

then by expanding $h_\beta(x) = h_\beta(0) + \sum_{i=1}^n g_{\beta,i}(x)x_i$ we get the Taylor expansion of g to order r. Apply induction and the $r = 0$ case to obtain the desired result. □

Let A_n^l be the vector space of polynomials of $\mathbf{R}^n \to \mathbf{R}$ of degree $\le l$.

Lemma 2.5. *Let $r \ge 0$ and $s > 0$ be integers and let K be a compact convex nbhd of 0 in $\mathbf{R}^n$. Let Z be an open nbhd of 0 in A_n^l where $l = (r + 1)^s$. Then there exists an $\varepsilon > 0$ so that if $p_1, \ldots, p_s$ are in K and if $g: \mathbf{R}^n \to \mathbf{R}$ is smooth and satisfies $\|g\|_{s(r+1)}^K < \varepsilon$, then there exists a polynomial v in Z such that*

$$\frac{\partial^{|\alpha|}v}{\partial x^\alpha}(p_i) = \frac{\partial^{|\alpha|}g}{\partial x^\alpha}(p_i) \quad \textit{for} \quad 1 \le i \le s \textit{ and } 0 \le |\alpha| \le r.$$

Proof. We prove the lemma by induction on s. Let $s = 1$ and let $p = p_1$. By Taylor's Theorem

$$g(x) = \sum_{0 \le |\alpha| \le r} a_\alpha(x - p)^\alpha + \sum_{|\alpha| = r+1} (x - p)^\alpha g_\alpha(x)$$

where each g_α is a smooth function. Let $v = \sum_{0 \le |\alpha| < r} a_\alpha(x - p)^\alpha$. Since p is assumed to vary within the compact set K, the coordinates of p are bounded by some constant. It is then easy to see that the coefficients of v are bounded by some constant multiple of ε (where the constant depends on K but not on p) since $|a_\alpha| \le |(\partial^{|\alpha|}g/\partial x^\alpha)(p)| < \varepsilon$. Thus by making ε small enough we can guarantee that v is in Z. (Note that $\deg v \le r$ so that v is in A_n^l.)

Assume that the lemma is true for $s - 1$ and let $p_1, \ldots, p_s$ be distinct points in K. Again apply Taylor's Theorem to g and obtain

$$g(x) = \sum_{0 \le |\alpha| \le r} a_\alpha(x - p_s)^\alpha + \sum_{|\alpha| = r+1} g_\alpha(x)(x - p_s)^\alpha.$$

If $\|g\|_{s(r+1)}^K < \varepsilon$, then $\|g_\alpha\|_{(s-1)(r+1)}^K < \varepsilon$ by Lemma 2.4. So by induction we may choose polynomials v_α with $\deg v_\alpha \le (r + 1)^{k-1}$ so that

$$\frac{\partial^{|\beta|}v_\alpha}{\partial x^\beta}(p_i) = \frac{\partial^{|\beta|}g_\alpha}{\partial x^\beta}(p_i) \quad \text{for} \quad 1 \le i \le s - 1 \text{ and } 0 \le |\beta| \le r.$$

Moreover we know that we may assume that the coefficients of v_α are smaller than some constant (depending on K) multiple of ε. Let

$$v = \sum_{0 \le |\alpha| \le r} a_\alpha(x - p_s)^\alpha + \sum_{|\alpha| = r+1} (x - p_s)^\alpha v_\alpha.$$

Note that $\deg v \le \deg v_\alpha \cdot (r + 1) \le (r + 1)^k$ so that v is in A_n^l. Clearly the coefficients of v are smaller than some constant multiple of ε since $p_1, \ldots, p_s$ are assumed to be in K, and the coefficients of v_α are bounded by a constant times ε. Thus by choosing ε small enough we can guarantee that v is in Z. Finally, we note that

$$\frac{\partial^{|\beta|}g}{\partial x^\beta}(p_i) = \frac{\partial^{|\beta|}}{\partial x^\beta}\left(\sum_{0 \le |\alpha| \le r} a_\alpha(x - p_s)^\alpha + \sum_{|\alpha| = r+1} (x - p_s)^\alpha g_\alpha(x)\right)$$
$$= \frac{\partial^{|\beta|}v}{\partial x^\beta}(p_i)$$

for $1 \le i \le s$ and $0 \le |\beta| \le r$ since the middle term of this equality depends only on $p_1, \ldots, p_s$, a_α ($|\alpha| \le r$), and $(\partial^{|\gamma|}g_\alpha/\partial x^\gamma)(p_i)$ ($\gamma \le \beta$) and the RHS of the equation depends in exactly the same way on these parameters. Thus the induction is proved. □

Proposition 2.6. *Suppose that f is infinitesimally stable and that f is stable under k-deformations for k large. Then f satisfies condition $\mathcal{O}$.*

Proof. Let p be in $X^r - X^{(r)}$, (the generalized diagonal of X^r). Assume $p = (p_1, \ldots, p_r)$ and let $q_1, \ldots, q_s$ be the distinct points among the p_i's as in Definition 1.11. Choose disjoint coordinate nbhds $\tilde{U}_i$ of q_i and coordinate nbhds V_i of $f(q_i)$ such that $\overline{F(\tilde{U}_i)} \subset V_i$. Let $\tilde{W}$ be a nbhd of f such that if g is in $\tilde{W}$, then $\overline{g(\tilde{U}_i)} \subset V_i$ for all i. Finally choose neighborhoods U_i' of q_i such that U_i' is convex and contained in $\tilde{U}_i$. Let $\rho\colon X \to \mathbf{R}$ be a smooth function which is one on a neighborhood of $U = \bigcup_{i=1}^t U_i'$ and zero off $\bigcup_{i=1}^t \tilde{U}_i$. If g is in $\tilde{W}$, then $g|\tilde{U}_i$ may be thought of as a mapping $g^i\colon \mathbf{R}^n \to \mathbf{R}^m$. Let $g^i = (g_1^i, \ldots, g_m^i)$ be the coordinate functions and define

$$\|g\|_\alpha^U = \max_{\substack{1 \le i \le t \\ 1 \le j \le m}} \|g_j\|_\alpha^{U_i}.$$

Let r and s in Lemma 2.5 both equal $m + 1$ so that $l = (m + 2)^{m+1}$. Let $B_{n,m}^l$ be the polynomial functions from $\mathbf{R}^n$ to $\mathbf{R}^m$ of degree $\le l$, i.e., $B_{n,m}^l = \bigoplus_{j=1}^m A_n$. Let $B = \bigoplus_{j=1}^t B_{n,m}^l$ and let $k = \dim B$. We may think of $v \in B$ as a map defined on U as follows. Let $v = (v_1, \ldots, v_t)$ and $v(x) = v_j(x)$ if $x \in U_j$. Then ρv is defined on X.

We now define a k-deformation, F, of f. For v in B, let $F_v(x) = f(x) + (\rho v)(x)$. Certainly $F\colon X \times B \to Y \times B$ is smooth and $F_0 = f$, so that F is a k-deformation. Since f is stable under k-deformations, there is a neighborhood Z of 0 in B in which F is trivial. Let

$$\tilde{W}_G = \{g \in \tilde{W} \mid \|g - f\|_{(m+1)(m+2)}^U < \varepsilon\}.$$

First note that $g - f$ makes sense since in each $\tilde{U}_i$ coordinate subtraction is defined. Also $\tilde{W}_G$ is an open neighborhood of f in the C^∞ topology. Choose $\varepsilon > 0$ by using Lemma 2.5 as follows: Let $p_1, \ldots, p_s$ be distinct points in $U = \bigcup_{j=1}^t U_j$ with $s \le m + 1$ and let g be in $\tilde{W}_G$. Then there exists a v in Z for which $j^{m+1}(g - f)(p_k) = j^{m+1}v(p_k)$ for $1 \le k \le s$.

Finally we shall show that if g is in W_ε, then g satisfies (†); that is, if $S = \{p_1, \ldots, p_s\} \subset U_p \cap g^{-1}(q)$, then

$$J^m(g^*TY)_S = (dg)J^m(TX)_S + g^*J^m(TY)_q.$$

Note that if this statement is true then f satisfies condition $\mathcal{O}$ for the choices U_p and W_ε. So let $g, p_1, \ldots, p_s$, and q be given satisfying the conditions of (†). Since g is in W_ε, there exists a v in Z so that $j^{m+1}g(p_i) = j^{m+1}(f + v)(p_i)$ for $1 \le i \le s$. Since $\rho \equiv 1$ on a nbhd of $\overline{U}_p$, $j^{m+1}(f + v)(p_i) = j^{m+1}F_v(p_i)$ for $1 \le i \le s$. Since F is trivial on Z, F_v is equivalent (as mappings of $X \to Y$) with f and since f is infinitesimally stable so is F_v. Thus (†) is satisfied by F_v at the points $p_1, \ldots, p_s, q$. Now the equations in (†) depend only on $j^{m+1}F_v(p_i)$ $(1 \le i \le s)$ so these same equations must be satisfied by g since $j^{m+1}F_v(p_i) = j^{m+1}g(p_i)$. ☐

To summarize this discussion we have:

Proposition 2.7. *If infinitesimal stability is equivalent to stability under k-deformations for k large (e.g., $k = \dim B^l_{n,m}$ where $l = (m + 2)^{m+1}$), then infinitesimal stability implies stability.*

Proof. By Proposition 2.6 all infinitesimally stable mappings satisfy condition $\mathcal{O}$. By Lemma 1.12 the set of infinitesimally stable mappings is an open set. Apply Lemma 2.2 and the hypothesis of this Proposition to see that infinitesimal stability implies stability. ☐

§3. A Characterization of Trivial Deformations

Let V be a nbhd of 0 in $\mathbf{R}^k$ and let $t_1, \ldots, t_k$ be the standard coordinates on $\mathbf{R}^k$.

Definition 3.1. *Let $f: X \to Y$ be smooth and let $F: X \times V \to Y \times V$ be a k-deformation of f. Define the vector field along F*

$$\tau_F{}^i = (dF)\left(\frac{\partial}{\partial t_i}\right) - F^*\left(\frac{\partial}{\partial t_i}\right)$$

where $\partial/\partial t_i$ is a vector field on $X \times V$ or $Y \times V$ as required.

We now establish some notation. Let $\pi: X \times V \to X$ and $\rho: X \times V \to V$ be the obvious projections. Then $T(X \times V) = \pi^*(TX) \oplus \rho^*(TV)$. So any vector ζ in $T(X \times V)$ can be written uniquely as $\zeta = \zeta_X + \zeta_V$ where ζ_X is in $\pi^*(TX)$ and ζ_V is in $\rho^*(TV)$. We call ζ_X the X-component of ζ and ζ_V the $\mathbf{R}^k$-component of ζ and denote ζ_X by $\pi(\zeta)$ and ζ_V by $\rho(\zeta)$.

Lemma 3.2. *Let F be a k-deformation of f. Then $F = f \times id_V$ iff $\tau_F{}^i \equiv 0$ for $1 \le i \le k$. In particular F is independent of t_i if $\tau_F{}^i \equiv 0$.*

Proof. If $F = f \times id_V$, then

$$(dF)_{(x,v)}\left(\left.\frac{\partial}{\partial t_i}\right|_{(x,v)}\right) = \left.\frac{\partial}{\partial t_i}\right|_{F(x,v)}$$

So that $\tau_F{}^i = 0$. So assume that $\tau_F{}^i = 0$ for all i. Fix (x_0, v_0) in $X \times V$ and choose coordinates $x_1, \ldots, x_n$ near x_0 in X and $y_1, \ldots, y_m$ near $F_{v_0}(x_0, v_0)$ in Y. In these coordinates we may write $F(x, v) = (F_1(x, v), \ldots, F_m(x, v), v)$. Then

$$(dF)_{(x,v)}\left(\frac{\partial}{\partial t_i}\bigg|_{(x,v)}\right) = \left(\frac{\partial F_1}{\partial t_i}(x, v), \ldots, \frac{\partial F_m}{\partial t_i}(x, v)\right) + \frac{\partial}{\partial t_i}\bigg|_{F(x,v)}.$$

Thus $\tau_F{}^i \equiv 0$ implies that $\partial F_j/\partial t_i(x, v) \equiv 0$ for $1 \le j \le m$. Thus $F_j(x, v)$ is independent of t_i for all i and $F_j(x, v) = F_j(x) = f_j(x)$ where $f = (f_1, \ldots, f_m)$ in these coordinates. Since (x_0, v_0) is arbitrary we find that globally $F(x, v) = f(x)$. □

***Theorem 3.3** (Thom-Levine). Let $f\colon X \to Y$ be smooth and let $F\colon X \times V \to Y \times V$ be a k-deformation of f. Then F is trivial iff there exists an open nbhd U of 0 in V and vector fields ζ^i on $X \times U$ and η^i on $Y \times U$ (for $1 \le i \le k$) satisfying*

(a) $\rho(\zeta^i) = 0 = \rho(\eta^i)$, *and*
(b) $\tau_F{}^i = (dF)(\zeta^i) + F^*(\eta^i)$ *on* $X \times U$.

Proof. Necessity. Assume that F is trivial, then there exists a nbhd U and diffeomorphisms $G\colon X \times U \to X \times U$ and $H\colon Y \times U \to Y \times U$ satisfying Definition 2.3(b). First we note that for any deformation K the $\mathbf{R}^k$-component of

$$(dK)_p\left(\frac{\partial}{\partial t_i}\bigg|_p\right) = \frac{\partial}{\partial t_i}\bigg|_{K(p)}$$

since $\rho\cdot K = \rho$. Thus

$$(dK)_p\left(\frac{\partial}{\partial t_i}\bigg|_p\right) = \frac{\partial}{\partial t_i}\bigg|_{K(p)} + \pi(dK)_p\left(\frac{\partial}{\partial t_i}\bigg|_p\right).$$

Now, by assumption, $F = H\cdot(f \times id_U)\cdot G^{-1}$ where all the mappings are k-deformations. Let p be in $X \times U$ and let $r = (f \times id_U)\cdot G^{-1}(p)$. Then compute

$$(*)\quad (dF)_p\left(\frac{\partial}{\partial t_i}\bigg|_p\right)$$
$$= \frac{\partial}{\partial t_i}\bigg|_p + \pi(dH)_r\left(\frac{\partial}{\partial t_i}\bigg|_r\right) + (dH)_r(df \times id_U)_{G^{-1}(p)}\pi(dG^{-1})_p\left(\frac{\partial}{\partial t_i}\bigg|_p\right).$$

Let $\zeta_p{}^i = (dG)_{G^{-1}(p)}\pi(dG^{-1})_p((\partial/\partial t_i)|_p)$. Thus ζ^i is a vector field on $X \times U$ and $\rho(\zeta^i) = \rho\cdot\pi(dG^{-1})(\partial/\partial t_i) = 0$ since G is a deformation and $\rho\cdot\pi = 0$.

Now insert $(dG)_p{}^{-1}\cdot(dG)_{G^{-1}(p)}$ before π in the last term of the RHS of (*) to obtain

$$(dF)_p\left(\frac{\partial}{\partial t_i}\bigg|_p\right) = \frac{\partial}{\partial t_i}\bigg|_{F(p)} + \pi(dH)_r\left(\frac{\partial}{\partial t_i}\bigg|_p\right) + (dF)_p(\zeta_p{}^i).$$

Thus

$$(**)\quad \tau_F{}^i(p) = (dF)_p\left(\frac{\partial}{\partial t_i}\bigg|_p\right) - \frac{\partial}{\partial t_i}\bigg|_{F(p)} = \pi(dH)_r\left(\frac{\partial}{\partial t_i}\bigg|_r\right) + (dF)_p(\zeta_p{}^i).$$

Define $\eta_q{}^i = \pi(dH)_{H^{-1}(q)}((\partial/\partial t_i)|_{H^{-1}(q)})$ where q is in $Y \times U$. Clearly η^i is a vector field of $Y \times U$ and the $\mathbf{R}^k$-component of η^i is zero. Substituting η in (**) we see that $\tau_F{}^i(p) = (dF)_p(\zeta_p{}^i) + \eta^i_{H(r)}$. But $H(r) = H\cdot(f\cdot id_U)\cdot G^{-1}(p) = F(p)$. So $\tau_F{}^i = (dF)(\zeta^i) + F^*(\eta^i)$. $\square$

Before proving the sufficiency part of the theorem we make some preparatory calculations.

Lemma 3.4. *Let ζ be a compactly supported vector field on $X \times \mathbf{R}^k$ such that the $\mathbf{R}^k$-component of ζ is zero. Then there is a diffeomorphism $g: X \times \mathbf{R}^k \to X \times \mathbf{R}^k$ which is a deformation of id_X satisfying*

$$(*) \qquad (dg)(g^{-1})^*(\partial/\partial t_k) = \zeta + \partial/\partial t_k.$$

Proof. Since ζ is compactly supported and $\partial/\partial t_k$ has a globally defined one parameter group, I, Corollary 6.5 guarantees that $\zeta + \partial/\partial t_k$ has a globally defined one parameter group $\phi: (X \times \mathbf{R}^k) \times \mathbf{R} \to X \times \mathbf{R}^k$.

Let $e_k = (0, \ldots, 0, 1)$ in $\mathbf{R}^k$. We claim that $\phi_s: X \times \{v\} \to X \times \{v + se_k\}$. Let $p = (x, v)$ be in $X \times \{v\}$. Then

$$(d\rho)_p\left(\left(\zeta + \frac{\partial}{\partial t_k}\right)\bigg|_p\right) = \frac{\partial}{\partial t_k}\bigg|_{\rho(p)}$$

since the $\mathbf{R}^k$-component of ζ is zero. Since $\zeta + (\partial/\partial t_k)$ is the infinitesimal generator of ϕ, the curve $s \mapsto \phi_s(p)$ is an integral curve for $\zeta + \partial/\partial t_k$ and thus represents the vector $(\zeta + \partial/\partial t_k)|_{\phi_s(p)}$ for each s. Thus $(d/ds)\rho(\phi_s(p)) = e_k$ and $\rho(\phi_s(p)) = se_k + \rho(\phi_0(p)) = v + se_k$. This proves the claim.

Next, define $g: X \times \mathbf{R}^k \to X \times \mathbf{R}^k$ by $g(x, v) = \phi_{v_k}(x, v - v_k e_k)$ where $v = (v_1, \ldots, v_k)$. Then $g: X \times \mathbf{R}^k \to X \times \mathbf{R}^k$ is a smooth mapping Since $g: X \times \{v\} \to X \times \{v\}$ and $g(x, 0) = \phi_0(x, 0) = (x, 0)$, g is a deformation of id_X. Note that $g|X \times \{v\}$ is a diffeomorphism since $h(x, v) = \phi_{-v_k}(x, v + v_k e_k)$ is the smooth inverse of $g|X \times \{v\}$. To see that g is a diffeomorphism we need only show that $\operatorname{Ker}(dg) \cap \rho^*(T\mathbf{R}^k) = \{0\}$. But this is clear since g is a deformation; i.e., $(dg)(\partial/\partial t_i) = \partial/\partial t_i + \pi(dg)(\partial/\partial t_i) \neq 0$.

Finally we compute (*).

The curve $s \mapsto (x, v + se_k)$ represents $(\partial/\partial t_k)|_{(x,v)}$ so the curve $s \mapsto g(x, v + se_k) = \phi_{v_k + s}(x, v - v_k e_k) = \phi_s(\phi_{v_k}(x, v - v_k e_k)) = \phi_s(g(x, v))$ represents

$$(dg)_{(x,v)}\left(\frac{\partial}{\partial t_k}\bigg|_{(x,v)}\right) = \left(\zeta + \frac{\partial}{\partial t_k}\right)\bigg|_{g(x,v)}. \qquad \square$$

Lemma 3.5. *Using the same notation as in Lemma* 3.4, *we have that*

(i) $\zeta = \pi(dg)(g^{-1})^*(\partial/\partial t_i)$, *and*

(ii) $\zeta = -(dg)\pi(dg^{-1})(\partial/\partial t_i)$.

Proof. (i) follows trivially from Lemma 3.4 since $\pi(\partial/\partial t_i) = 0$. Applying $(dg)_p{}^{-1}$ to both sides of Lemma 3.4 we have that

$$(dg)_p{}^{-1}\left(\left(\zeta + \frac{\partial}{\partial t_i}\right)\bigg|_p\right) = \frac{\partial}{\partial t_i}\bigg|_{g^{-1}(p)}.$$

Thus

$$0 = \pi\left(\frac{\partial}{\partial t_i}\bigg|_{g^{-1}(p)}\right) = \pi(dg)_p^{-1}(\zeta_p) + \pi(dg)_p^{-1}\left(\frac{\partial}{\partial t_i}\bigg|_p\right).$$

Since the $\mathbf{R}^k$-component of $\zeta_p = 0$ so does the $\mathbf{R}^k$-component of $(dg^{-1})_p(\zeta_p)$. So $\pi(dg)_p^{-1}(\zeta_p) = (dg)_p^{-1}(\zeta_p)$. Apply $(dg)_{g^{-1}(p)}$ to obtain (ii). □

Proof of Theorem 3.3. *Sufficiency.* Let $F: X \times V \to Y \times V$ be a k-deformation of f and let ζ^i and η^i be vector fields on $X \times V$ and $Y \times V$ respectively, such that the $\mathbf{R}^k$-components of ζ^i and η^i are zero and $\tau_F^i = (dF)(\zeta^i) + F^*(\eta^i)$ on $X \times V$. We must show that F is trivial. Since X is compact ζ^i is trivially compactly supported. By shrinking V we may assume that $\overline{V}$ is compact and that $\tau_F^i = (df)(\zeta^i) + F^*\eta^i$ on $X \times \overline{V}$. We can then damp η^i to zero off a compact nbhd of $F(X \times \overline{V})$ and assume that η^i is compactly supported.

Apply Lemmas 3.4 and 3.5 to show the existence of diffeomorphisms $G: X \times V \to X \times V$ and $H: Y \times V \to Y \times V$ so that

$$-\zeta^k = \pi(dG)(G^{-1})^*\left(\frac{\partial}{\partial t_k}\right)$$

and

$$\eta^k = -(dH)\pi(dH^{-1})\left(\frac{\partial}{\partial t_k}\right).$$

Let $M = H^{-1}\cdot F\cdot G$ and let p be in $X \times V$ with $q = G(p)$ and $r = F\cdot G(p)$. Then

$$(dM)_p\left(\frac{\partial}{\partial t_k}\bigg|_p\right) = \frac{\partial}{\partial t_k}\bigg|_{M(p)} + \pi(dH^{-1})_r\left(\frac{\partial}{\partial t_k}\bigg|_r\right)$$
$$+ (dH^{-1})_r\pi(dF)_q\left(\frac{\partial}{\partial t_k}\bigg|_q\right) + (dH^{-1})_r(dF)_q\pi(dG)_p\left(\frac{\partial}{\partial t_k}\bigg|_p\right)$$

using the fact that H, F, and G are deformations.

So

$$(dM)_p\left(\frac{\partial}{\partial t_k}\bigg|_p\right) = \frac{\partial}{\partial t_k}\bigg|_{M(p)} + (dH^{-1})_r\left(-\eta_r^k + \pi(dF)_q\left(\frac{\partial}{\partial t_k}\bigg|_q\right) - (dF)_q(\zeta_q^k)\right)$$

Now $\tau_F^k(q) = (dF)_q(\zeta_q^k) + \eta_r^k$ by assumption. Hence

$$\begin{aligned}\tau_M^k(p) &= (dM)_p\left(\frac{\partial}{\partial t_k}\bigg|_p\right) - M^*\left(\frac{\partial}{\partial t_k}\right)\\ &= (dH^{-1})_r\left(-\tau_F^k(q) + \pi(dF)_q\left(\frac{\partial}{\partial t_k}\bigg|_q\right)\right)\\ &= 0\end{aligned}$$

since

$$\tau_F^k = (dF)\left(\frac{\partial}{\partial t_k}\right) - F^*\left(\frac{\partial}{\partial t_k}\right) = \pi(dF)\left(\frac{\partial}{\partial t_k}\right).$$

Applying Lemma 3.2 we see that M is, in reality, a $(k-1)$ deformation trivially extended to a k-deformation. Thus if we can show that $\tau_M{}^i$ $(1 \le i \le k-1)$ can be written in the form $\tau_M{}^i = (dM)(\bar{\zeta}^i) + M^*(\bar{\eta}^i)$ where $\bar{\zeta}^i$ and $\bar{\eta}^i$ are vector fields whose R^k-components are zero, then we will be able to use induction to conclude that F is trivial.

Now note that

$$(*) \qquad \tau_M{}^i \cdot G = (dH)\tau_F{}^i - (dM)\pi(dG)(\partial/\partial t).$$

For $(dH)(\tau_F{}^i) = (dH)\pi(dF)(\partial/\partial t_i) = \pi(dH)(dF)(\partial/\partial t_i)$ since H and F are deformations. So

$$(dH)(\tau_F{}^i) = \pi(dM)(dG)\left(\frac{\partial}{\partial t_i}\right) = \pi(dM)\left(\frac{\partial}{\partial t_i}\bigg|_G\right) + (dG)\left(\frac{\partial}{\partial t_i}\right)$$

$$= \pi(dM)\left(\frac{\partial}{\partial t_i}\bigg|_G\right) + (dM)\pi(dG)\left(\frac{\partial}{\partial t_i}\right).$$

This proves (*) since $\pi(dM)(\partial/\partial t_i)|_G) = \tau_M{}^i \cdot G$. From (*) we see that to show that $\tau_M{}^i$ has the desired form it is sufficient to show that $(dH)\tau_F{}^i$ has the desired form.

Finally we compute $(dH)(\tau_F{}^i) = (dH)(dF)(\zeta^i) + (dH)(\eta^i|_F)$. Define $\bar{\zeta}_{G(p)} = (dG)_p(\zeta_p{}^i)$ and $\bar{\eta}_q{}^i = (dH)_{H^{-1}(q)}(\eta^i_{H^{-1}(q)})$. Then $\bar{\zeta}$ and $\bar{\eta}$ are vector fields whose $\mathbf{R}^k$-components equal zero. Moreover,

$$\tau_M{}^i(G(p)) = (dM)_{G(p)}(\bar{\zeta}_{G(p)}) + \bar{\eta}^i_{M(G(p))}. \qquad \square$$

Exercises

It is possible to use the Thom-Levine Theorem to prove that certain mappings are homotopically stable. For example, show that:

(1) Submersions are stable under deformations and

(2) 1:1 immersions are stable under deformations.

§4. Infinitesimal Stability ⇒ Stability

Proposition 4.1. *Let $f: X \to Y$ be stable under k-deformations; then f is infinitesimally stable.*

Proof. Since f is stable under k-deformations, f is homotopically stable. To show that f is infinitesimally stable we must produce for each vector field τ along f vector fields ζ on X and η on Y so that $\tau = (df)(\zeta) + f^*\eta$. Consider $X_f = \text{graph} f$ in $X \times Y$. We can view τ as a vector field on X_f pointing in the Y-direction as follows: $\bar{\tau}_{(p,f(p))} = \tau_p$ in $\{0\} \oplus T_{f(p)}Y \subset T_{(p,f(p))}(X \times Y)$. Extend $\bar{\tau}$ to a compactly supported vector field on $X \times Y$. (This is possible since X_f is closed in $X \times Y$ and thus has a tubular nbhd Z. Trivially translate $\bar{\tau}_{(p,f(p))}$ along the vector space fiber of Z at $(p, f(p))$ and damp-out off a compact nbhd of 0 in this vector space. This can clearly be done smoothly.)

Next let $\phi_t: X \times Y \to X \times Y$ be the one parameter group whose infinitesimal generator is $\bar{\tau}$. Finally define $F: X \times \mathbf{R} \to Y \times \mathbf{R}$ by $F(x, t) = (\pi_Y \cdot \phi_t(x, f(x)), t)$ where $\pi_Y: X \times Y \to Y$ is the obvious projection. Clearly F is smooth and is a deformation of f since $F(x, 0) = (\pi_Y(x, f(x)), 0) = (f(x), 0)$. Since f is stable under deformations there exist vector fields $\bar{\zeta}$ on $X \times I_\delta$ and $\bar{\eta}$ on $Y \times I_\delta$ (where $\delta > 0$) whose $\mathbf{R}$-components are zero satisfying $\tau_F = (dF)(\bar{\zeta}) + F^*\bar{\eta}$ on $X \times I_\delta$ by Theorem 3.3. Restrict this equation to $X \times \{0\}$ to obtain $\tau_F|_{X \times \{0\}} = (df)(\zeta) + f^*\eta$ where $\zeta_p = \bar{\zeta}_{(p,0)}$ and $\eta_q = \bar{\eta}_{(q,0)}$ define vector fields on X and Y respectively since the $\mathbf{R}$-components of $\bar{\zeta}$ and $\bar{\eta}$ are zero. Finally we compute $\tau_F|_{(p,0)} = \pi(dF)_{(p,0)}((\partial/\partial t)|_{(p,0)})$. The curve $t \mapsto (p, t)$ represents $(\partial/\partial t)|_{(p,0)}$ so that $t \mapsto \pi_Y \cdot \phi_t(p, f(p))$ represents $\tau_F|_{(p,0)}$. Now

$$\frac{d}{dt} \pi_Y \cdot \phi_t(p, f(p))|_{t=0} = (d\pi_Y)_{(p,f(p))}(\bar{\tau}|_{(p,f(p))}) = \tau_p$$

since $\bar{\tau}|_{(p,f(p))}$ points in the Y-direction. Thus $\tau_F|_{(p,0)} = \tau_p$. □

The proof of the fact that stability under deformations implies infinitesimal stability is a calculation involving nothing deeper than the global integration of certain vector fields. This is not true for the converse statement. As we shall see, the proof of this implication uses the generalized Malgrange Preparation Theorem and is quite similar in spirit to the proof of the formulation of infinitesimal stability given in §1.

Theorem 4.2. *Let $f: X \to Y$ be infinitesimally stable, then f is stable under k-deformations for all k.*

Let $F: X \times V \to Y \times V$ be a k-deformation of f. We need to show that F is trivial. By applying Theorem 3.3 we see that it is enough to find a nbhd U of 0 with $U \subset V$ and vector fields ζ on $X \times U$ and η on $Y \times U$ such that the $\mathbf{R}^k$-components of ζ and η are zero and $\tau_F = (dF)(\zeta) + F^*\eta$ on $X \times V$.

First we prove that ζ and η exist locally.

Proposition 4.3. *If f is infinitesimally stable at p, then there exists germs of vector fields ζ and η with $\mathbf{R}^k$-components equal to zero such that*

$$[\tau_F]_{(p,0)} = (dF)[\zeta]_{(p,0)} + [\eta]_{(f(p),0)}.$$

Proof. Let $N_F{}^p \equiv N \equiv$ {germs of vector fields $\tau: X \times \mathbf{R}^k \to T(Y \times \mathbf{R}^k)$ along F at $(p, 0)$ | $\mathbf{R}^k$-component of $\tau = 0$}, and let $A_F{}^p \equiv A \equiv N/K$ where $K = \{(df)[\zeta]_{(p,0)} \mid \zeta$ is a vector field on $X \times \mathbf{R}^k$ with $\mathbf{R}^k$ component $= 0\}$. There is an obvious action of $C^\infty_{(p,0)}(X \times \mathbf{R}^k)$ on N given by multiplication. Thus N is a module over $C^\infty_{(p,0)}(X \times \mathbf{R}^k)$ and is finitely generated. For if we choose coordinates $x_1, \ldots, x_n$ based at p on X and $y_1, \ldots, y_m$ based at $f(p)$ on Y, then every vector field along F whose $\mathbf{R}^k$-component is zero can be written as

$$\sum_{i=1}^m \tau_i(x, t) \frac{\partial}{\partial y_i}\bigg|_{F(x,t)}.$$

Thus the vector fields along F, $F^*(\partial/\partial y_i)$ are generators of the module N. Thus A is a finitely generated module over $C^\infty_{(p,0)}(X \times \mathbf{R}^k)$. Finally we note that via F^*, A is a module over $C^\infty_{(q,0)}(Y \times \mathbf{R}^k)$ where $q = f(p)$. We claim that A is also a finitely generated $C^\infty_{(q,0)}(Y \times \mathbf{R}^k)$ module with a set of generators given by e_i = projection of $F^*(\partial/\partial y_i)$ in A. It is in proving the claim that we shall use the fact that f is infinitesimally stable at p.

First we show that the claim is sufficient to prove the Proposition. In A,

$$[\tau_F]_{(p,0)} = \sum_{i=1}^{m} \left[\eta_i F^*\left(\frac{\partial}{\partial y_i}\right)\right]_{(p,0)}.$$

Thus in N,

$$[\tau_F]_{(p,0)} = (dF)[\zeta]_{(p,0)} + F^*\left[\sum_{i=1}^{m} \eta_i \frac{\partial}{\partial y_i}\right]_{(p,0)}$$

where ζ has $\mathbf{R}^k$-component equal to zero since $(dF)[\zeta]_{(p,0)}$ is in K.

Now $\eta = \sum_{i=1}^{m} \eta_i(\partial/\partial y_i)$ clearly has $\mathbf{R}^k$-component equal to zero so $\tau_F = (dF)(\zeta) + F^*\eta$ on the germ level near $(p, 0)$. Apply the obvious local form of Theorem 3.3 to prove the proposition.

To prove the claim we shall use the Malgrange Preparation Theorem (IV, Theorem 3.6). Using Taylor's Theorem write

$$(*) \qquad \tau(x, t) = \tau_0(x) + \sum_{i=1}^{k} t_i \tau_i(x, t).$$

Since τ is a vector field along F, τ_0 is a vector field along f. Since f is infinitesimally stable, there exist vector fields ζ on X and η on Y such that $\tau_0 = (df)(\zeta) + f^*\eta$. Extend ζ and η trivially to be vector fields on $X \times \mathbf{R}^k$ and $Y \times \mathbf{R}^k$ and apply Taylor's Theorem again to obtain

$$(**) \qquad \tau_0(x) - ((dF)(\zeta) + F^*\eta)(x, t) = \sum_{i=1}^{k} t_i \tau_i'(x, t).$$

Substituting (**) in (*) we obtain

$$(***) \qquad \tau(x, t) = [df(\zeta) + F^*\eta](x, t) + \sum_{i=1}^{k} t_i \tau_i''(x, t).$$

Next we consider the vector space $A/(t_1, \ldots, t_k)A$. The equivalence class of τ in $A/(t_1, \ldots, t_k)A$ is $F^*\eta$. Now $F^*\eta = \sum_{i=1}^{m} (\eta_i \cdot f)(\partial/\partial y_i)|_F$ since η is the trivial extension of a vector field on Y to $Y \times \mathbf{R}^k$. So the projections of $F^*(\partial/\partial y_1), \ldots, F^*(\partial/\partial y_m)$ generate the vector space $A/(t_1, \ldots, t_k)A$.

Finally consider the vector space $A/\mathscr{M}_{(q,0)}(Y \times \mathbf{R}^k)A$. Since $(t_1, \ldots, t_k) \subset \mathscr{M}_{(q,0)}(Y \times \mathbf{R}^k)$ there is a natural projection of $A/(t_1, \ldots, t_k)A$ onto $A/\mathscr{M}_{(q,0)}(Y \times \mathbf{R}^k)A$ so that $e_1, \ldots, e_m$ generate this last vector space. Now apply the Malgrange Theorem to obtain the desired result. □

Corollary 4.4. *Let f be infinitesimally stable and let $F: X \times V \to Y \times V$ be a k-deformation of f. Let $S = \{p_1, \ldots, p_s\} \subset f^{-1}(q)$. Then there exists a nbhd U of $S \times \{0\}$ in $X \times \mathbf{R}^k$ and vector fields ζ on $X \times \mathbf{R}^k$ and η on $Y \times \mathbf{R}^k$ such that the $\mathbf{R}^k$-components of ζ and η are zero and $\tau_F = (dF)(\zeta) + F^*\eta$ on $X \times U$.*

Proof. In the case that S is a single point, this is just Proposition 4.3. The proof proceeds precisely as the proof of Proposition 4.3 substituting $A_F{}^S = A_F{}^{p_1} \oplus \cdots \oplus A_F{}^{p_k}$ for $A_F{}^p$ and using Lemma 1.4 rather than the Malgrange Preparation Theorem. ▯

Our next step is to show that τ_F can be written in the desired form on a nbhd of the critical point set of f. Let Σ_q denote the critical points of f in $f^{-1}(q)$ and let $\Sigma = \bigcup_{q \in Y} \Sigma_q$ denote the critical point set of f.

Note. p is a *critical point* of f if $(df)_p : T_pX \to T_{f(p)}Y$ is not onto. Recall Lemma 1.8 which showed that Σ_q is a finite set.

Proposition 4.5. *Let X be compact, let $f: X \to Y$ be infinitesimally stable, and let F be a k-deformation of f. Then there exist vector fields ζ on $X \times \mathbf{R}^k$ and η on $Y \times \mathbf{R}^k$ with the $\mathbf{R}^k$-component of ζ and η equal to zero on a nbhd B of $\Sigma \times \{0\}$ in $X \times \mathbf{R}^k$ such that $\tau_F = (dF)(\zeta) + F^*\eta$ on B.*

Proof. We claim that there exist open sets $U_1, \ldots, U_N$ in X; $V_1, \ldots, V_N$ in Y; and $W_1, \ldots, W_N$ in Y; and $\varepsilon > 0$ satisfying

(a) $f(\Sigma) \subset \bigcup_{i=1}^N W_i$

(b) $\overline{W}_i \subset V_i$

(c) for all $v \in R^k$ with $|v| < \varepsilon$, $F_v{}^{-1}(\overline{W}_i) \cap \Sigma \subset U_i$ where $F(x, v) = (F_v(x), v)$

(d) for $|v| < \varepsilon$, $U_i \subset F_v{}^{-1}(V_i)$

(e) there exist vector fields ζ_i on $U_i \times B_\varepsilon$ and η_i on $V_i \times B_\varepsilon$ with $\mathbf{R}^k$-components equal to zero (where $B_\varepsilon = \{v \in \mathbf{R}^k \mid |v| < \varepsilon\}$ so that

$$\tau_F = (dF)(\zeta_i) + F^*\eta_i \quad \text{on} \quad U_i \times B_\varepsilon.$$

We need only verify the choices at each point q in $f(\Sigma)$. The finiteness then follows since $f(\Sigma)$ is compact. Since Σ_q is a finite set we may apply Corollary 4.4 for $S = \Sigma_q$ and gain the existence of U, V, ζ, η, and ε satisfying (e). Shrink U so that $f(\overline{U}) \subset V$. Now choose W satisfying (b) and (c) for $f = F_0$. By taking ε smaller if necessary we may assume that (c) and (d) hold.

Next choose a partition of unity $\rho_1, \ldots, \rho_N$ on $\bigcup_{i=1}^N W_i$ with supp $\rho_i \subset W_i$ and extend ρ_i to be $\equiv 0$ off W_i. Now there is a nbhd U of Σ such that $f^{-1}(\overline{W}_i) \cap U \subset U_i$ for each i since Σ is compact and U_i is open. Choose a smooth function ρ on X such that supp $\rho \subset U$ and $\rho \equiv 1$ on a nbhd of Σ. Let $\zeta = \sum_{i=1}^N \rho F^*\rho_i\zeta_i$. (This is globally defined and smooth since

$$\operatorname{supp}(\rho F^*\rho_i) \subset \operatorname{supp}\rho \cap \operatorname{supp} F^*\rho_i \subset U \cap F^{-1}(\overline{W}_i \times B_\varepsilon) \subset U_i \times B_\varepsilon.$$

Hence $\rho F^*\rho_i\zeta_i$ can be extended to all of $X \times \mathbf{R}^k$ trivially.) Let $\eta = \sum_{i=1}^N \rho_i\eta_i$. ($\eta$ is globally defined on all of $Y \times \mathbf{R}^k$.) Then calculate

$$\begin{aligned} (dF)(\zeta) + F^*\eta &= (dF)\left(\sum_{i=1}^N \rho F^*\rho_i\zeta_i\right) + \sum_{i=1}^N F^*(\rho_i\eta_i) \\ &= \sum_{i=1}^N F^*\rho_i[(dF)(\zeta_i) + F^*\eta_i] \text{ on a nbhd of } \Sigma \times \{0\} \\ &= \tau_F. \end{aligned}$$

Let B be that nbhd. ▯

Proof of Theorem 4.2. If $\dim X < \dim Y$, then $\Sigma = X$ and the last proposition proves the theorem. So assume $\dim X \geq \dim Y$ and let B, ζ, and η be given by Proposition 4.5. Choose a nbhd Z of $\Sigma \times \{0\}$ so that $\bar{Z} \subset B$. Let $\sigma = \tau_F - (df)(\zeta) - F^*\eta$. Then σ is a vector field along F whose $\mathbf{R}^k$-component is zero and which is zero on B. Now $f|(X - Z)$ is a submersion. Thus for v in $\mathbf{R}^k$ small enough $F_v|(X - Z)$ is a submersion. Since F is a deformation, F is a submersion on a nbhd D of $X \times \mathbf{R}^k - Z$ in $X \times \mathbf{R}^k$. Thus $T(X \times \mathbf{R}^k) = T\mathbf{R}^k \oplus \operatorname{Ker}(dF) \oplus G$ on D where G is some complementary subbundle. Moreover $(dF): G \to TY$ is an isomorphism on D. Since σ is a vector field along F whose $\mathbf{R}^k$-component is zero, there exists a vector field ζ' on D so that $(dF)(\zeta') = \sigma$ on D. Moreover, we can assume that ζ' is a section of G so that the $\mathbf{R}^k$-component of ζ' is zero. Since $\sigma \equiv 0$ on B, we can extend ζ' to be $\equiv 0$ near the boundary of D and thus extend ζ' to a vector field on $X \times \mathbf{R}^k$ whose $\mathbf{R}^k$-component is zero. Then $\tau_F = (dF)(\zeta + \zeta') + F^*\eta$ on a nbhd of $X \times \{0\}$. □

Theorem 4.6. *Suppose $f: X \to Y$ is smooth and X is compact. If f is infinitesimally stable, then f is stable.*

Proof. A trivial consequence of Proposition 4.1, Theorem 4.2, and Proposition 2.6. □ Q.E.D.

§5. Local Transverse Stability

We will show that local infinitesimal stability (see Definition 1.1) is equivalent to a certain transversality condition. First we must construct the submanifolds which will appear in this transversality statement. To do this, consider the action of $\operatorname{Diff}(X) \times \operatorname{Diff}(Y)$ on $J^k(X, Y)$ given by $(g, h)\cdot\sigma = j^k h(q)\cdot\sigma\cdot j^k(g^{-1})(g(p))$ where σ is in $J^k(X, Y)_{p,q}$. Let $\mathscr{D}_\sigma$ be the orbit of the action thru the k-jet σ. It is true that $\mathscr{D}_\sigma$ is a submanifold of $J^k(X, Y)$ but, for our purposes, we shall not need this fact. We shall only prove the following.

Theorem 5.1. *$\mathscr{D}_\sigma$ is an immersed submanifold of $J^k(X, Y)$.*

Before proving this result, we need some facts about extending diffeomorphisms and damping translations.

Lemma 5.2. *Let $\eta: \mathbf{R}^n \to \mathbf{R}^n$ be an immersion such that η is a diffeomorphism outside of some compact set K. Then η is a diffeomorphism.*

Proof. We need only show that η is 1:1 and onto as the Inverse Function Theorem will imply the result. To show that η is onto we note first that η is a submersion and so $\operatorname{Im}\eta$ is open. Let L be a compact set with $K \subset \operatorname{Int} L$. Then $\operatorname{Im}\eta = \eta(L) \cup \eta(\mathbf{R}^n - \operatorname{Int} L)$. Both sets in the union are closed so $\operatorname{Im}\eta$ is closed. Thus $\operatorname{Im}\eta = \mathbf{R}^n$.

To show that η is 1:1, define $S = \{x \in \mathbf{R}^n \mid \exists y \in \mathbf{R}^n, y \neq x, \text{ with } \eta(y) = \eta(x)\}$. Since η is a diffeomorphism off K, $\mathbf{R}^n - S \neq \varnothing$. Thus it is enough to show that S is both open and closed; for then $S = \varnothing$ and η is 1:1. Let x

be in S and y be in $\mathbf{R}^n - \{x\}$ such that $\eta(x) = \eta(y) = q$. Choose nbhds U of x, V of y, and W of q so that $U \cap V = \varnothing$ and $\eta|U: U \to W$ and $\eta|V: V \to W$ are diffeomorphisms. (This is possible since η is an immersion.) Then $U \subset S$ for if a is in U, then $b = (\eta|V)^{-1} \cdot (\eta|U)(a)$ satisfies $b \neq a$ and $\eta(b) = \eta(a)$. Thus S is open. To see that S is closed let $x_1, x_2, \ldots$ be a sequence of points in S converging to x. Choose an open nbhd W of x so that $\eta|W$ is a diffeomorphism. We may assume that each x_i is in W. Choose $y_i \neq x_i$ so that $\eta(y_i) = \eta(x_i)$. Clearly the y_i's are not in W since $\eta|W$ is 1:1. Also, the y_i's are contained in the compact set $\eta^{-1}(\eta(K))$. Thus we may assume that the y_i's converge to y in $\eta^{-1}(\eta(K)) - W$. Clearly $y \neq x$ and the continuity of η guarantees that $\eta(x) = \eta(y)$. Thus x is in S and S is closed. ☐

Proposition 5.3. *Let $T_a: \mathbf{R}^n \to \mathbf{R}^n$ be translation by a in $\mathbf{R}^n$—i.e., $T_a(x) = x + a$. Given an open set B in $\mathbf{R}^n$, there exists a diffeomorphism $\eta: \mathbf{R}^n \to \mathbf{R}^n$ such that $\eta = T_a$ on B and $\eta = id_{\mathbf{R}^n}$ outside of some compact set.*

Proof. Choose a smooth function $\sigma: \mathbf{R}^n \to \mathbf{R}$ which is 1 on a ball centered at 0 containing B and which has compact support. Let $\rho(x) = \sigma(tx)$ for some t. Choose t so small that $|d\rho| = t|d\sigma| < 1/|a|$. By also demanding that $t \leq 1$ we see that $\rho \equiv 1$ on B. Now consider $\eta(x) = x + \rho(x)a$ and observe that $\eta = T_a$ on B and $\eta = id_{\mathbf{R}^n}$ off of some compact set. By applying Lemma 5.2 it is enough to show that η is an immersion in order to show that η is a diffeomorphism. Now

$$(d\eta)_x = I_n + \begin{pmatrix} a_1 \\ \vdots \\ a_n \end{pmatrix} \left(\frac{\partial \rho}{\partial x_1}, \ldots, \frac{\partial \rho}{\partial x_n} \right) \quad \text{where} \quad a = (a_1, \ldots, a_n).$$

(A short computation is necessary here.) Thus for $v \neq 0$, $|(d\eta)_x(v)| \geq |v| - |a| \cdot |d\rho| \cdot |v| > 0$ by the choice of ρ. Hence η is an immersion. ☐

The following is left as an exercise.

Lemma 5.4. *The connected component of the identity in $GL(n, \mathbf{R})$ is the set of matrices with positive determinant.*

Proposition 5.5. *Let ϕ be a local diffeomorphism on $\mathbf{R}^n$ defined near 0 satisfying $\phi(0) = 0$ and $\det(d\phi)_0 > 0$. Then there is a diffeomorphism $\eta: \mathbf{R}^n \to \mathbf{R}^n$ such that $\eta = id_{\mathbf{R}^n}$ outside of some compact set K and $\eta = \phi$ on some nbhd of 0.*

Proof. Since $\phi(0) = 0$, $\phi = (d\phi)_0 + \beta$ where β is $0(|x|^2)$ near 0. We first show that β can be damped out off K. Let $\rho: \mathbf{R}^n \to \mathbf{R}$ be a smooth function such that $\rho \equiv 1$ on a nbhd of 0 and $\rho \equiv 0$ off K. Consider $\tau = (d\phi)_0 + \rho\beta$. Clearly $\tau = \phi$ on a nbhd of 0 and $\tau = (d\phi)_0$ off K. We wish to choose ρ so that τ will be a diffeomorphism. By Lemma 5.2 it is enough to show that τ is an immersion. Now for v in $T_x\mathbf{R}^n = \mathbf{R}^n$, $|(d\tau)_x(v)| \geq |d(d\phi)_0(v)| - |(d\rho\beta)_x| \cdot |v| \geq (c - |(d\rho\,\beta)_x|)|v|$ where $c = |(d\phi)_0{}^{-1}|$. Thus if we choose ρ so that $|(d\rho\,\beta)| < c$, then τ will be an immersion. Choose $\sigma: \mathbf{R}^n \to \mathbf{R}$ such that $\sigma \equiv 1$ on a nbhd of 0 and $\operatorname{supp} \sigma \subset B(1) =$ ball of radius 1 centered at the origin. Let $\rho(x) = \sigma(rx)$ for some positive constant r. Then $\rho \equiv 1$ on a nbhd of 0 and supp $\rho \subset B(1/r)$ = ball of radius $1/r$ centered at 0. By choosing r large enough supp $p \subset K$. Let $M = \sup |d\sigma|$. Then $rM \geq |d\rho|$. Since β is $0(|x|^2)$ and supp $\rho \subset B(1/r)$ there exist constants e and f so that $|\beta(x)| \leq e|x|^2$ and $|(d\beta)_x| \leq f|x|$ on supp ρ. Thus $|(d\rho\beta)| \leq |d\rho|\cdot|\beta| + |\rho|\cdot|d\beta| \leq (eM + f)/r$. Choose r large enough so that $(eM + f)/r < c$.

Next we show that given a linear map α with $\det \alpha > 0$ there exists a diffeomorphism g so that $g = \alpha$ on a nbhd of 0 and $g = id_{\mathbf{R}^n}$ outside of K. If g exists, then $\eta = \alpha^{-1}\cdot g\cdot\tau$ is the desired diffeomorphism where $\alpha = (d\phi)_0$. (We use the hypothesis that $\det (d\phi)_0 > 0$ here.) Moreover, it is sufficient to show that there exists a $\delta > 0$ so that g exists whenever $|\alpha - I_n| < \delta$. For we may choose a curve $c: R \to Gl(n, R)$ so that $c(0) = I_n$ and $c(1) = \alpha$ using Lemma 5.4. Also, since $[0, 1]$ is compact there exist points $t_0 = 0 < t_1 < \cdots < t_k = 1$ such that $|c(t_i)\cdot c(t_{i-1})^{-1} - I_n| < \delta$ for $1 \leq i \leq k$. Let g_i be the diffeomorphism associated with $c(t_i)\cdot c(t_{i-1})^{-1}$, then $g = g_k\cdot\cdots\cdot g_1$ is the desired diffeomorphism. Let $\rho: \mathbf{R}^n \to \mathbf{R}$ be a smooth function such that $\rho \equiv 1$ on a nbhd of 0 and $\rho \equiv 0$ off K, then consider $g = I_n + \rho(\alpha - I_n)$. Clearly $g = \alpha$ near 0 and $g = I_n$ off K. Using Lemma 5.2 again, we need only show that g is an immersion to see that g is a diffeomorphism. Indeed

$$|(dg)_x(v)| \geq |v| - (|(d\rho)_x| + |\rho(x)|)\cdot|\alpha - I_n|\cdot|v|.$$

Thus if we choose $\delta < 1/\sup(|d\rho| + |\rho|)$, then whenever $|\alpha - I_n| < \delta$, the associated g will be an immersion. □

At this point we shall need some standard facts about Lie groups. We refer the reader who is not familiar with this topic to the appendix where we give definitions, examples, and sketch the results that are used here.

Now let p be in X and q be in Y. Let $G^k(X)_p$ and $G^k(Y)_q$ be the invertible k-jets in $J^k(X, X)_{p,p}$ and $J^k(Y, Y)_{q,q}$ respectively and let $G = G^k(X)_p \times G^k(Y)_q$. Then G is a Lie group (see Example (4) after Definition A.1) and there is an obvious action of G on $J^k(X, Y)_{p,q}$ given as follows: $(\alpha, \beta)(\sigma) = \beta\cdot\sigma\cdot\alpha^{-1}$ where (α, β) is in G and σ is in $J^k(X, Y)_{p,q}$. Let $\mathcal{O}_\sigma$ be the orbit in $J^k(X, Y)_{p,q}$ thru the k-jet σ. Applying Theorem A.13 we see that $\mathcal{O}_\sigma$ is an immersed submanifold. (In fact, $\mathcal{O}_\sigma$ is a submanifold as it is the orbit of an algebraic group acting algebraically on a manifold. See Borel, Linear Algebraic Groups—Proposition 6, 7, p. 180. For our purposes we shall not need this fact. We also note that the knowledge that $\mathcal{O}_\sigma$ is a submanifold would be enough to prove that $\mathcal{D}_\sigma$ is actually a submanifold of $J^k(X, Y)$.) Now let $\mathring{\mathcal{O}}_\sigma$ be the connected component of $\mathcal{O}_\sigma$ containing σ. Clearly $\mathring{\mathcal{O}}_\sigma$ is also an immersed submanifold of $J^k(X, Y)_{p,q}$.

Proof of Theorem 5.1. Suppose that $\dim X = n$ and $\dim Y = m$. Choose p in X and q in Y with chart nbhds U of p and V of q. Via charts we may identify U with $\mathbf{R}^n$, V with $\mathbf{R}^m$, p with 0, and q with 0. Consider $T: U \times V \times J^k(U, V)_{p,q} \to J^k(U, V)$ defined by $(p', q', \tau) \mapsto j^k T_{q'}\cdot\tau\cdot j^k(T_{p'}{}^{-1})$ where T_c is translation by the vector c. (This makes sense using the identifications above.) T is a diffeomorphism as it is essentially the inverse of a chart in the manifold structure of $J^k(X, Y)$. (Identify the domain with $\mathbf{R}^n \times \mathbf{R}^m \times J^k(\mathbf{R}^n, \mathbf{R}^m)_{0,0}$ and see II, Theorem 2.6.) We claim that $T(U \times V \times \mathring{\mathcal{O}}_\sigma) = \mathcal{D}_\sigma^{U,V} \equiv$ connected component of $\mathcal{D}_\sigma \cap J^k(U, V)$ containing σ. This will imply the Theorem as $\mathring{\mathcal{O}}_\sigma$ is an immersed submanifold since T is a diffeomorphism; as long as we also show that $\mathcal{D}_\sigma \cap J^k(U, V)$ has at most a countable number of components. First we show that $T(U \times V \times \mathring{\mathcal{O}}_\sigma) \subset \mathcal{D}_\sigma^{U,V}$. Since $U \times V \times \mathring{\mathcal{O}}_\sigma$ is connected, it is enough to show that $T(U \times V \times \mathring{\mathcal{O}}_\sigma) \subset \mathcal{D}_\sigma \cap J^k(U, V)$. Let (p', q', τ) be in $U \times V \times \mathring{\mathcal{O}}_\sigma$. Now $\mathring{\mathcal{O}}_\sigma = \mathring{G} \cdot \sigma$ (using Lemma A.14) and $\mathring{G} = \mathring{G}^k(X)_p \times \mathring{G}^k(Y)_q$ where $\mathring{G}^k(X)_p = \{\alpha \in G^k(X)_p \mid \det (d\alpha)_p > 0\}$. Thus $\tau = \beta \cdot \sigma \cdot \alpha^{-1}$ where $\alpha \in \mathring{G}^k(X)_p$ and $\beta \in \mathring{G}^k(Y)_q$. Now we can represent α and β by mappings $\bar{\alpha}: X \to X$ and $\bar{\beta}: Y \to Y$ which are diffeomorphism nbhds of p and q respectively. Using Proposition 5.5 we can insure that $\bar{\alpha}$ and $\bar{\beta}$ are globally defined diffeomorphisms. So $\tau = j^k(\bar{\beta})(q) \cdot \sigma \cdot j^k(\bar{\alpha}^{-1})(p)$. Now $T(p', q', \tau) = j^k(T_{q'})(q) \cdot \tau \cdot j^k(T_{p'}{}^{-1})(p')$. By Proposition 5.3, we may assume that $T_{q'}: Y \to Y$ and $T_{p'}: X \to X$ are globally defined diffeomorphisms. Thus $T(p', q', \tau) = (T_{p'} \cdot \bar{\alpha}, T_{q'} \cdot \bar{\beta}) \cdot \sigma \in \mathcal{D}_\sigma$. For the reverse inclusion, let τ be in $\mathcal{D}_\sigma^{U,V}$. Let the source of τ be p' and the target be q'. Consider $\rho = j^k(T_{q'}{}^{-1})(q') \cdot \tau \cdot j^k(T_{p'})(p)$. Since $T_{p'}$ and $T_{q'}$ are in Diff(X) and Diff(Y) respectively (Proposition 5.3 again), we see that ρ is still in $\mathcal{D}_\sigma$. Thus there exist (γ, δ) in Diff(X) $\times$ Diff(Y) such that $\rho = j^k(\delta)(q) \cdot \sigma \cdot j^k(\gamma^{-1})(p)$ and so ρ is in $\mathcal{O}_\sigma$. So we have shown that $\mathcal{D}_\sigma^{U,V} \subset T(U \times V \times \mathcal{O}_\sigma)$. Also we have shown that $\mathcal{D}_\sigma \cap J^k(U, V) \subset T(U \times V \times \mathcal{O}_\sigma)$ so that $\mathcal{D}_\sigma \bigcap J^k(U, V)$ has at most a countable number of components. Since T is a diffeomorphism and $\mathcal{D}_\sigma^{U,V}$ is connected we must have that $\mathcal{D}_\sigma^{U,V} \subset T(U \times V \times$ (connected component of $\mathcal{O}_\sigma$)). But certainly σ is in $\mathcal{D}_\sigma^{U,V} \cap T(U \times V \times \mathring{\mathcal{O}}_\sigma)$ so that $\mathcal{D}_\sigma^{U,V} \subset T(U \times V \times \mathring{\mathcal{O}}_\sigma)$. Thus the components of $\mathcal{D}_\sigma \cap J^k(U, V)$ are a subset of the components of $T(U \times V \times \mathcal{O}_\sigma)$ and this last set is at most countable. □

Definition 5.6. *Let $f: X \to Y$ be smooth and let p be in X. Let $m = \dim Y$ and let $\sigma = j^m f(p)$. Then f is* locally transverse stable *at p if $j^m f \pitchfork \mathcal{D}_\sigma$ at p.*

Notes. (1) The concept of intersecting transversely an immersed submanifold makes sense in an obvious way, since the tangent space to an immersed submanifold at a point is well-defined. More precisely, let W be an immersed submanifold of Y and let $f: X \to Y$. Then $f \pitchfork W$ at p if either $f(p) \notin W$ or $f(p) \in W$ and $T_{f(p)} Y = (df)_p(T_p X) + T_{f(p)} W$.

(2) Another way of phrasing this definition is that an immersed submanifold, W, is the countable union of submanifolds $W_1, W_2, \ldots$ which are open subsets of W. Thus $f \pitchfork W$ iff $f \pitchfork W_i$ for each i.

(3) Applying (2) we see that the Thom Transversality Theorem still applies to immersed submanifolds; i.e., if W is an immersed submanifold of $J^k(X, Y)$, then $T_W = \{f \in C^\infty(X, Y) \mid j^k f \pitchfork W\}$ is a residual subset of $C^\infty(X, Y)$. (Trivial, since $T_W = \bigcap_{i=1}^\infty T_{W_i}$ where T_{W_i} is defined in the obvious fashion.)

A corollary of all this is the following:

Lemma 5.7. *Let $f: X \to Y$ be stable, then f is locally transverse stable at p for all p in X.*

Proof. Choose an open nbhd U of f in $C^\infty(X, Y)$ such that all g in U are equivalent to f. Let $\sigma = j^m f(p)$. By the Thom Transversality Theorem (and Note (3) above), there is a g in U such that $j^m g \pitchfork \mathscr{D}_\sigma$. Choose (α, β) in $\text{Diff}(X) \times \text{Diff}(Y)$ such that $f = \beta \cdot g \cdot \alpha^{-1}$. Then $j^m f \pitchfork \mathscr{D}_\sigma$ at p since $j^m g \pitchfork \mathscr{D}_\sigma$ at $\alpha(p)$. □

Let $f: X \to Y$ be smooth with p in X and $\sigma = j^k f(p)$. Our next goal is to compute $T_\sigma \mathscr{D}_\sigma$ as well as a normal subspace in $T_\sigma J^k(X, Y)$. The idea is to compute a jet version of $T_f C^\infty(X, Y) = C_f^\infty(X, TY) = C^\infty(f^*TY)$.

Let ω be in $J^k(f^*TY)_p$ and let $\tau: X \to TY$ be a vector field along f representing ω. (We will constantly use the identification of $C_f^\infty(X, TY)$ with $C^\infty(f^*TY)$.) Let F be a deformation of f satisfying $dF_t/dt|_{t=0} = \tau$. (The existence of such a deformation can be shown as follows; Consider graph f as a submanifold of $X \times Y$ and identify τ as a vector field along graph f always pointing in the Y directions. Since X is compact we may extend τ to a compactly supported vector field $\tilde{\tau}$ on $X \times Y$. Let ϕ_t be the corresponding one parameter group and let $\pi_Y : X \times Y \to Y$ be the obvious projection. Define $F_t(x) = \pi_Y \cdot \phi_t(x, f(x))$. Then the associated F is a deformation of f with the desired property.) Consider the path $t \mapsto j^k F_t(p)$ in $J^k(X, Y)$ based at σ and define $\lambda(\omega)$ to be the tangent vector to this path at $t = 0$.

Proposition 5.8. *$\lambda: J^k(f^*TY)_p \to T_\sigma J^k(X, Y)$ is a well-defined linear injection.*

Proof. To see that λ is well-defined we shall compute a formula for $\lambda(\omega)$ which just depends on the k-jet of τ. Choose coordinates $x_1, \ldots, x_n$ based at p in X and coordinates $y_1, \ldots, y_m$ based at $q = f(p)$ in Y. Let $f_1, \ldots, f_m$ be the coordinate functions of f in these coordinates. We may write $\tau = \sum_{i=1}^m g_i f^*(\partial/\partial y_i)$ where g_i is a smooth function on X and $F_t = (F_t^1, \ldots, F_t^m)$ where $F_t^i = f_i + tg_i + O(t^2)$. This last equality follows since $(dF_t/dt)|_{t=0} = \tau$. Thus

$$j^k F_t^i(0) = \sum_{|\alpha| \le k} \frac{x^\alpha}{\alpha!} \frac{\partial^{|\alpha|}}{\partial x^\alpha} F_t^i(0) = \sum_{|\alpha| \le k} \frac{x^\alpha}{\alpha!} \left(\frac{\partial^{|\alpha|}}{\partial x^\alpha} f_i(0) + t \frac{\partial^{|\alpha|}}{\partial x^\alpha} g_i(0) + O(t^2) \right)$$

Hence

$$(*) \qquad \frac{d}{dt} j^k F_t^i(0)|_{t=0} = \sum_{|\alpha| \le k} \frac{x^\alpha}{\alpha!} \frac{\partial^{|\alpha|}}{\partial x^\alpha} g_i(0).$$

This sum is completely determined by $j^k g(0)$; that is, the k-jet of τ at p. Now suppose that $\lambda(\omega) = 0$. Then by (*) $(\partial^{|\alpha|}/\partial x^\alpha) g_i(0) = 0$ for $|\alpha| \le k$ and thus $\omega = 0$ in $J^k(f^*TY)_p$; so λ is injective. The linearity of λ also follows from (*). □

Proposition 5.9. *Let $\alpha: J^k(X, Y) \to X$ be the source map; then the sequence $0 \to J^k(f^*TY)_p \xrightarrow{\lambda} T_\sigma J^k(X, Y) \xrightarrow{(d\alpha)_\sigma} T_p X \to 0$ is exact.*

Proof. Since λ is injective and α is a submersion we need only show that $\text{Im}\,\lambda = \text{Ker}\,(d\alpha)_\sigma$. First we note that $(d\alpha)_\sigma \cdot \lambda = 0$, since $t \mapsto \alpha \cdot j^k F_t(p) = p$ is a curve representing $(d\alpha)_\sigma \cdot \lambda(\omega)$ when F_t is a deformation defining $\lambda(\omega)$. Since this curve is constant $(d\alpha)_\sigma \cdot \lambda(\omega) = 0$. To finish the proof we show that dim Im λ = dim Ker $(d\alpha)_\sigma$. Now dim Im λ = dim $J^k(f^*TY)$ = dim (polynomial mappings of $\mathbf{R}^n$ into $\mathbf{R}^m$ of degree $\leq k$) where n = dim X and m = dim Y. On the other hand dim Ker $(d\alpha)_\sigma$ = dim $J^k(X, Y)$ − dim X = dim (polynomial mappings of $\mathbf{R}^n$ into $\mathbf{R}^m$ of degree $\leq k$). □

There is a natural mapping of $C^\infty(TY)_q \to T_\sigma\mathcal{D}_\sigma$ given by the action of Diff(Y) on $J^k(X, Y)$ where q = target of σ. Let η be a vector field on Y representing $[\eta]_q$. We may assume that η has compact support. Let ϕ_t be the one parameter group whose infinitesimal generator is η. Consider the curve $c(t) = j^k\phi_t(q)\cdot\sigma$. Clearly this curve lies in $\mathcal{D}_\sigma$. Let $(d\gamma_1)(\eta) = (dc/dt)|_{t=0}$. Thus $(d\gamma_1): C^\infty(TY)_q \to T_\sigma\mathcal{D}_\sigma$ and is just the k-jet version of

$$(d\gamma_f): T_{\mathrm{id}_Y} \operatorname{Diff}(Y) \to T_f C^\infty(X, Y).$$

Proposition 5.10. *The diagram*

$$\begin{array}{ccc} C^\infty(TY)_q & \xrightarrow{(d\gamma_1)} & T_\sigma J^k(X, Y) \\ \Big\downarrow \pi_k^\infty & & \Big\uparrow \lambda \\ J^k(TY)_q & \xrightarrow{f^*} & J^k(f^*TY)_p \end{array}$$

commutes where $\pi_k^\infty : C^\infty(TY)_q \to J^k(TY)_q$ is the obvious projection.

Proof. Let η and ϕ be as above. Then

$$\lambda\cdot f^*\cdot\pi_k^\infty(\eta) = \frac{d}{dt} j^k(\phi_t\cdot f)(p)|_{t=0} = \frac{d}{dt} j^k\phi_t(q)\cdot\sigma|_{t=0} = (d\gamma_1)(\eta)$$

since a deformation of f corresponding to $f^*\eta$ is given by $F_t = \phi_t\cdot f$.

There is also a natural mapping of $C^\infty(TX)_p \to T_\sigma\mathcal{D}_\sigma$ given by the action of Diff(X) on $J^k(X, Y)$ where p = source of σ. Let ζ be a vector field on X and let ψ_t be the one parameter group whose infinitesimal generator is ζ. Consider the curve $c(t) = \sigma\cdot j^k\psi_t^{-1}(\psi_t(p))$. Clearly this curve lies in $\mathcal{D}_\sigma$. Let $(d\gamma_2)(\zeta) = (dc/dt)|_{t=0}$. Thus $(d\gamma_2): C^\infty(TX)_p \to T_\sigma\mathcal{D}_\sigma$ and is the k-jet version of $(d\gamma_f): T_{\mathrm{id}_X} \operatorname{Diff}(X) \to T_f C^\infty(X, Y)$. □

Proposition 5.11. *Let $\mathcal{K} = -\lambda\cdot(df) + (dj^kf)_p\cdot\pi_0{}^k$ where $\pi_0{}^k : J^k(TX)_p \to J^0(TX)_p = T_pX$ is the obvious projection. Then the diagram*

$$\begin{array}{ccc} C^\infty(TX)_p & \xrightarrow{(d\gamma_2)} & T_\sigma\mathcal{D}_\sigma \\ \Big\downarrow \pi_k^\infty & \nearrow \mathcal{K} & \\ J^k(TX)_p & & \end{array}$$

commutes where π_k^∞ is the obvious projection.

Proof. Let ζ and ψ be as above. Let $\omega = \pi_k^\infty(\zeta)$. We compute $\lambda\cdot(df)(\omega)$. Note that $F_t = f\cdot\psi_t$ is a deformation of f satisfying $(dF_t/dt)|_{t=0} = (df)(\zeta)$.

Thus we may use F_t to compute $\lambda((df)(\omega))$. Now

$$\lambda\cdot(df)(\omega) = \frac{d}{dt}(j^k F_t)(p)|_{t=0} = \frac{d}{dt}[j^k f(\psi_t(p))\cdot j^k\psi_t(p)]|_{t=0}$$

$$= \frac{d}{dt} j^k f(\psi_t(p))|_{t=0} + \frac{d}{dt}\sigma\cdot j^k\psi_t(p)|_{t=0}.$$

Thus

$$\lambda\cdot(df)(\omega) = (dj^k f)_p \pi_0^\infty(\zeta) - \frac{d}{dt}\sigma\cdot j^k\psi_{-t}(p)|_{t=0}$$

where $\pi_0^\infty : C^\infty(TX)_p \to T_pX = J^0(TX)_p$ is the obvious projection. Since $\pi_0^\infty = \pi_0{}^k\cdot\pi_k$ and $\psi_{-t} = \psi_t{}^{-1}$ we have that

$$\lambda\cdot(df)(\omega) = (dj^k f)_p\pi_0{}^k(\omega) - (d\gamma_2)(\zeta)$$

or

$$(d\gamma_2)(\zeta) = \mathscr{K}\cdot\pi_k^\infty(\zeta) \qquad \square$$

***Lemma* 5.12.** *$(d\gamma_2)\oplus(d\gamma_1) : C^\infty(TX)_p \oplus C^\infty(TY)_q \to T_\sigma\mathscr{D}_\sigma$ is onto.*

Proof. Let v be in $T_\sigma\mathscr{D}_\sigma$ and $c(t)$ a curve representing v in $\mathscr{D}_\sigma$. Suppose there exists a curve of diffeomorphisms $t \mapsto (g_t, h_t)$ in $\mathrm{Diff}(X) \times \mathrm{Diff}(Y)$ such that $c(t) = j^k h_t(q)\cdot\sigma\cdot j^k g_t{}^{-1}(g_t(p))$ then v is in the image $(d\gamma_2)\oplus(d\gamma_1)$. Let

$$\zeta_p = \frac{dg_t}{dt}(p)|_{t=0} \quad \text{and} \quad \eta_q = \frac{dh_t}{dt}(q)|_{t=0}.$$

Then ζ and η are vector fields on X and Y respectively. (Since we are only interested in the germ of η at q we may assume that η has compact support.) Let ϕ_t and ψ_t be the one parameter groups associated with ζ and η respectively. Then the curve $\tilde{c}(t) = j^k\psi_t(q)\cdot\sigma\cdot j^k\phi_{-t}(\phi_t(p))$ satisfies

$$\left.\frac{d\tilde{c}}{dt}\right|_{t=0} = \left.\frac{dc}{dt}\right|_{t=0} = v$$

since $(dc/dt)|_{t=0}$ only depends on $(dg_t/dt)|_{t=0}$ and $(dh_t/dt)|_{t=0}$ but

$$\left.\frac{dg_t}{dt}\right|_{t=0} = \left.\frac{d\phi_t}{dt}\right|_{t=0} \quad \text{and} \quad \left.\frac{dh_t}{dt}\right|_{t=0} = \left.\frac{d\psi_t}{dt}\right|_{t=0}.$$

Thus

$$(d\gamma_2)\oplus(d\gamma_1)(\zeta, \eta) = \left.\frac{d\tilde{c}}{dt}\right|_{t=0} = v.$$

Recall that when we proved that $\mathscr{D}_\sigma$ is an immersed submanifold (see the proof of Proposition 5.1) we showed that $\mathscr{D}_\sigma^{U,V}$ = connected component of $\mathscr{D}_\sigma \cap J^k(U, V)$ containing σ is equal to $T(U \times V \times \mathring{\mathcal{O}}_\sigma)$ where $\mathring{\mathcal{O}}_\sigma$ is the orbit thru σ of the action of the Lie group $\mathring{G} = \mathring{G}^k(X)_p \times \mathring{G}^k(Y)_q$ on $J^k(X, Y)_{p,q}$. Since $T_\sigma\mathscr{D}_\sigma = T_\sigma\mathscr{D}_\sigma^{U,V}$ we may assume that the curve $c(t)$ is in $\mathscr{D}_\sigma^{U,V}$. Since T is a diffeomorphism there is a curve $a(t) = (p(t), q(t), \tau(t))$ in $U \times V \times \mathring{\mathcal{O}}_\sigma$

such that $T(a(t)) = c(t)$. Thus $c(t) = j^k(T_{q(t)})\cdot\tau(t)\cdot j^k(T_{p(t)}^{-1})$. So if we can show that $\tau(t) = j^k(h(t))\cdot\sigma\cdot j^k(g(t)^{-1})$ then we will be finished by the last paragraph. But $\tau(t)$ is a curve in $\mathring{\mathcal{O}}_\sigma$. Thus there is a curve $(\tilde{g}(t), \tilde{h}(t))$ in $\mathring{G}^k(X)_p \times \mathring{G}^k(Y)_q$ such that $\tilde{h}(t)\cdot\sigma\cdot\tilde{g}(t)^{-1} = \tau(t)$ since $\mathring{G}$ is a Lie group.

In local coordinates $\mathring{G}^k(X)$ is just polynomial mappings of degree $\leq k$ on $\mathbf{R}^n$ which are diffeomorphisms on a nbhd of 0. Let $g(t)$ be the unique polynomial of degree $\leq k$ such that $j^kg(t) = \tilde{g}(t)$. Similarly for $h(t)$ and $\tilde{h}(t)$. Since $\det(dg_0)_p > 0$ and $\det(dh_0)_q > 0$ we may assume that g_t and h_t are globally defined diffeomorphisms on X and Y respectively. (Apply Proposition 5.5.) Thus we obtain the desired g_t and h_t. □

We now state and prove the main Theorem of this section.

Theorem 5.13. *Let $f: X \to Y$ be smooth and let $m = \dim Y$. If f is locally transverse stable at p, then f is locally infinitesimally stable at p.*

Proof. Assume that f is locally transverse stable at p. By Corollary 1.3 it is sufficient to show that f satisfies the conditions of infinitesimal stability to order m. In particular, we need to show that if τ is in $J^m(f^*TY)_p$, then there is a v_1 in $J^m(TX)_p$ and a v_2 in $J^m(TY)_q$ so that $\tau = (df)(v_1) + f^*(v_2)$. Consider $\lambda(\tau)$ in $T_\sigma J^m(X, Y)$. Since f is locally transverse stable at p there exists w in $T_\sigma\mathscr{D}_\sigma$ and v in T_pX so that $\lambda(\tau) = w + (dj^mf)_p(v)$. By Lemma 5.12 there exists ζ in $C^\infty(TX)_p$ and η in $C^\infty(TY)_q$ so that $w = -(d\gamma_2)(\zeta) + (d\gamma_1)(\eta)$. Let $v_1 = \pi_m^\infty(\zeta)$ and $v_2 = \pi_m^\infty(\eta)$. Applying Propositions 5.10 and 5.11 we have that $w = \lambda\cdot f^*(v_2) - \mathscr{K}(v_1)$. Thus

$$\lambda(\tau) = \lambda\cdot(df)(v_1) - (dj^mf)_p\cdot\pi_0{}^m(v_1) + \lambda\cdot f^*(v_2) + (dj^mf)_p(v).$$

Apply $(d\alpha)_\sigma$ to both sides and apply Proposition 5.9 to obtain

$$0 = v - \pi_0{}^m(v_1). \qquad (\text{Note } (d\alpha)_\sigma\cdot(dj^mf)_p = id.)$$

So $\lambda(\tau) = \lambda\cdot(df)(v_1) + \lambda\cdot f^*(v_2)$. But λ is injective (Proposition 5.8), so $\tau = (df)(v_1) + f^*v_2$. □

§6. Transverse Stability

The problem of transferring the result that local transverse stability implies local infinitesimal stability to a global result is, as usual, at the self-intersections of the mapping in question. Multijet transversality is again the tool used to solve the problem.

Let $\mathscr{D}_\sigma{}^s$ be the orbit through the s-fold multijet σ under the action of $\mathrm{Diff}(X) \times \mathrm{Diff}(Y)$ on $J_s{}^m(X, Y)$.

Proposition 6.1. *$\mathscr{D}_\sigma{}^s$ is an immersed submanifold of $J_s{}^m(X, Y)$ for each s-fold multijet σ.*

Remarks. (1) $\mathscr{D}_\sigma{}^s$ is, in fact, a submanifold for the same reasons that $\mathscr{D}_\sigma$ is a submanifold of $J^k(X, Y)$.

(2) We are only interested in s-fold m-jets $\sigma = (\sigma_1, \ldots, \sigma_s)$ where target $\sigma_1 = \cdots =$ target σ_s since these jets reflect the problem of self-intersection. Such multijets are called *diagonal elements.*

Proof. The proof splits into two cases: namely, whether or not σ is a diagonal element. Suppose $\sigma = (\sigma_1, \ldots, \sigma_s)$ is a diagonal element. Recall from Proposition 6.1 the diffeomorphism $T\colon U \times V \times \mathring{\mathcal{O}}_\sigma \to J^k(U, V)$. Now we extend this to an immersion as follows: Let $U_1, \ldots, U_s$ be disjoint chart nbhds of source $(\sigma_1) = p_1, \ldots,$ source $(\sigma_s) = p_s$ respectively and V a chart nbhd of $q =$ target $(\sigma_1) = \cdots =$ target (σ_s). Define $\tilde{T}\colon U_1 \times \cdots \times U_s \times V \times \mathring{\mathcal{O}}_{\sigma_1} \times \cdots \times \mathring{\mathcal{O}}_{\sigma_s} \to J^k(U_1, V) \times \cdots \times J^k(U_s, V) \subset J_s{}^k(X, Y)$ as follows:

$$\tilde{T}(x_1, \ldots, x_s, y, \tau_1, \ldots, \tau_s) = (T(x_1, y, \tau_1), \ldots, T(x_s, y, \tau_s)).$$

Define $\mathscr{D}_\sigma{}^s(U_1, \ldots, U_s, V) =$ connected component of σ in $\mathscr{D}_\sigma{}^s \cap (J^k(U_1, V) \times \cdots \times J^k(U_s, V))$. With arguments similar to those in Proposition 5.1 one shows that $\operatorname{Im} \tilde{T} = \mathscr{D}_\sigma{}^s(U_1, \ldots, U_s, V)$ so that $\mathscr{D}_\sigma{}^s$ is an immersed submanifold. The only catch is that if we have s diffeomorphisms one each defined on U_i that there is a global diffeomorphism on X which is equal to each on a nbhd of σ_i. This is possible as long as the diffeomorphisms are given by translations or have a fixed point where the diffeomorphism has a Jacobian with positive determinant. But these are the only diffeomorphisms that are considered in the proof. In other words, the diffeomorphisms on U_i can be damped near the boundary to extend smoothly to the identity off U_i.

Now suppose σ is not a diagonal element. For simplicity suppose that $s = 2$ so that $\sigma = (\sigma_1, \sigma_2)$ and target $\sigma_1 \neq$ target σ_2. Then we claim that $\mathscr{D}_\sigma{}^2 = (\mathscr{D}_{\sigma_1} \times \mathscr{D}_{\sigma_2}) \cap Z$ where Z is the open set

$$\{(\sigma_1, \sigma_2) \in J_2{}^m(X, Y) \mid \text{target } \sigma_1 \neq \text{target } \sigma_2\}$$

(at least locally). If this is true, then certainly $\mathscr{D}_\sigma{}^2$ is an immersed submanifold. We shall leave the proof of this claim to the reader as it is not difficult and we shall not make further use of this fact. Note the case for general s is similar to $s = 2$; the details of this observation are also left as an exercise. □

Definition 6.2. *Let $f\colon X \to Y$ be smooth and let $m = \dim Y$. Then f is* transverse stable *if for every s with $1 \leq s \leq m + 1$ and diagonal element σ, $j_s{}^m f \pitchfork \mathscr{D}_\sigma{}^s$.*

Lemma 6.3. *Let $f\colon X \to Y$ be smooth. If f is stable, then f is transverse stable.*

Proof. The proof is the same as the proof of Lemma 5.7 except that we substitute the Multijet Transversality Theorem for the Thom Transversality Theorem. □

The main result is the following:

Theorem 6.4. *Let $f\colon X \to Y$ be smooth. If f is transverse stable, then f is infinitesimally stable.*

The proof of Theorem 6.4 is almost identical to that of Theorem 5.13. We will just give a sketch as the important ideas have already been presented.

Sketch of Proof. We assume that f is transverse stable; we need only show that condition (†) of Theorem 1.6 is satisfied by f. So let $S = \{p_1, \ldots, p_s\} \subset f^{-1}(q)$, with $1 \leq s \leq m + 1$. Define $\lambda^S : J^m(f^*TY)_S \to T_\sigma J_s^m(X, Y)$ as follows: $\lambda : J^m(f^*TY)_{p_i} \to T_{\sigma_i} J^m(X, Y)$ has already been defined in the discussion before Proposition 5.8. Since $J^m(f^*TY)_S = \bigoplus_{i=1}^s J^m(f^*TY)_{p_i}$ and $T_\sigma J_s^m(X, Y) = \bigoplus_{i=1}^s T_{\sigma_i} J^m(X, Y)$ it makes sense to define $\lambda^S = \bigoplus_{i=1}^s \lambda$. Proposition 5.8 still applies so that λ^S is a linear injection. Let $\alpha^{(s)} : J_s^m(X, Y) \to X^{(s)}$ be the source map. Just as in Proposition 5.9 the sequence

$$0 \longrightarrow J^m(f^*TY)_S \xrightarrow{\lambda^S} T_\sigma J_s^m(X, Y) \xrightarrow{(d\alpha^{(s)})_\sigma} T_{(p_1, \ldots, p_s)} X^{(s)} \longrightarrow 0$$

is exact.

Next let η be a vector field on Y with compact support represent $[\eta]_q$ and let ϕ_t be the associated one parameter group. Define $(d\gamma_1^S)(\eta)$ by $(dc/dt)|_{t=0}$ where $c(t) = j_s^k \phi_t(q) \cdot \sigma$. Then $(d\gamma_1^S)$ is a mapping of $C^\infty(TY)_q \to T_\sigma \mathcal{D}_\sigma^s$. Just as in Proposition 5.10 the diagram

$$\begin{array}{ccc} C^\infty(TY)_q & \xrightarrow{(d\gamma_1^S)} & T_\sigma \mathcal{D}_\sigma^s \\ \Big\downarrow \pi_m^\infty & & \Big\uparrow \lambda^S \\ J^m(TY)_q & \xrightarrow{f^*} & J^m(f^*TY)_S \end{array}$$

commutes.

Similarly, we can define $(d\gamma_2^S) : C^\infty(TX)_S \to T_\sigma \mathcal{D}_\sigma^s$. Let $[\zeta_1]_{p_1}, \ldots, [\zeta_s]_{p_s}$ be germs of vector fields on X. Since $p_1, \ldots, p_s$ are all distinct, there is one vector field ζ on X such that $[\zeta]_{p_i} = [\zeta_i]_{p_i}$. Let ψ_t be the one parameter group associated with ζ. Then define $(d\gamma_2^S)([\zeta_1]_{p_1}, \ldots, [\zeta_s]_{p_s}) = (dc/dt)|_{t=0}$ where $c(t) = \sigma \cdot j_s^m \psi_t^{-1} \psi_t(S)$. Next define $\mathcal{K}^S = -\lambda^S \cdot (df) + (dj_s^m f)_S \cdot \pi_0^m$ where $\pi_0^m : J^m(TX)_S \to T_S X^{(s)}$ is the obvious projection. Then just as in Proposition 5.11 the diagram

$$\begin{array}{ccc} C^\infty(TX)_S & \xrightarrow{(d\gamma_2^S)} & T_\sigma \mathcal{D}_\sigma^s \\ \Big\downarrow \pi_m^\infty & \nearrow \mathcal{K}^S & \\ J^m(TX)_S & & \end{array}$$

commutes where π_m^∞ is the obvious projection.

Finally we note that $(d\gamma_2^S) \oplus (d\gamma_1^S) : C^\infty(TX)_S \oplus C^\infty(TY)_q \to T_\sigma \mathcal{D}_\sigma^s$ is onto as in Lemma 5.12.

The calculations to show that condition (†) holds for the set S proceed in an entirely analogous way as the calculations in the proof of Theorem 5.13. □

§7. Summary

The following summarizes the work of this Chapter.

Theorem 7.1. *Let $f: X \to Y$ be smooth and let X be compact. Then the following are equivalent:*

(a) *f is stable;*
(b) *f is transverse stable;*
(c) *f is infinitesimally stable;*
(d) *f is homotopically stable;*
(e) *f is stable under k-parameter families of deformations.*

Proof

(a) $\Rightarrow$ (b) Lemma 6.3;
(b) $\Rightarrow$ (c) Theorem 6.4;
(c) $\Leftrightarrow$ (d) $\Leftrightarrow$ (e) Theorem 4.2 and Proposition 4.1.
(d) $\Rightarrow$ (a) Proposition 2.6. ∎

At this point a few comments about the heuristics of stable mappings seem in order. It should be clear that "generic" properties have something to do with stable mappings. This can be made precise as follows. A property P of smooth mappings of $X \to Y$ is *generic* if it satisfies the following two conditions. Let $W_P = \{f \in C^\infty(X, Y) \mid f \text{ satisfies property } P\}$:

(1) W_P contains a residual subset of $C^\infty(X, Y)$. (Preferably W_P is open and dense.)

(2) If f is in W_P, then any mapping equivalent to f is in W_P; that is, W_P is an invariant subset under the action of $\text{Diff}(X) \times \text{Diff}(Y)$ on $C^\infty(X, Y)$.

A simple argument shows that with this definition of generic a stable mapping does satisfy every generic property. Examples of generic properties were developed in Chapters II and III. (e.g., Morse functions with distinct critical values, 1:1 immersions when $2 \dim X < \dim Y$, and immersions with normal crossings when $2 \dim X = \dim Y$). In these cases we used the Thom Transversality and Multijet Transversality Theorems to show that the property in question is valid for a residual set of mappings and is thus a generic property. John Mather's Theorem which states that infinitesimally stable mappings are stable enabled us to show (in Chapter III) that (for the relative dimensions of X and Y under consideration) there are no other interesting generic properties; i.e., if there were other generic properties they would be satisfied automatically if the function in question satisfied the generic properties listed above. The subsequent chapters will be devoted to finding generic properties for mappings between manifolds of arbitrary dimensions. One might hope for a list of "interesting" generic properties along with a result which states that if a mapping satisfies the properties on this list then it is stable. This turns out not to be possible!

The proof that infinitesimal stability implies stability—in particular, the notion of transverse stability—allows us to describe what all of the generic properties are (at least those generic properties which depend only on the m-jet of the function for some m). Since the m-jet of a function determines whether or not the function is stable (Theorem 1.6 where $m = \dim Y$), this means that we can describe all of the "interesting" generic properties. Said more precisely—to each orbit, $\mathcal{O}$, in $J^m_{m+1}(X, Y)$ under the action of $\mathrm{Diff}(X) \times \mathrm{Diff}(Y)$ there is associated a generic property, $P_{\mathcal{O}}$ of mappings of $X \to Y$; namely, the property that the m-jet extension of the function intersects $\mathcal{O}$ transversely. Theorem 7.1 states that a mapping is stable iff it satisfies all generic properties $P_{\mathcal{O}}$ constructed in this fashion. Since the number of orbits is uncountably infinite it is by no means clear that the set of stable mappings is dense. Quite the reverse, it seems remarkable that a large subset of mappings could satisfy all of these properties simultaneously.

The problem with the generic properties $P_{\mathcal{O}}$ described in the last paragraph is that they do not translate easily into more familiar properties of smooth mappings such as the structure of the singular sets. In the next chapter we will show how to construct submanifolds of $J^k(X, Y)$ (which are unions of orbits) and which do translate into nice geometric notions. In doing so, we shall also be able to show that certain of these properties are contradictory so that stable mappings are not always dense. In fact, there exist manifolds X and Y for which there are no stable mappings in $C^\infty(X, Y)$.

Chapter VI

Classification of Singularities
Part I: The Thom-Boardman Invariants

§1. The S_r Classification

For a mapping $f: X \to Y$ we can make the following rudimentary classification of singularities. We say that f has a singularity of type S_r at x in X if $(df)_x$ drops rank by r; i.e., if $\operatorname{rank}(df)_x = \min(\dim X, \dim Y) - r$. Denote by $S_r(f)$ the singularities of f of type S_r. Recall that in the proof of the Whitney Immersion Theorem we introduced the submanifolds S_r of $J^1(X, Y)$ consisting of jets of corank r. (See II, Theorem 5.4.) Clearly $S_r(f) = (j^1f)^{-1}(S_r)$. To prove the Whitney Theorem we showed that if X and Y have the "right" relative dimensions then generically $S_r(f) = \varnothing$ $(r > 0)$ and f has no singularities; i.e., f is an immersion. Without restricting the relative dimensions of X and Y we can still say that generically $S_r(f)$ is a submanifold of X and $\operatorname{codim} S_r(f) = \operatorname{codim} S_r = r^2 + er$ where $e = |\dim X - \dim Y|$. This statement follows immediately from the Thom Transversality Theorem and II, Theorem 4.4. In particular, the set of mappings for which $j^1f \pitchfork S_r$ (for all r) is residual. Besides the Transversality Theorem, the main fact used in the proof of this statement is that S_r is actually a submanifold of $J^1(X, Y)$. We shall sketch a different proof of this fact in order to motivate the material in §4.

Given a pair of vector spaces V and W, let $L^r(V, W)$ be the set of linear maps of V into W which drop rank by r. The main fact needed is:

Proposition 1.1. *$L^r(V, W)$ is a submanifold of* $\operatorname{Hom}(V, W)$ *of codimension $r^2 + er$ where $e = |\dim V - \dim W|$.*

This is just II, Proposition 5.3, which we shall reprove here using a trick involving Grassmann manifolds.

Let $s = r + \max(0, \dim V - \dim W)$ and let $G(s, V)$ be the Grassmannian of s planes in V. Let E be the canonical bundle over $G(s, V)$. (See Example (4) after I, Proposition 5.4.) We will denote by Q the vector bundle with fiber V/E_p at $p \in G(s, V)$ and by $\operatorname{Hom}(Q, W)$ the vector bundle over $G(s, V)$ whose fiber at p is $\operatorname{Hom}(Q_p, W)$. (The construction is functorial and thus yields a smooth vector bundle. See I, Proposition 5.4.) This fiber contains $L^0(Q_p, W)$ as an open subset, and $L^0(Q, W) = \bigcup_p L^0(Q_p, W)$ is an open submanifold of $\operatorname{Hom}(Q, W)$.

Now we claim there is a natural identification (as sets)

$$L^0(Q, W) \cong L^r(V, W). \tag{1.2}$$

In fact, for each $p \in G(s, V)$ we have a projection $\pi: V \to Q_p$ and this induces a transpose map backwards.

$$\pi^* : \mathrm{Hom}\,(Q, W) \to \mathrm{Hom}\,(V, W). \tag{1.3}$$

The image of this map is the set of elements in Hom (V, W) of corank $\geq r$. Moreover, π^* maps $L^0(Q, W)$ bijectively onto the elements in Hom (V, W) which are *precisely of corank* r; so π^* gives us the identification (1.2) whose existence we asserted.

We now use (1.2) to provide $L^r(V, W)$ with a manifold structure. We will let the reader compute the dimension of this manifold and check that it gives the same answer as our computation in Chapter II. (Hint: It is the same as the dimension of Hom (Q, W) regarded as a manifold.) This does not yet show that $L^r(V, W)$ is a submanifold of Hom (V, W). Let $\pi^* : \mathrm{Hom}\,(Q, W) \to \mathrm{Hom}\,(V, W)$ be the map described above and let M be the subset of Hom (V, W) consisting of elements of corank $> r$. To conclude the proof one has to show

Proposition 1.4. *The map* $\pi^* : L^0(Q, W) \to \mathrm{Hom}\,(V, W) - M$ *is a* 1:1 *proper immersion.*

The proof is straightforward but a little tedious. We will not include it here. (See exercises.) Anyway, given Proposition 1.4, it is easy to prove our assertion about $S_r(f)$. Namely, we first observe that Proposition 1.1 is true for vector bundles as well as vector spaces. Given two vector bundles $E \to X$ and $F \to Y$ denote by $L^r(E, F)$ the fiber bundle over $X \times Y$ with typical fiber $L^r(E_x, F_y)$. Then by Proposition 1.1 $L^r(E, F)$ is a subfiber-bundle of Hom (E, F) of codimension $r^2 + er$.

Recall now the canonical identification $J^1(X, Y) = \mathrm{Hom}\,(TX, TY)$. (See Remark (2) after II, Theorem 2.7.) Just note that S_r is the subfiber-bundle of $J^1(X, Y)$ corresponding to $L^r(TX, TY)$ and we have the desired conclusion that S_r is a submanifold of $J^1(X, Y)$.

Definition 1.5. *We will say that a mapping* $f: X \to Y$ *is* one generic *if* $j^1 f \pitchfork S_r$ *for all* r.

From now on, we will assume all maps are one-generic.

Exercises

(The purpose of these exercises is to supply the reader with an outline of the proof of Proposition 1.4.)

(1) Show that the map (1.2) is a homeomorphism; i.e., give $L^r(V, W)$ the induced topology and show that the inverse of the map (1.2) is continuous.

(2) Let $Z \xrightarrow{\sigma} X$ be a fiber bundle and let $\rho : Z \to Y$ be a map. Let $z \in Z$, $x = \sigma(z)$, and $y = \rho(z)$. Show that

$$0 \to T_z Z_x \to T_z Z \to T_x X \to 0$$

is exact, and deduce the existence of a mapping

$$T_xX \to T_yY/(d\rho)_z(T_zZ_x). \tag{1.6}$$

(3) Apply Exercise (2) with the following data: $X = G(s, V)$, x = an s dimensional subspace K of V, $Y = \operatorname{Hom}(V, W)$, y = a linear map A in $\operatorname{Hom}(V, W)$ with $\operatorname{Ker} A = K$, $Z = L^0(Q, W)$, and $z = A \cdot \pi$ where π is as in Proposition 1.4. (Note that V/K is the fiber of Q above x and $L^0(V/K, W)$ is the fiber of Z above x.) Let $\rho = \pi^*: L^0(Q, W) \to \operatorname{Hom}(V, W)$ and let $\sigma: L^0(Q, W) \to G(s, V)$ be the canonical projection. Note that $T_yY = T_A \operatorname{Hom}(V, W) \cong \operatorname{Hom}(V, W)$ and

$$T_zZ_x = T_{A\cdot\pi} \operatorname{Hom}(V/K, W) \cong \operatorname{Hom}(V/K, W)$$

since Hom's are vector spaces.

(i) Show that $(d\rho)_z: T_zZ_x \to T_yY$ is an injection. In particular show that the image is all maps containing K in their kernels.

(ii) From (i) conclude that $(d\rho)_z(T_zZ_x) \cong \operatorname{Hom}(V/K, W)$.

(iii) Finally conclude that $T_yY/(d\rho)_z(T_zZ_x) \cong \operatorname{Hom}(K, W)$.

Moreover, show all these identifications are canonical.

(4) Let $x \in G(s, V)$ and let K be the s-dimensional subspace it represents in V. Recall there is a canonical identification $T_xG(s, V) \cong \operatorname{Hom}(K, V/K)$. (See Note (2) at the end of I, §3.) Show that the map (1.6) of Exercise 2 is just the map $A_*: \operatorname{Hom}(K, V/K) \to \operatorname{Hom}(K, W)$ given by composition on the last factor where A is given in Exercise 3.

(5) From 2–4 deduce that the map π^* in Proposition 1.4 is an immersion.

(6) Using Exercise 4, prove that $N_A = T_A \operatorname{Hom}(V, W)/T_AL^r(V, W)$ the normal space to $L^r(V, W)$ in $\operatorname{Hom}(V, W)$ at A is canonically isomorphic to $\operatorname{Hom}(\operatorname{Ker} A, \operatorname{coker} A)$.

(7) Let $f: \mathbf{R}^n \to \mathbf{R}^n$ be given by $f(x_1, \ldots, x_n) = (f_1(x), \ldots, f_r(x), x_{r+1}, \ldots, x_n)$. Suppose that $f(0) = 0$ and that f has an S_r singularity at 0. Show that $j^1f \pitchfork S_r$ at 0 iff the r^2 vectors $\left(d\dfrac{\partial f_i}{\partial x_j}\right)_0$ where $1 \le i, j \le r$ are all linearly independent. Hint: Use the proof of II, Proposition 5.3 to identify $T_\sigma L^r(\mathbf{R}^n, \mathbf{R}^n)$ in coordinates where $\sigma = j^1f(0)$.

§2. The Whitney Theorem for Generic Mappings between 2-Manifolds

The Thom-Boardman Theory has to do with the behavior of maps restricted to their singular sets; i.e., if $f: X \to Y$ is one-generic, $S_r(f)$ is a submanifold, so $f|S_r(f)$ is again a map between manifolds; and we can, for example, ask whether it has singularities generically. It is this type of question to which the Thom-Boardman Theory addresses itself. Before we outline this theory, we will discuss one example in detail—the Whitney Theory for maps between 2-manifolds.

Let X and Y be 2-dimensional manifolds and let $f: X \to Y$ be a one-generic mapping. By our computation in §1 $S_1(f)$ is of codimension 1 in X

and $S_2(f)$ does not occur since it would have to be of codimension 4. Let p be in $S_1(f)$ and $q = f(p)$. One of the following two situations can occur.

$$\text{(2.1)} \qquad \begin{cases} \text{(a)} & T_pS_1(f) \oplus \operatorname{Ker}(df)_p = T_pX; \\ \text{(b)} & T_pS_1(f) = \operatorname{Ker}(df)_p. \end{cases}$$

Note that if p is a S_1 singularity satisfying (a), then p is a fold point. (See III, Definition 4.1.) The first theorem of Whitney for maps between 2-manifold gives the normal form for fold points.

Theorem 2.2. *If* (a) *occurs then one can choose a system of coordinates* (x_1, x_2) *centered at* p *and* (y_1, y_2) *centered at* q *such that* f *is the map*

$$(*) \qquad (x_1, x_2) \mapsto (x_1, x_2^2).$$

This theorem is just a special case of the normal form that we derived for submersions with folds. (See III, Theorem 4.5.)

Now we will suppose that condition (b) holds; i.e., $\ker(df)_p = T_pS_1(f)$. This situation is considerably more complicated. Let us choose a smooth nonvanishing vector field, ξ, along $S_1(f)$ such that at each point of $S_1(f)$ ξ is in the kernel of (df). (Locally this is always possible.) By assumption, ξ is tangent to $S_1(f)$ at p. The nature of the singularity at p obviously depends on *what order of contact* ξ has with $S_1(f)$ at p. Let us make this statement more precise. Let k be a smooth function on X, such that $k = 0$ on $S_1(f)$ and $(dk)_p \neq 0$. Consider $(dk)(\xi)$ as a function on $S_1(f)$. By assumption this has a zero at p. We let the reader check as an exercise that the order of this zero does not depend on the choice of ξ or k. (Hint: for another choice (ξ', k') show that ξ' and k' are nonzero multiples of ξ and k.)

Definition 2.3. *We will say* p *is a* simple cusp *if this zero is a simple zero.*

The second main theorem of Whitney states

Theorem 2.4. *If* p *is a simple cusp then one can find coordinates* (x_1, x_2) *centered at* p *and* (y_1, y_2) *centered at* q *such that*

$$\begin{cases} f^*y_1 = x_1 \\ f^*y_2 = x_1x_2 + x_2^3. \end{cases}$$

A picture of this map is sketched in Figure 3. Let X be the graph of $x_3 = x_1x_2 + x_2^3$. This graph can be viewed as a family of cubic curves in (x_2, x_3) depending on the parameter x_1. For x_1 positive these curves are without critical points. For $x_1 = 0$ there is a critical point which is a point of inflection and for $x_1 < 0$ there are max and min's. Let $f\colon X \to \mathbf{R}^2$ be the projection of X onto the x_1x_3-plane. There is a natural set of global coordinates on X given by $(x_1, x_2) \mapsto (x_1, x_2, x_1x_2 + x_2^3)$. In these coordinates on X f has the form of Theorem 2.4. The fold curve is the locus of extrema and the cusp is the inflection point. Note that $S_1(f)$ is a parabola twisted in $\mathbf{R}^3$ so that any vector tangent to $S_1(f)$ at $(0, 0, 0)$ is killed by f. Also note that the image of $S_1(f)$ under f is the cusped plane curve $t \mapsto (-3t^2, -2t^3)$.

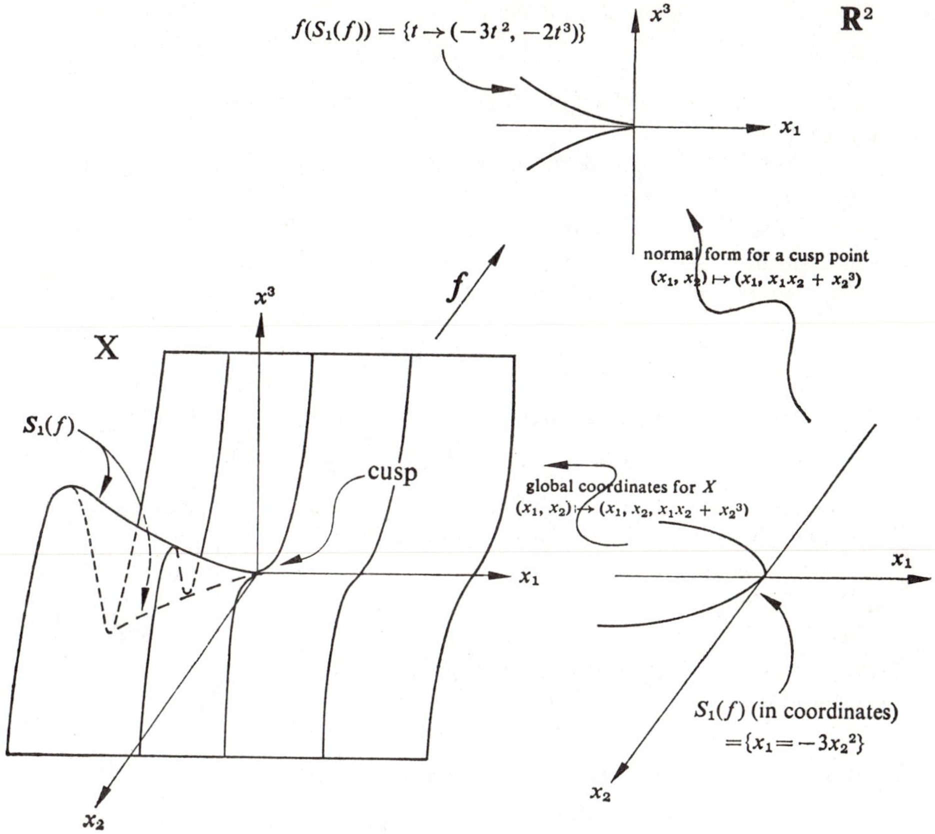

Figure 3: The Simple Cusp

The proof we will give of Theorem 2.4 is due to Morin and uses the Malgrange Preparation Theorem. Whitney gives a more elementary but slightly more complicated proof in [58].

Proof. Let us choose coordinates (x_1, x_2) centered at p and (y_1, y_2) centered at q such that f has the form

$$(\dagger) \qquad \begin{aligned} f^*y_1 &= x_1 \\ f^*y_2 &= h(x_1, x_2). \end{aligned}$$

Since f has rank 1 at p this is possible. We can also assume that at the origin $(df)_0 = \begin{bmatrix} 1 & 0 \\ 0 & 0 \end{bmatrix}$ in this coordinate system; i.e., $(\partial h/\partial x_1)(0) = (\partial h/\partial x_2)(0) = 0$. We note, however, $d(\partial h/\partial x_2)_0 \neq 0$, otherwise f would not be one-generic. (*Proof.* Suppose

$$\frac{\partial}{\partial x_1}\left(\frac{\partial h}{\partial x_2}\right) = \frac{\partial}{\partial x_2}\left(\frac{\partial h}{\partial x_2}\right) = 0$$

at 0. Let $\alpha = \frac{1}{2}(\partial^2 h/\partial x_1^2)(0)$ and compare f with the map

(††) $$(x_1, x_2) \mapsto (x_1, \alpha x_1^2).$$

They have the same 2-jet at 0, but (††) is of rank 1 everywhere, so it is not one-generic.)

The set $S_1(f)$ is defined by the equation $\partial h/\partial x_2 = 0$; so at each point of $S_1(f)$ the kernel of (df) is spanned by $\partial/\partial x_2$. This means that we can take $\partial h/\partial x_2$ to be the function k and $\partial/\partial x_2$ to be the vector field ξ of Definition 2.3. The condition for the origin to be a cusp is that $(\partial^2 h/\partial x_2^2)(0) = 0$; and for it to be a simple cusp, $(\partial^3 h/\partial x_2^3)(0) \neq 0$. Therefore, at the origin, we have

$$h = \frac{\partial h}{\partial x_2} = \frac{\partial^2 h}{\partial x_2^2} = 0 \quad \text{and} \quad \frac{\partial^3 h}{\partial x_2^3} \neq 0.$$

The Generalized Malgrange Preparation Theorem allows us to write

$$x_2^3 = 3a_2(x_1, h)x_2^2 + a_1(x_1, h)x_2 + a_0(x_1, h)$$

where a_0, a_1, and a_2 are smooth functions of y_1 and y_2 vanishing at 0. (To see this recall that f is given by $f(x_1, x_2) = (x_1, h(x_1, x_2))$. Then via f $C_0^\infty(\mathbf{R}^2)$ becomes a module over itself; i.e., $a \cdot b(x_1, x_2) = a(f(x_1, x_2))b(x_1, x_2)$ where a is in the ring $C_0^\infty(\mathbf{R}^2)$ and b is in the module $C_0^\infty(\mathbf{R}^2)$. By the Malgrange Theorem (IV, Corollary 3.11) this module is generated by 1, x_2, and x_2^2 if the vector space $C_0^\infty(\mathbf{R}^2)/((x_1, h) + \mathcal{M}_0(\mathbf{R}^2)^4)$ is generated by 1, x_2, x_2^2. The assumptions on h guarantee that this is so.)

Now the equation above can be written in the form

(*) $$(x_2 - a)^3 + b(x_2 - a) = c$$

(with $a = a_2$, and b and c new functions of (y_1, y_2) vanishing at 0.) If we set $x_1 = 0$ in (*) we see that the left hand side is of the form $x_2^3 + \cdots$, the dots indicating terms of order >3 in x_2. Since $h(0, x_2) = x_2^3 + \cdots$, the right and left hand sides of (*) can be equal only if $(\partial c/\partial y_2)(y_1, y_2) \neq 0$ at 0. The leading term in the Taylor series of h is a nonzero multiple of $x_1 x_2$; so, comparing the linear and quadratic terms on the right and left hand sides of (*), one easily sees that $\partial c/\partial y_1 = 0$ and $\partial b/\partial y_1 \neq 0$ at the origin. This means that the following are legitimate coordinate changes:

$$\begin{cases} \bar{x}_1 = b(x_1, h) \\ \bar{x}_2 = x_2 - a(x_1, h) \end{cases}$$

$$\begin{cases} \bar{y}_1 = b(y_1, y_2) \\ \bar{y}_2 = c(y_1, y_2). \end{cases}$$

In these coordinates we have

$$f^*\bar{y}_1 = \bar{x}_1$$
$$f^*\bar{y}_2 = \bar{x}_2^3 + \bar{x}_1\bar{x}_2$$

which is Whitney's canonical form. □

Finally, Whitney proved that the singularities described above are generically the only singularities that can occur for maps between 2-manifolds.

Theorem 2.5. *Let X and Y be 2-manifolds. Then there is a residual set in $C^\infty(X, Y)$ such that if f belongs to this set, its only singularities are folds and simple cusps.*

This is not hard to show directly, but we prefer to deduce it from a more general result. We defer the proof to §4 below. (See §4, Exercise 4.)

§3. The Intrinsic Derivative

In this section we will develop a technique due to Porteous for differentiating maps between vector bundles. This is the intrinsic derivative, and it plays a rather important role in the Thom-Boardman theory (a special case of the intrinsic derivative was introduced in Chapter II, Definition 6.5).

We will first of all give a pedestrian (and uninvariant) definition of this notion, then later a more sophisticated (invariant) definition.

Let X be a manifold and let $E = X \times \mathbf{R}^k$ and $F = X \times \mathbf{R}^l$ be product bundles over X. Let $\rho : E \to F$ be a vector bundle homomorphism. We may view ρ as a mapping of $X \to \operatorname{Hom}(\mathbf{R}^k, \mathbf{R}^l)$. Then for p in X, $(d\rho)_p : T_pX \to T_{\rho(p)} \operatorname{Hom}(\mathbf{R}^k, \mathbf{R}^l) = \operatorname{Hom}(\mathbf{R}^k, \mathbf{R}^l)$ makes sense. Let $K_p = \operatorname{Ker} \rho(p)$ and let $L_p = \operatorname{Coker} \rho(p)$. Then we define the *intrinsic derivative*, $(D\rho)_p$, in this local situation by the composite map

$$T_pX \to \operatorname{Hom}(\mathbf{R}^k, \mathbf{R}^l) \to \operatorname{Hom}(K_p, L_p)$$

where this second arrow is given by "restricting and projecting".

We claim that the intrinsic derivative does not depend on which choices of trivializations of E and F are made. More precisely, let $A : E \to E$ and $B : F \to F$ be vector bundle isomorphisms. We may view A and B as mappings of $X \to \operatorname{Hom}(\mathbf{R}^k, \mathbf{R}^k)$ and $X \to \operatorname{Hom}(\mathbf{R}^l, \mathbf{R}^l)$ respectively. With these trivializations ρ has the form $\bar{\rho}$ where $\bar{\rho}(x) = B(x)\cdot\rho(x)\cdot A(x)^{-1}$. Let $\bar{K}_p = \operatorname{Ker} \bar{\rho}(p)$ and $\bar{L}_p = \operatorname{Coker} \bar{\rho}(p)$. Clearly $A(p)$ and $B(p)$ induce isomorphisms of $K_p \to \bar{K}_p$ and $L_p \to \bar{L}_p$ respectively. Thus the different trivializations give a natural mapping $\phi: \operatorname{Hom}(K_p, L_p) \to \operatorname{Hom}(\bar{K}_p, \bar{L}_p)$ defined by $\phi(C) = B(p)\cdot C\cdot A(p)^{-1}$. Our statement of invariance reduces to showing that the diagram

$$\begin{array}{ccc} T_pX & \xrightarrow{(D\rho)_p} & \operatorname{Hom}(K_p, L_p) \\ & {\scriptstyle (D\bar{\rho})_p}\searrow & \downarrow{\scriptstyle \phi} \\ & & \operatorname{Hom}(\bar{K}_p, \bar{L}_p) \end{array}$$

commutes. First note that if A and B are linear changes of trivializations—i.e., $A = id_X \times a$ and $B = id_X \times b$ where $a : \mathbf{R}^k \to \mathbf{R}^k$ and $b : \mathbf{R}^l \to \mathbf{R}^l$ are linear isomorphisms—then the computation that the diagram commutes is trivial. In the general case, we may, by using linear changes of trivializations, assume that $A(p) = id_{\mathbf{R}^n}$ and $B(p) = id_{\mathbf{R}^l}$. By doing so we see that $\bar{K}_p = K_p$

and $\bar{L}_p = L_p$; so we need only show that $(D\bar{\rho})_p = (D\rho)_p$ as mappings of $T_pX \to \text{Hom}(K_p, L_p)$. First compute

$$\begin{aligned} d(\bar{\rho})_p &= d(B\cdot\rho\cdot A^{-1})_p \\ &= (dB)_p\cdot\rho(p)\cdot A(p)^{-1} + B(p)\cdot(d\rho)_p\cdot A(p)^{-1} + B(p)\cdot\rho(p)\cdot(dA^{-1})_p \\ &= (dB)_p\cdot\rho(p) + (d\rho)_p + \rho(p)\cdot(dA^{-1})_p. \end{aligned}$$

(Note that we use the product rule and not the chain rule in the computation since we are differentiating the product of matrices whose coefficients depend smoothly on the parameters x.) When we "restrict and project" the first and third terms vanish. Thus, $(D\rho)_p = (D\bar{\rho})_p$.

In general, let E and F be vector bundles over X and let $\rho : E \to F$ be a vector bundle homomorphism. Fixing p in X and defining K_p and L_p as above, we may define the *intrinsic derivative of* ρ *at* p, $(D\rho)_p : T_pX \to \text{Hom}(K_p, L_p)$, by choosing trivializations of E and F on a nbhd of p and computing as in the local situation. The last paragraph implies that this mapping is independent of the choice of trivializations.

To prepare the reader for our other definition of the intrinsic derivative, we need to look again at some elementary properties of the manifold $L^r(V, W) \subset \text{Hom}(V, W)$ discussed in §1.

Let A be in $L^r(V, W)$ and let K_A = kernel A, L_A = cokernel A. Let N_A be the normal space to $L^r(V, W)$ in $\text{Hom}(V, W)$ at the point represented by A, i.e., $N_A = T_A \text{Hom}(V, W)/T_A L^r(V, W)$. We will show that there is a canonical identification $\text{Hom}(K_A, L_A) \cong N_A$.

Since $T_A \text{Hom}(V, W)$ is canonically isomorphic with $\text{Hom}(V, W)$ there is a canonical surjective linear map

$$T_A \text{Hom}(V, W) \to \text{Hom}(K_A, L_A) \tag{3.1}$$

given by "restricting and projecting."

Lemma 3.2. *The kernel of the mapping* (3.1) *is the tangent space to* $L^r(V, W)$ *at* A.

Proof. We can choose linear coordinates in V and W so that A has the form

$$\left(\begin{array}{c|c} I_m & 0 \\ \hline 0 & 0 \end{array}\right) \quad \text{where} \quad m = \text{rank } A.$$

If we define $L^r(V, W)$ in a nbhd of A as the pre-image of 0 with respect to the map

$$\left(\begin{array}{c|c} S & T \\ \hline U & V \end{array}\right) \mapsto V - US^{-1}T, \tag{3.3}$$

(see II, Proposition 5.3), then $T = U = 0$ at A, so the derivative of (3.3) at A is just the map (3.1). ∎

Corollary 3.4. *The map* (3.1) *induces an isomorphism* $N_A \cong \text{Hom}(K_A, L_A)$. (*For a more elegant proof, see Exercise* 6 *at the end of* §1.)

This result is true of vector bundles as well as vector spaces. Let $E \to X$ and $F \to X$ be vector bundles, and let σ be in the fiber of $L^r(E, F)$ above x. Since we are dealing with fiber bundles, the normal space to $L^r(E_x, F_x)$ in $\mathrm{Hom}(E_x, F_x)$ at σ is identical with the normal space to $L^r(E, F)$ in $\mathrm{Hom}(E, F)$; so just as in the case of vector spaces we have a canonical identification $N_\sigma \cong \mathrm{Hom}(K_\sigma, L_\sigma)$.

Now let $\rho: E \to F$ be a vector bundle map. For the moment we will view ρ as a map $\rho: X \to \mathrm{Hom}(E, F)$. Let $\sigma = \rho_x$, assume σ is of corank r (i.e., $\sigma \in L^r(E, F)$), and let N_σ be the normal space to $L^r(E, F)$ in $\mathrm{Hom}(E, F)$. Then we get a sequence of maps

$$(3.5) \qquad T_xX \xrightarrow{(d\rho)_x} T_\sigma \mathrm{Hom}(E, F) \to N_\sigma \cong \mathrm{Hom}(K_\sigma, L_\sigma).$$

An easy result, whose proof we leave to the reader, is:

Proposition 3.6. *The composite of the maps* (3.5) *is identical with the intrinsic derivative.*

(Hint: The intrinsic derivative was defined in terms of a trivialization, so prove the assertion for trivial bundles.)

This proposition immediately provides us with the following:

Proposition 3.7. *Suppose $\sigma = \rho_x$ is of corank r. Then the following two assertions are equivalent.*

(a) *$(D\rho)_x : T_xX \to \mathrm{Hom}(K_\sigma, L_\sigma)$ is surjective.*
(b) *$\rho \pitchfork L^r(E, F)$ at x where ρ is viewed as a mapping of $X \to \mathrm{Hom}(E, F)$.*

Exercises

To become familiar with this definition, the reader ought to try computing some special cases. One good case to look at is the following. Let $f: X \to Y$ be a smooth map. Let $E = TX$ and $F = f^*TY$ and let $\rho: E \to F$ be (df). (Note that we use f^*TY which is a bundle over X rather than TY which is not.)

(1) Compute the intrinsic derivative

$$D(df)_x : T_xX \to \mathrm{Hom}(K_x, L_x)$$

where $\quad K_x = \ker(df)_x \quad$ and $\quad L_x = \mathrm{coker}(df)_x$.

(Hint: Trivialize the tangent bundles by choosing coordinate systems $(x_1, \ldots, x_n)$ centered at x and $(y_1, \ldots, y_m)$ centered at y. Moreover, choose these coordinates so that $(df)_x$ has the form

$$\left(\begin{array}{c|c} I_s & 0 \\ \hline 0 & 0 \end{array}\right).$$

In terms of these coordinates the last $m - s$ coordinate functions of f can be written in the form:

$$(3.8) \qquad f_\alpha = \sum_{i,j=1}^{s} a_\alpha^{ij} x_i x_j + \sum_{j=1}^{s} \sum_{\beta=s+1}^{n} b_\alpha^{ij} x_\beta x_j + \sum_{\beta,\gamma=s+1}^{n} c_\alpha^{\beta\gamma} x_\beta x_\gamma$$

plus terms of order $0(|x|^3)$. Show that the intrinsic derivative can be viewed as a map

$$(3.9) \qquad D(df)_x : T_xX \otimes K_x \to L_x.$$

Moreover, show that this map is just the map defined by the last two terms of (3.8) (providing one uses the linear coordinates $x_{s+1}, \ldots, x_n$ in K_x and the linear coordinates $y_{s+1}, \ldots, y_m$ in L_x).)

(2) Show that the intrinsic derivative $D(df)_x$ is determined by the 2-jet of f at x (Hint: Use Exercise 1.)

(3) Show that the map (3.9) when restricted to $K_x \otimes K_x$ is in fact a symmetric map; i.e., $D(df)_x(k_1, k_2) = D(df)_x(k_2, k_1)$. Thus $D(df)_x$ induces a mapping

$$(3.10) \qquad \delta^2 f_x : K_x \circ K_x \to L_x$$

(where $K_x \circ K_x$ denotes the symmetric product of K_x with itself). (Hint: Use Exercise 1.)

(4) Let f be a real valued function and let x be a critical point of f. Then $K_x = T_xX$ and $L_x = \mathbf{R}$. Show that

$$\delta^2 f_x : T_xX \circ T_xX \to \mathbf{R}$$

is just the Hessian of f at x. (In Chapter II, Definition 6.5.)

§4. The $S_{r,s}$ Singularities

Let $f: X \to Y$ be one-generic. We will denote by $S_{r,s}(f)$ the set of points where the map $f: S_r(f) \to Y$ drops rank s. Note that codim $S_r(f) > \dim X - \dim Y$ by Proposition 1.1, so $\dim S_r(f) < \dim Y$. Therefore, $x \in S_{r,s}(f)$ if and only if $x \in S_r(f)$ and the kernel of $(df)_x$ intersects the tangent space to $S_r(f)$ in an s dimensional subspace. For example, for maps between 2-manifolds the points $S_{1,0}(f)$ are fold points and the points $S_{1,1}(f)$ are cusps. (See §2.)

Our goal in this section is to show that the $S_{r,s}(f)$ are generically manifolds (just like the sets $S_r(f)$) and to compute their dimensions. The idea of the proof will be to construct universal $S_{r,s}$'s in $J^2(X, Y)$ (analogous to the S_r's described in §1) such that

$$x \in S_{r,s}(f) \Leftrightarrow j^2f(x) \in S_{r,s}.$$

To begin, recall the identification: $S_r \cong L^r(TX, TY)$. Given $\sigma \in S_r$ with source at x and target at y we now can attach to σ the vector spaces $K_\sigma = \ker \sigma$ and $L_\sigma = \operatorname{coker} \sigma$ in T_xX and T_yY respectively. This defines for us vector bundles on S_r which we denote by K and L. As we saw in §3, the normal bundle to S_r in $J^1(X, Y)$ is canonically isomorphic to Hom (K, L).

Now let $S_r^{(2)}$ be the pre-image of S_r in $J^2(X, Y)$. By Exercise 2 at the end of §3 the intrinsic derivative gives us a map of fiber bundles:

$$\begin{array}{ccc} S_r^{(2)} & \longrightarrow & \operatorname{Hom}(K \circ K, L) \\ & \searrow \quad \swarrow & \\ & S_r & \end{array} \tag{4.1}$$

Moreover, the top arrow is surjective. (This is clear from Exercise 1 in §3. The last two terms on the RHS of (3.8) are completely arbitrary except for the symmetry condition.) We will construct our universal $S_{r,s}$'s from the diagram (4.1). The main step in the construction is a theorem about vector spaces similar to Proposition 1.1 of §1. Let V and W be vector spaces, let $V \circ V$ be the symmetric product of V with itself, and let $V \wedge V$ be the space of alternating tensors. Recall the standard algebraic fact that $V \otimes V = (V \circ V) \oplus (V \wedge V)$, so that there is a canonical projection $\pi : V \otimes V \to V \circ V$ whose kernel is $V \wedge V$. Consider the map

$$\operatorname{Hom}(V \circ V, W) \to \operatorname{Hom}(V \otimes V, W) \to \operatorname{Hom}(V, \operatorname{Hom}(V, W)) \tag{4.2}$$

where the first arrow is given by $A \mapsto A \cdot \pi$ and the second arrow is given by $B \mapsto \phi_B$ where $\phi_B(v)(v') = B(v \otimes v')$. Let $\operatorname{Hom}(V \circ V, W)_s$ be the pre-image under (4.2) of $L^s(V, \operatorname{Hom}(V, W))$. (Note that $\operatorname{Hom}(V \circ V, W)_s$ is not the same as $L^s(V \circ V, W)$.)

Proposition 4.3. $\operatorname{Hom}(V \circ V, W)_s$ *is a submanifold of* $\operatorname{Hom}(V \circ V, W)$ *of codimension*

$$\frac{l}{2} k(k + 1) - \frac{l}{2}(k - s)(k - s + 1) - s(k - s) \tag{4.4}$$

where $k = \dim V$ *and* $l = \dim W$.

Proof. We use the "Grassmannian trick" used to prove Proposition 1.1. Let $G(s, V)$ be the Grassmannian of s planes in V. Let E be the canonical bundle over $G(s, V)$ and let Q be the vector bundle over $G(s, V)$ whose fiber at p is V/E_p. Let $\operatorname{Hom}(Q \circ Q, W)$ be the vector bundle over $G(s, V)$ whose fiber at p is $\operatorname{Hom}(Q_p \circ Q_p, W)$. The set $\operatorname{Hom}(Q_p \circ Q_p, W)_0$ is an open subset of this fiber (being the inverse image under the continuous map (4.2) of an open set), and

$$\operatorname{Hom}(Q \circ Q, W)_0 = \bigcup_p \operatorname{Hom}(Q_p \circ Q_p, W)_0$$

is an open subfiber-bundle of $\operatorname{Hom}(Q \circ Q, W)$.

The map $\pi_p : V \to Q_p$ induces a map $\pi_p \otimes \pi_p : V \otimes V \to Q_p \otimes Q_p$. It is easy to see that $\pi_p \otimes \pi_p : V \circ V \to Q_p \circ Q_p$ and is onto. Just as in §1, this map induces a transpose map $\pi^* : \operatorname{Hom}(Q \circ Q, W) \to \operatorname{Hom}(V \circ V, W)$ whose image is the set $\bigcup_{t \geq s} \operatorname{Hom}(V \circ V, W)_t$. Moreover, its restriction to $\operatorname{Hom}(Q \circ Q, W)_0$ maps this set bijectively onto $\operatorname{Hom}(V \circ V, W)_s$; so there is a canonical isomorphism (of sets)

$$\operatorname{Hom}(Q \circ Q, W)_0 \cong \operatorname{Hom}(V \circ V, W)_s. \tag{4.5}$$

The left hand side is a manifold; so we can define a manifold structure on $\operatorname{Hom}(V \circ V, W)_s$ by requiring (4.5) to be a diffeomorphism. To prove that $\operatorname{Hom}(V \circ V, W)_s$ is a submanifold of $\operatorname{Hom}(V \circ V, W)$ requires a little more work. We will need, in fact, the following proposition.

Proposition 4.6. *Let* $M_s = \bigcup_{t>s} \operatorname{Hom}(V \circ V, W)_t$. *The map*

$$\pi^* : \operatorname{Hom}(Q \circ Q, W)_0 \to \operatorname{Hom}(V \circ V, W) - M_s$$

is a 1:1 *proper immersion.*

The proof of this is almost identical with the proof of Proposition 1.2. For the details of that proof see the Exercises at the end of §1.

Finally we have to compute the dimension of $\operatorname{Hom}(V \circ V, W)_s$. This is the same as the dimension of $\operatorname{Hom}(Q \circ Q, W)$. The fiber dimension of Q is $k - s$; so the fiber dimension of $\operatorname{Hom}(Q \circ Q, W)$ is $(k - s)(k - s + 1)l/2$. The dimension of the base space (i.e., $G(s, V)$) is $s(k - s)$; therefore, the total dimension is $(k - s)(k - s + 1)l/2 + s(k - s)$; and the codimension is as asserted in Proposition 4.3. □

Proposition 4.3 is valid for vector bundles as well as vector spaces. Given two vector bundles $E \to X$ and $F \to X$ let $\operatorname{Hom}(E \circ E, F)_s$ be the fiber bundle whose fiber at p in X is $\operatorname{Hom}(E_p \circ E_p, F_p)_s$. Then by Proposition 4.3 $\operatorname{Hom}(E \circ E, F)_s$ is a fiber subbundle of $\operatorname{Hom}(E \circ E, F)$, and its codimension is given by (4.4) with k the fiber dimension of E and l the fiber dimension of F.

Let us now go back to the map (4.1) described earlier in this section. $\operatorname{Hom}(K \circ K, L)_s$ is a submanifold of $\operatorname{Hom}(K \circ K, L)$, so its pre-image is a submanifold (of the same codimension) in $S_r^{(2)}$ (since (4.1) is a submersion). We will denote this manifold by $S_{r,s}$. Our main result of this section is

Theorem 4.7. *Let* $f: X \to Y$ *be one-generic. Then* $x \in S_{r,s}(f) \Leftrightarrow j^2f(x) \in S_{r,s}$.

Proof. Let $j^1f(x) = \sigma$ in S_r. The normal space to S_r in $J^1(X, Y)$ at σ is $\operatorname{Hom}(K_\sigma, L_\sigma)$; and the map

$$(dj^1f)_x : T_xX \to T_\sigma J^1(X, Y)$$

induces a map:

$$T_xX \to \operatorname{Hom}(K_\sigma, L_\sigma) \tag{4.8}$$

which is, as we saw in the last section, the intrinsic derivative of (df). This map is surjective since $j^1f \pitchfork S_r$ by assumption, and its kernel is the tangent space to $S_r(f)$ at x. If x is in $S_{r,s}(f)$ the kernel of (4.8) intersects the kernel, K_σ, of $(df)_x$ in an s dimensional subspace; that is, the restriction of (4.8) to K_σ has an s dimensional kernel. This means $j^2f(x)$ is in $\operatorname{Hom}(K_\sigma \circ K_\sigma, L_\sigma)_s$. The converse is equally easy to see, and this proves Theorem 4.7. □

Corollary 4.9. *Let* $f: X \to Y$ *be smooth. If* $j^2f \pitchfork S_{r,s}$, *then* $S_{r,s}(f)$ *is a submanifold of* $S_r(f)$ *whose codimension is given by the formula* (4.4) *where* $l = \dim Y - \dim X + k$ *and* $k = r + \max(\dim X - \dim Y, 0)$.

Thus $S_{r,s}(f)$ *is a submanifold of* X *and* $\dim S_{r,s}(f) = \dim X - r^2 - er -$ (codim $S_{r,s}(f)$ *in* $S_r(f)$) *where* $e = |\dim X - \dim Y|$.

The Transversality Theorem says that the condition $j^2 f \pitchfork S_{r,s}$ is satisfied by a residual set of mappings. A mapping for which this condition is satisfied for all r, s will be called *2-generic*.

Exercises

(1) Show that the condition $j^2 f \pitchfork S_{r,s}$ at x is a condition on the three jet of f at x.

(2) Let X and Y be 2-manifolds and $f\colon X \to Y$ a 2-generic map. Show that $\dim S_1(f) = 1$ and $\dim S_{1,1}(f) = 0$.

(3) Let X and Y be two manifolds and let $f\colon X \to Y$ be one-generic. Let x in X be a cusp point. Show that if x is *not* a simple cusp there exist coordinates (x_1, x_2) centered at x, coordinates (y_1, y_2) centered at $y = f(x)$, and a mapping of the form

$$(x_1, x_2) \mapsto (x_1, x_1\alpha(x_1, x_2)) \tag{4.10}$$

with the same 3-jet at 0 as f. Hint: Use the coordinates in the proof of Theorem 2.2. In these coordinates f is of the form $(x_1, x_2) \mapsto (x_1, h(x_1, x_2))$; and at a cusp point which is not simple

$$h = \frac{\partial h}{\partial x_2} = \frac{\partial^2 h}{\partial x_2{}^2} = \frac{\partial^3 h}{\partial x_2{}^3} = 0.$$

(4) Let X and Y be 2-manifolds and $f\colon X \to Y$ be a 2-generic mapping. Show that its only singularities are folds and simple cusps. (Hint: Show that for the map (4.10) $S_{1,1}$ is the whole x_2 axis. Now use Exercise 1 and Exercise 2.)

(5) Show that for 2-generic maps of n manifolds into n manifolds $S_{2,1}$ occurs for the first time in dimension 7 and $S_{2,2}$ occurs for the first time in dimension 10.

(6) Let A be an element of Hom $(V \circ V, W)_s$. Using (4.2) A is associated with a linear mapping $\tilde{A}$ of $V \to$ Hom (V, W) of corank s. Let $K = \ker \tilde{A}$ and let $L = \operatorname{coker} \tilde{A}$. Since $K \subset V$, there is a natural mapping

$$K^* \otimes V^* \to K^* \otimes K^* \tag{4.11}$$

given by restriction; i.e., $k^* \otimes v^* \mapsto k^* \otimes (v^*|K)$. Since Hom (V, W) is naturally identified with $V^* \otimes W$, we can regard L as a quotient space of $V^* \otimes W$ and obtain a natural map

$$K^* \otimes V^* \otimes W \to K^* \otimes L \tag{4.12}$$

given by $id \otimes \pi$ where π is the obvious projection of $V^* \otimes W \to L$. Let $K^* \circ V^*$ be the pre-image of $K^* \circ K^*$ with respect to (4.11), and let $K^* \circ L$ be the image of $(K^* \circ V^*) \otimes W$ with respect to (4.12). Finally let N_A be the normal space of the point A to $\mathrm{Hom}\,(V \circ V, W)_s$ in $\mathrm{Hom}\,(V \circ V, W)$. Show that there is a canonical identification

$$N_A \cong K^* \circ L.$$

(Hint: Use the same argument as in Exercise 6 of §1.)

(7) Let $f\colon X \to R^2$ be a 2-generic mapping where X is a compact 2-manifold. Show that $S_1(f)$ is a disjoint union of a finite number of circles and that $S_{1,1}(f)$ is a finite collection of points. Also show that no other types of singularities aside from fold points and simple cusp points occur.

(8) Let $f\colon X \to X$ be 2-generic. How large must dim X be to allow the existence of an S_3 singularity? Similarly for an $S_{3,1}$ singularity?

(9) Show that $S_{1,2}$ singularities occur generally when $\dim X = 5$ and $\dim Y = 3$.

(10) Show that for all $i \geq 1$ $\dim S_i \leq \dim Y$.

§5. The Thom-Boardman Stratification

It is clear, in principle anyway, how to define higher order versions of the $S_{i,j}$'s. If $f\colon X \to Y$ is 2-generic, $S_{i,j,k}(f)$ is defined to be the set of points in $S_{i,j}(f)$ where the map

$$f\colon S_{i,j}(f) \to Y$$

drops rank by k. This definition makes sense because, as we know from the previous section, $S_{i,j}(f)$ is a submanifold of X. If, by chance, $S_{i,j,k}(f)$ turns out to be a manifold, we can define $S_{i,j,k,l}(f)$ similarly. Thom conjectured that for a residual set of maps this process could be continued indefinitely. This conjecture was proved by Boardman in his I.H.E.S. paper [6]. Specifically what Boardman proved is the following.

Theorem 5.1. *For every sequence of integers $r_1, \ldots, r_k$ such that $r_1 +$* max (0, dim X − dim Y) $\geq r_2 \geq \cdots \geq r_k \geq 0$, *one can define a fiber subbundle, $S_{r_1,\ldots,r_k}$ of $J^k(X, Y)$ (relative to the fibration $J^k(X, Y) \to X \times Y$) such that if $j^l f$ is transversal to all manifolds $S_{t_1,\ldots,t_l}$ where $l < k$, then $S_{r_1,\ldots,r_k}(f)$ is well-defined and*

$$x \in S_{r_1,\ldots,r_k}(f) \Leftrightarrow j^k f(x) \in S_{r_1,\ldots,r_k}.$$

Boardman's proof of this depends on characterizing the $S_{r_1,\ldots,r_k}$'s by "Jacobean extensions" (A short description of Jacobean extensions can be found in Arnold's survey article [4].) An alternative proof was given by Michael Menn in [34]. His proof is based on the Grassmannian trick of propositions 1.1 and 4.3. (The enterprising reader might try, as an exercise, to construct the $S_{i,j,k}$'s by the Grassmannian trick, taking as his starting point Exercise 6 of the previous section.)

Remark. Though we won't attempt to prove the Boardman theorem here, we will prove in the next chapter that for maps $f\colon X \to Y$ between equidimensional manifolds the $S_{1,\ldots,1}$ singularities occur generically as codimension k submanifolds of X. These singularities are in some sense the most frequently encountered of the Thom-Boardman singularities. For example, they are the only singularities that occur in dimensions ≤ 6 except for $S_{2,0}$ ($S_{2,1}$ occurs for the first time in dimension 7.)

Assuming the Boardman Theorem we will call a mapping $f: X \to Y$ whose k jet extension $j^k f$ is, for all k, transversal to the $S_{r_1,\ldots,r_k}$'s a *Boardman* map. It is clear from the Thom transversality theorem that Boardman maps are a residual subset of $C^\infty(X, Y)$. In particular if a map is stable it is a Boardman map. We might ask whether the converse is true? It turns out the converse is false, even locally (for rather subtle reasons which we will go into in the next section.) However, it is easy to see why it can't be true globally: Consider the submersion with folds depicted in Figure 1 of III, §4. It is clear from this example that for a Boardman map $f: X \to Y$ to be stable the restriction

$$f: S_{r_1,\ldots,r_k}(f) \to Y$$

must have "normal crossing" properties similar to those described in Chapter III §§3–4. We will now make this condition precise.

Condition NC. Let $f: X \to Y$ be a Boardman map, and let $I_1, \ldots, I_s$ be multi-indices (not necessarily distinct). Let $x_1, \ldots, x_s$ be distinct points of X with x_j in $S_{I_j}(f)$ and

$$f(x_1) = \cdots = f(x_s) = y.$$

Let H_j be the tangent space to $S_{I_j}(f)$ at x_j. Then the subspaces

$$(df)_{x_1}H_1, \ldots, (df)_{x_s}H_s$$

are in general position in $T_y Y$.

The condition NC implies among other things that the maps

$$f: S_{r_1,\ldots,r_{k-1},0}(f) \to Y$$

are immersions with normal crossings and that the images of these immersions intersect transversally as immersed submanifolds.

We will now prove

Theorem 5.2. *The set of Boardman maps satisfying the condition NC is residual in $C^\infty(X, Y)$ (so, in particular, stable maps have to satisfy this condition).*

For the proof we will need:

Lemma 5.3. *Let X and Y be manifolds and let Z_1 and Z_2 be submanifolds of Y with $Z_2 \subset Z_1$. Let $f: X \to Y$ be transversal to Z_1, and let $X_1 = f^{-1}(Z_1)$. (Because of the transversality, this is a submanifold of X). Then $f: X \to Y$ is transversal to Z_2 iff $f: X_1 \to Z_1$ is transversal to Z_2.*

Lemma 5.4. *Let the diagram*

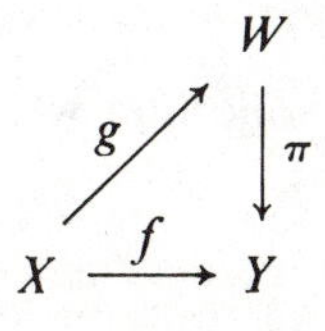

commute, and let π be a submersion. Then f is transversal to $Z \subset Y$ iff g is transversal to $\pi^{-1}(Z)$.

The proofs of both these lemmas are elementary, and are left to the reader.

With some future applications in view, we will formulate the NC condition a little more generally: Let $T_1, \ldots, T_s$ be submanifolds of $J^k(X, Y)$ (not necessarily distinct). Suppose each T_j has the property that the target map: $T_j \to Y$ is a submersion. By the transversality theorem there is a residual set of maps, $f: X \to Y$, such that $j^k f \pitchfork T_j$ for $j = 1, \ldots, s$. For the moment just consider such maps, and let $T_j(f) = (j^k f)^{-1}(T_j)$. (The transversality assumption assures that the $T_j(f)$'s are submanifolds of X.) Consider the following normal crossing condition, relative to $T_1, \ldots, T_s$:

Let $x_1, \ldots, x_s$ be distinct points of X with x_j in $T_j(f)$ and

$$f(x_1) = \cdots = f(x_s) = y.$$

Let H_j be the tangent space to $T_j(f)$ at x_j. Then

$$(df)_{x_1} H_1, \ldots, (df)_{x_s} H_s \tag{5.5}$$

are in general position in $T_y Y$.

We will prove:

Proposition 5.6. *The above normal crossing condition relative to the T_j's is satisfied for a residual set of maps $f: X \to Y$.*

(Note that Theorem 5.2 is a corollary of this proposition: just take the T_j's to be the S_{I_j}'s.)

Proof. The target map $\beta: T_1 \times \cdots \times T_s \to Y \times \cdots \times Y$ is a submersion, so the pre-image $\beta^{-1}(\Delta Y)$ where ΔY is the diagonal in $Y \times \cdots \times Y$ is a submanifold of $T_1 \times \cdots \times T_s$. By the multi-jet transversality theorem there exists a residual set of maps $f: X \to Y$ such that the multi-jet extension $j_s^k f$ is transversal to $\beta^{-1}(\Delta Y)$. It is easy to see that if the usual k jet extension $j^k f$ is transversal to T_j for $j = 1, \ldots, s$ then $j_s^k f$ is transversal to $T_1 \times \cdots \times T_s$ and vice-versa. Suppose both these transversality conditions are satisfied. The pre-image of $T_1 \times \cdots \times T_s$ with respect to $j_s^k f$ is just the set $T_1(f) \times \cdots \times T_s(f)$ with the generalized diagonal $X^{(s)}$ deleted. Call this set W. By Lemma 5.3 the map:

$$j_s^k f: W \to T_1 \times \cdots \times T_s$$

is transversal to $\beta^{-1}(\Delta Y)$; so, by Lemma 5.4, the map

$$f \times \cdots \times f: W \to Y \times \cdots \times Y$$

is transversal to ΔY. We showed, however, in Chapter III, §3 that this is the same as the normal crossing condition. $\square$

Let's now go back to the conjecture we made earlier (with the normal crossing condition added).

Conjecture. $f: X \to Y$ is stable $\Leftrightarrow$ f is a Boardman map satisfying NC.

We will see in the next section that this conjecture is false for maps between 9-manifolds. (In fact it is false for maps between n manifolds of dim > 3 though in dimensions 4–7 it is "nearly right".) Nevertheless, let's attempt to give a proof of it. If $f: X \to Y$ is a Boardman map we can partition X into a disjoint union of subsets consisting of the nonsingular points: $X - \bigcup S_i(f)$ and the Boardman sets $S_{r_1,\ldots,r_k}(f)$ with $r_k = 0$.

The map

$$f: X - \bigcup_{i \neq 0} S_i(f) \to Y$$

is an immersion or a submersion depending on whether dim $X \leq$ dim Y or dim $X \geq$ dim Y; and if $r_k = 0$ the map

$$f: S_{r_1,\ldots,r_k}(f) \to Y$$

is an immersion. This partition of X that we have just described is called the *Thom-Boardman stratification.* It has the property that f, restricted to each "stratum," is a particularly simple kind of stable map (either a submersion or an immersion with normal crossings). How do the various strata fit together, i.e., how do the closures of the higher dimensional strata intersect the lower dimensional strata? Obviously the story is quite complicated; but because of the transversality theorem, they might be expected to fit together in the same way that the universal strata $S_{r_1,\ldots,r_k}$ fit together in $J^k(X, Y)$. Hence, if we perturb f, the Thom-Boardman stratification of the perturbed map should look like the Thom-Boardman stratification of the unperturbed map. This suggests a way to prove the conjecture: construct an isotopy of X carrying the first stratification into the second, and then adjust it so that it conjugates the first mapping into the second.

This "proof" is unfortunately based on an erroneous assumption, namely that if we know the stratification and know that on each stratum f is either an immersion or submersion, then we have enough data at our disposal to describe f in the large. In fact, this data doesn't even describe the C^0 structure of the mapping; e.g., compare the two Morse functions $(x, y) \mapsto x^2 + y^2$ and $(x, y) \mapsto x^2 - y^2$.

One might suppose that if we know f, the Thom-Boardman data give us enough information to determine the structure of small perturbations of f; but even this isn't true as we will see in the next section.

Our "proof" does however have an intriguing air of plausibility about it, and we might ask whether some refined conjecture is true. It turns out that if we just restrict ourselves to the C^0 stability problem (two maps being equivalent if they can be conjugated, one into the other, by *homeomorphisms* of the source space and target space) then there is a finer stratification of the jet space than the Thom-Boardman stratification for which the "proof" above can be made rigorous. Thom is able to conclude from this that for all X and Y the C^0 stable maps form a residual subset of $C^\infty(X, Y)$. A careful proof of this can be found in a forthcoming book of John Mather [33]. See also [48]. In the next section we will see that the usual stable maps don't always form a residual subset of $C^\infty(X, Y)$ or even a dense subset.

§6. Stable Maps Are Not Dense

We have seen that immersions with normal crossings are stable; so stable maps are dense in $C^\infty(X, Y)$ if $\dim Y \geq 2 \dim X$. Stable maps are also dense in $C^\infty(X, Y)$ if $\dim Y = 1$ (Morse theory) and if $\dim X = \dim Y = 2$ (the Whitney theory sketched in §2). For a time it was conjectured that stable maps are always dense in $C^\infty(X, Y)$. Thom and Levine proved in [18] that this is not the case. In fact, they showed that for maps of 9 manifolds into 9 manifolds, stable maps are not dense. In this section we will give their demonstration of this fact. We will first prove:

Proposition 6.1. *Let X and Y be manifolds of dimension n^2. Then there exists a one-generic map $f: X \to Y$ such that $S_n(f)$ is nonempty.*

Proof. It is enough to prove that a map f exists taking on an S_n singularity transversely at a single point, say x_0, because we can always find a nearby mapping which is one-generic. If this mapping is close enough to f it must also take on an S_n singularity at a point close to x_0 by Exercise 1 of II, §4.

Let $N = \dim X = \dim Y$. With choices of coordinate on X and Y we can identify $C^\infty(X, Y)_{p,q}$ with $C^\infty(\mathbf{R}^N, \mathbf{R}^N)_{0,0}$; so to exhibit a map of X into Y taking on an S_n singularity transversely it is enough to construct a map germ $f: (\mathbf{R}^N, 0) \to (\mathbf{R}^N, 0)$ with this property. To do this we will need the following lemma.

Lemma 6.2 *Let V and W be vector spaces and let K be a subspace of V. If $\dim V \geq \dim K \cdot \dim W$, there exists an element A in $(V^* \circ K^*) \otimes W$ such that the map*

$$V \to K^* \otimes W \tag{6.3}$$

associated to A is onto. (For notation, see Exercise 6 of §4.)

Proof. The requirement that A be in $(V^* \circ K^*) \otimes W$ means simply that A, regarded as a map of $K \otimes K$ into W, is symmetric, otherwise A can be arbitrary. In particular, if H is a complement to K in V we can define A so that (6.3) is completely arbitrary on H. Since $\dim H \geq \dim (K^* \otimes W) - \dim K$ it is enough to show that there exists a B in $(K^* \circ K^*) \otimes W$ such that the map

$$K \to K^* \otimes W \tag{6.3$'$}$$

associated to B is injective. Let W_1 be a one dimensional subspace of W. One can already construct a B in $(K^* \circ K^*) \otimes W$ such that the induced map $K \to K^* \otimes W_1$ is injective. (Just take a nondegenerate bilinear form.) □

To prove Proposition 6.1, choose f as in Exercise 1 of §3 and identify A with the last two terms of expression (3.8). If A is chosen as in the lemma, then $j^1 f \pitchfork S_n$ at 0 by Proposition 3.7. □

We will now prove

Proposition 6.4. *Let X and Y be manifolds of dimension n^2. Let $f: X \to Y$ be one-generic. If $S_n(f)$ is nonempty and $n > 2$ then f is not stable.*

Combined with Proposition 6.1 this shows that stable maps are not dense in $C^\infty(X, Y)$ when $\dim X = \dim Y = n^2$, $n > 2$.

For the proof of Proposition 6.4 we will need

Lemma 6.5. *Let X and Y be manifolds and W an open subset of $C^\infty(X, Y)$. Then the set $A_W = \{\sigma \in J^k(X, Y) \mid \exists f \in W$ and $x \in X$ with $\sigma = j^k f(x)\}$ is open in $J^k(X, Y)$.*

Proof. Let $\sigma = j^k f(x)$ be in A_W. Choose coordinate nbhds U and V about x and $f(x)$ such that $f(\bar{U}) \subset V$ and let ρ be a function which is 1 near x and has support in U. Consider the set of maps g defined by

$$g = \begin{cases} f + \rho h & \text{in } U \\ f & \text{in } X - U \end{cases}$$

where h is a polynomial map of degree $\leq k$. If the coefficients of h are sufficiently small this is well-defined (i.e., $g(\bar{U}) \subset V$) and is in W. The set $j^k g(y)$ for y near x and h with coefficients small define an open nbhd of $J^k(X, Y)$ in A_W. □

Now let us go back to the diagram (4.1) of §4.

$$\begin{array}{ccc} S_r^{(2)} & \longrightarrow & \mathrm{Hom}\,(K \circ K, L) \\ & \searrow \qquad \swarrow & \\ & S_r & \end{array} \tag{6.6}$$

The groups Diff(X) and Diff(Y) act on S_r and on $S_r^{(2)}$ in obvious ways as subsets of $J^1(X, Y)$ and $J^2(X, Y)$. They also act on the bundles K and L in the following simple way. If σ in S_r has source at x and target at y we can think of σ as an element of $\mathrm{Hom}\,(T_x X, T_y Y)$. Then by definition K_σ = kernel σ and L_σ = cokernel σ. If g is in Diff(Y) then g acts on σ by sending it to $\sigma' = (dg)_y \cdot \sigma$; so $(dg)_y$ maps the cokernel of σ onto the cokernel of σ' and leaves the kernels fixed. Thus Diff(Y) acts on K trivially and on L by the tangent bundle action. Diff(X) acts similarly; i.e., by the tangent bundle action on K and trivially on L.

Lemma 6.7. *The mappings in the diagram* 6.6 *commute with the respective actions of* Diff(X) *and* Diff(Y).

A proof is sketched in the exercises.

We will now prove Proposition 6.4. Let $f: X \to Y$ be one-generic. By the dimension formula in Proposition 1.1 $\dim S_n(f) = 0$, so $S_n(f)$ consists of a countable number of isolated points, say $x_0, x_1, x_2, \ldots$. Let $\alpha_0 = j^2 f(x_0)$, $\alpha_1 = j^2 f(x_1)$, etc.

Now suppose f is stable. Then the orbit, W_f, of f in $C^\infty(X, Y)$ is open; so by Lemma 6.5, the set

$$A_{W_f} = \{\sigma \in J^k(X, Y) \mid \exists g \in W_f \text{ and } x \in X \text{ with } \sigma = j^k g(x)\} \tag{6.8}$$

is open in $J^2(X, Y)$. If α is in the set A_{W_f} and is also in $S_n^{(2)}$ then it must be conjugate to one of the α_i's. This means that the open set $A_{W_r} \cap S_n^{(2)}$ in $S_n^{(2)}$ is covered by a countable number of orbits of the action of the group $\text{Diff}(X) \times \text{Diff}(Y)$. Let $\sigma = j^1f(x_0)$. Since the top arrow in (6.6) is surjective, this means there is an open set in $\text{Hom}(K_\sigma \circ K_\sigma, L_\sigma)$ covered by a countable number of orbits of the group $GL(K_\sigma) \times GL(L_\sigma)$. Suppose all these orbits were of dimension less than the dimension of $\text{Hom}(K_\sigma \circ K_\sigma, L_\sigma)$. Since they are immersed submanifolds they would have to be of measure zero by II, Lemma 1.5; and so would their union. We conclude that one of the orbits has to be open. We will show this is impossible by showing:

Lemma 6.9. *Let V and W be vector spaces of dimension n. Let G be the group $GL(V) \times GL(W)$ acting on the space* $\text{Hom}(V \circ V, W)$ *by its standard representation. (The action is given as follows: let C be in* $\text{Hom}(V \circ V, W)$ *and let (A, B) be in G, then $(A, B)\cdot C(\bar{v}) = C\cdot B\cdot A^{-1}(\bar{v})$ for $\bar{v}$ in $V \circ V$. Note that $A(v_1 \otimes v_2) = Av_1 \otimes Av_2$ and $V \circ V$ is an invariant subspace of $V \otimes V$ under this representation.) Then G has an open orbit only when $n < 3$.*

Proof. The dimension of $\text{Hom}(V \circ V, W)$ is $n^2(n + 1)/2$ and the dimension of G is $2n^2$. Note that an orbit of G is diffeomorphic to some homogeneous space G/H. So the dim (orbit) $= \dim G - \dim H \leq \dim G$. (See Theorem A.13).

Clearly $n^2(n + 1)/2 \leq 2n^2$ only when $n \leq 3$; so if $n > 3$, the dimension of G is less than the dimension of $\text{Hom}(V \circ V, W)$, and the assertion is trivial. When $n = 3$ the dimensions are equal; but G contains a one-dimensional subgroup which acts trivially on $\text{Hom}(V \circ V, W)$ namely $\{(c\, id_V, c^2\, id_W)\}$, c a nonzero real number; so the orbits are at least one dimension less than dim G. □

This concludes the proof of Proposition 6.4.

One can prove a slightly stronger result by the same technique; namely, one can show that stable maps are not dense in $C^\infty(X, Y)$ when $\dim X = \dim Y \geq 9$. John Mather has recently proved a much more striking result using his theory of stable mappings. He has shown that if the two-tuple ($\dim X$, $\dim Y$) occurs inside the region A of Figure 4 or on its boundary then stable maps are not dense in $C^\infty(X, Y)$; if it occurs outside this region they are. For the proof of this result see [31].

Let X be a compact n manifold and $f\colon X \to \mathbf{R}^n$ a one-generic mapping. Let Z be a submanifold of X of dimension $=$ codim $S_i(f) = i^2$, such that $S_i(f) \pitchfork Z$. One can show that the number of points in the intersection $S_i(f) \cap Z$ is, modulo 2, a topological invariant of the pair (X, Z). (In fact this invariant doesn't even depend on f.) It can be computed using the theory of characteristic classes. (See [46] and [13]).

Now one can considerably weaken the hypotheses of Proposition 6.4. One can show that if X and Y are n manifolds, then in the dimension range, $i^2 \leq n \leq i^2(i - 1)/2$, a stable map, $f\colon X \to Y$ cannot take on an S_i singularity. This means that if in this dimension range one can find a pair of manifolds (X, Z) for which the topological invariant described above is nonzero there will exist *no stable maps* of X into $\mathbf{R}^n$. (A case when this happens is $X = \mathbf{P}^{19}$ and $Z = \mathbf{P}^{16}$, viewed as a subspace of $\mathbf{P}^{19}$. Therefore, there are no stable maps of $\mathbf{P}^{19}$ into $\mathbf{R}^{19}$!)

In fact, the stable maps of $X^n \to Y^m$ are dense iff the pair (n, m) satisfies any of the following where $q = m - n$

(a) $m < 7q + 8$ when $q \geq 4$
(b) $m < 7q + 9$ when $3 \geq q \geq 0$
(c) $m < 8$ when $q = -1$
(d) $m < 6$ when $q = -2$
(e) $m < 7$ when $q \leq -3$

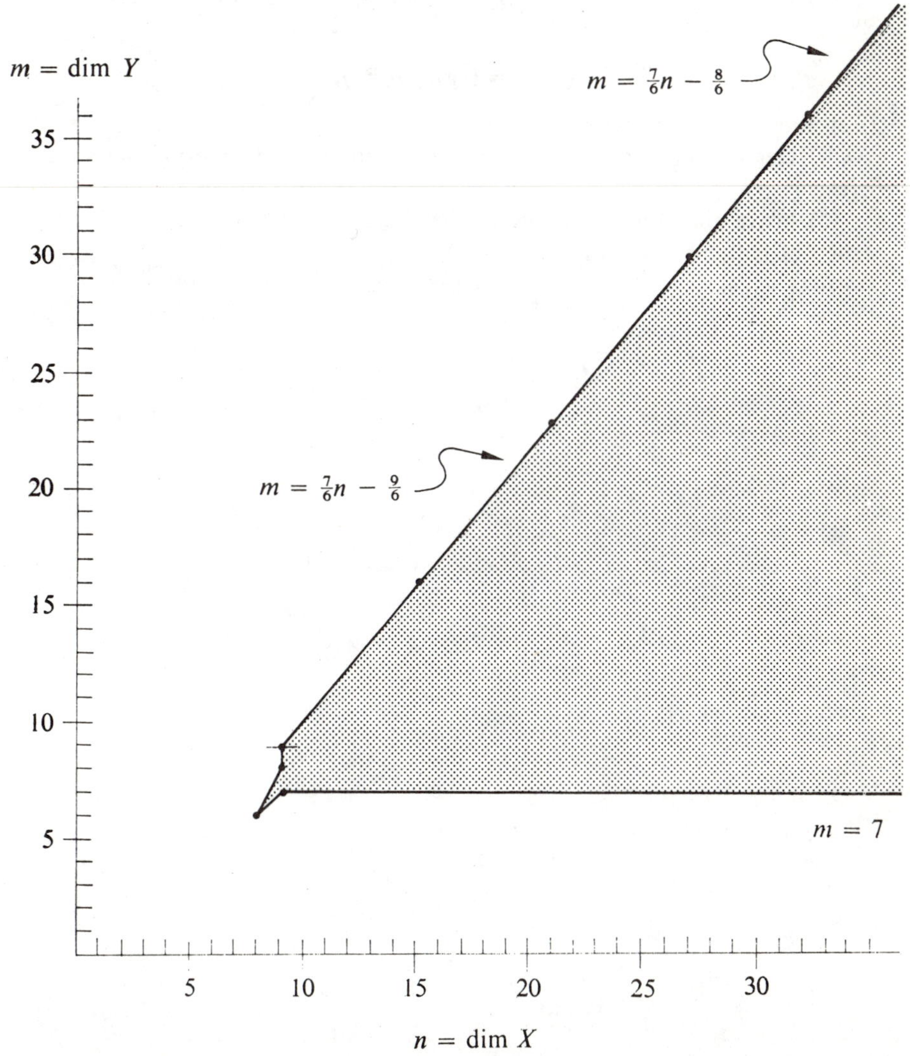

Figure 4: The region where stable maps are *not* dense is the shaded region including the boundary.

Exercises

(1) Let E, F, and G be vector bundles. Let $\rho: E \to F$ be a vector bundle morphism, and let $\tau: F \to G$ be a vector bundle isomorphism. Prove the *chain rule formula* for the intrinsic derivative:

$$D(\tau \cdot \rho)_x = \hat{\tau} \cdot (D\rho)_x$$

where $\hat{\tau}$: Hom (ker ρ, coker ρ) $\to$ Hom (ker ρ, coker $\tau\rho$) is the obvious map induced by τ.

(2) Determine the corresponding formula for a vector bundle isomorphism acting on the left.

(3) Let $f: X \to Y$ be a smooth map, and $g: Y \to Y$ a diffeomorphism. Show:

$$\delta^2(g \cdot f) = (dg)_{f(x)} \cdot (\delta^2 f)_x.$$

(For notation, see Exercise 3 in §3.)

(4) Derive a similar formula for composition by a diffeomorphism on the left.

(5) Prove that the mappings in the diagram (6.6) commute with the action of Diff(X) $\times$ Diff(Y).

Chapter VII

Classification of Singularities
Part II: The Local Ring of a Singularity

§1. Introduction

The Thom-Boardman theory gives us a way of breaking up a map into simple constituent pieces; however, from the Thom-Boardman data alone we usually cannot reassemble the constituent pieces and see what the map itself looks like. Consider for example the maps

$$f: \mathbf{R}^2 \to \mathbf{R}, \qquad (x_1, x_2) \mapsto x_1^2 + x_2^2$$

and

$$g: \mathbf{R}^2 \to \mathbf{R}, \qquad (x_1, x_2) \mapsto x_1^2 - x_2^2.$$

f and g have isolated S_1 singularities at the origin and are regular everywhere else. However, their map germs at the origin are not equivalent, even under homeomorphisms of $\mathbf{R}^2$ and $\mathbf{R}$, since f has an extremum at 0 and g does not. From the Thom-Boardman data alone there is no way of computing the Hessian of f at 0; and, of course, it is the signature of the Hessian which distinguishes f from g. (See II, Theorem 6.9.)

In this chapter, we will be concerned with a more subtle invariant of a singularity—its *local ring*. To define this we recall some notation from Chapter IV, §3. If x is a point in a manifold X then $C_x^\infty = C_x^\infty(X)$ denotes the ring of germs of smooth functions at x. This is a local ring and its maximal ideal, $\mathcal{M}_x = \mathcal{M}_x(X)$, is the ideal of germs of functions vanishing at x. A map germ $f: (X, x) \to (Y, y)$ induces a map $f^*: C_y^\infty \to C_x^\infty$ by pull-back; and this is a morphism of local rings.

Definition 1.1. *Let $f: (X, x) \to (Y, y)$ be a map germ. The* local ring *of f is the quotient ring: $C_x^\infty / C_x^\infty f^* \mathcal{M}_y$.*

This ring will be denoted $\mathcal{R}_f$ and its maximal ideal m_f. If $f: X \to Y$ is a map, not just a map germ, then at each point x in X we get the local ring of the germ *of f which we will denote $\mathcal{R}_f(x)$ and call the* local ring of f at x.

Example 1 (Immersions). If $f: (X, x) \to (Y, y)$ is the germ of an immersion, we can choose coordinates centered at x and y so that f is the immersion

$$f^*y_i = x_i \qquad i = 1, \ldots, n$$
$$f^*y_i = 0 \qquad i = n + 1, \ldots, m$$

where $n = \dim X$ and $m = \dim Y$. Then $\mathcal{M}_x = f^*\mathcal{M}_y$ and $\mathcal{R}_f = \mathbf{R}$.

Example 2 (Submersions). If $f: (X, x) \to (Y, y)$ is the germ of a submersion we can choose coordinates so that f is the canonical submersion

$$f^*y_i = x_i \qquad i = i, \ldots, m.$$

Then $C_x^\infty f^*\mathcal{M}_y$ is the ideal generated by $(x_1, \ldots, x_m)$ and $\mathcal{R}_f$ is the ring of germs of smooth functions in the remaining variables $x_{m+1}, \ldots, x_n$.

Example 3 (Morse functions). If $f: X \to \mathbf{R}$ is a Morse function with critical point at x, the product operation in the local ring $\mathcal{R}_f$ induces a linear map

(1.2) $$\mathfrak{m}_f/\mathfrak{m}_f^2 \otimes \mathfrak{m}_f/\mathfrak{m}_f^2 \mapsto \mathfrak{m}_f^2/\mathfrak{m}_f^3.$$

The kernel of the map (1.2) turns out to be one-dimensional, and is, in fact, all scalar multiples of the Hessian. (See Exercise 5 below.) Therefore up to scalar multiple the Hessian can be computed from $\mathcal{R}_f$.

Example 4 (Generic maps between 2 manifolds). At a fold, we can write such a map in the form $(x_1, x_2) \mapsto (x_1, x_2^2)$. If we divide C_0^∞ by the ideal (x_1, x_2^2) we get the truncated polynomial ring $\mathbf{R}[x_2]/(x_2^2)$, so $\mathcal{R}_f \cong \mathbf{R}[t]/(t^2)$.

At a simple cusp we can choose coordinates so that the mapping is $(x_1, x_2) \mapsto (x_1, x_1x_2 + x_2^3)$. If we divide C_0^∞ by the ideal $(x_1, x_1x_2 + x_2^3)$ (which is just the ideal (x_1, x_2^3)), we get the truncated polynomial ring $\mathbf{R}[x_2]/(x_2^3)$. Hence $\mathcal{R}_f \cong \mathbf{R}[t]/(t^3)$.

We recommend that you try a few other examples on your own. The ring $\mathcal{R}_f$ has been around a long time in algebraic geometry. For example, if $f: X \to Y$ is a morphism between schemes†, $\mathcal{R}_f$ is in a natural sense the "fiber" of f at x. From the algebraic geometer's point of view, $\mathcal{R}_f(x)$ is a much more natural invariant to attach to a singularity than, say, its Boardman data. Its use in differential geometry is fairly recent, however; and is mainly due to Malgrange, Mather, and Tougeron. Its importance is indicated by the following theorem of Mather. (See [29].)

Theorem 1.3. *Let $f, g: (X, x) \to (Y, y)$ be germs of stable maps. Then f and g are equivalent if and only if* $\mathcal{R}_f$ and $\mathcal{R}_g$ *are isomorphic as rings. Note that f and g are* equivalent *if there exist germs of diffeomorphisms $h: (X, x) \to (X, x)$ and $k: (Y, y) \to (Y, y)$ such that $g = k \cdot f \cdot h^{-1}$ near x.*

The proof of this theorem is beyond the scope of this book. We will, however, see it corroborated by the simple examples we are going to discuss in the following sections.

Exercises

(1) Let $f: (X, x) \to (Y, y)$ be map germs. Then f induces a map $f^*: C_y^\infty/\mathcal{M}_y^{k+1} \to C_x^\infty/\mathcal{M}_x^{k+1}$ for each k. Let $g: (X, x) \to (Y, y)$ be another map germ. Show that $j^kf(x) = j^kg(x)$ iff $f^* = g^*$. Thus f^* is the k-jet of f at x in algebraic disguise. (Hint: Let $(y_1, \ldots, y_m)$ be coordinates centered at y and $\bar{y}_1, \ldots, \bar{y}_m$ be the associated elements of the quotient ring. Show that $f^*\bar{y}_i$ is the k-jet of the ith coordinate function of f.)

† Whatever that means.

(2) Let $\gamma: C_y^\infty/\mathcal{M}_y^{k+1} \to C_x^\infty/\mathcal{M}_x^{k+1}$ be a ring homomorphism. Show that $\gamma = f^*$ for some map germ f.

(3) From Exercises 1 and 2, deduce an isomorphism (of sets)

$$J^k(X, Y)_{x,y} \cong \mathrm{Hom}\,(C_y^\infty/\mathcal{M}_y^{k+1}, C_x^\infty/\mathcal{M}_x^{k+1})$$

(Hom meaning homomorphisms of rings).

(4) Identify $\mathcal{M}_y/\mathcal{M}_y^2$ with the cotangent space of Y at y. Compare with IV, Lemma 3.3. Show that $f^*: \mathcal{M}_y/\mathcal{M}_y^2 \to \mathcal{M}_x/\mathcal{M}_x^2$ is just the transpose of $(df)_x: T_xX \to T_yY$.

(5) Let $f: \mathbf{R}^n \to \mathbf{R}$ be the Morse function given by $(x_1, \ldots, x_n) \mapsto x_1^2 \pm x_2^2 \pm \cdots \pm x_n^2$. Show that at the origin m_f/m_f^2 is just the space of linear functions in $x_1, \ldots, x_n$; and m_f^2/m_f^3 is the space of quadratic functions with $x_1^2 \pm x_2^2 \pm \cdots \pm x_n^2$ identified to zero. Conclude that the kernel of the map (1.2) is the one-dimensional space spanned by $x_1 \otimes x_1 \pm \cdots \pm x_n \otimes x_n$.

(6) Verify that nonconjugate map germs can have the same local ring. (Hint: Try $(x_1, x_2) \mapsto (x_1, x_1x_2 + x_2^3)$ and $(x_1, x_2) \mapsto (x_1, x_2^3)$.) Why doesn't this contradict Mather's theorem?

(7) Prove that the dimension of $m_f(x)/m_f^2(x)$ is equal to the dimension of $\mathrm{Ker}\,(df)_x: T_xX \to T_{f(x)}Y$. (Hint: Use Exercise 4.)

§2. Finite Mappings

Let $f: X \to Y$ be a smooth map. We say that f is *finite at x* if:

(2.1) $$\dim_{\mathbf{R}} \mathcal{R}_f(x) < \infty.$$

f is *finite* if it is finite at every point. Note that if $\dim X > \dim Y$, then $\dim_{\mathbf{R}} \mathcal{R}_f(x) = \infty$, (i.e., even *formally* the functions $f_1, \ldots, f_n$ cannot generate an ideal of finite codimension in the formal power series ring $\mathbf{R}[[x_1, \ldots, x_m]]$ if $m > n$. See [61].) Therefore, in talking about finite maps, we are implicitly assuming $\dim X \leq \dim Y$. "Finiteness" implies among other things that the map f is locally "finite-to-one" at x. In fact, we have

Proposition 2.2. *If f is finite at x, and $a = f(x)$ then there exists an open nbhd U of x such that x is the only point in U mapping onto a.* (*In particular, if f is finite, it is "finite-to-one" on compact subsets of X.*)

Proof. We can assume that X and Y are $\mathbf{R}^n$ and $\mathbf{R}^m$ respectively and that $x = a = 0$. Let $f_1, \ldots, f_m$ be the coordinate functions of f. The assumption (2.1) means that some power of the maximal ideal in C_0^∞ is contained in the ideal $(f_1, \ldots, f_m)$. Therefore, there exists an open set U and an integer N such that on U, $x_i^N = \sum_{j=1}^m a_{ij}f_j$, the a_{ij}'s being smooth functions on U. Therefore, if the f_i's vanish on U so do the coordinate functions x_i. In other words, zero is the only pre-image of zero. □

Remark. The converse is not true. The map $f: \mathbf{R} \to \mathbf{R}$, given by $t \to \exp(-1/t^2)$ is "finite-to-one" but $\dim \mathcal{R}_f(0) = \infty$.

Suppose f is finite at x. Let $p_1, \ldots, p_k$ be elements of C_x^∞ projecting onto a basis of $\mathscr{R}_f(x)$ over **R**. Then the Malgrange Preparation Theorem says that $p_1, \ldots, p_k$ generate C_x^∞ as a module over C_y^∞. Conversely, if $p_1, \ldots, p_k$ generate C_x^∞ as a module over C_y^∞ it is clear that they project onto a spanning set of vectors for $\mathscr{R}_f(x)$ viewed as a vector space over **R**. We will use this to prove

Proposition 2.3. $\dim_{\mathbf{R}} \mathscr{R}_f(x)$ *is an upper semi-continuous function of* x. *In particular, if* f *is finite at* x, *it is finite near* x.

Proof. Let $x_1, \ldots, x_n$ be a coordinate system centered at x, and $y_1, \ldots, y_m$ a coordinate system centered at the image point. We can assume that the functions $p_1, \ldots, p_k$ above are polynomials in x_i's of degree $< N$. We will show that there exists a fixed open set, U, about x such that on U every polynomial in the x_i's can be written as a linear combination of the p_i's with smooth functions of the y_i's as coefficients. First note that this statement is true for some U and for all polynomials of degree $\leq N$. (Note that for any monomial we can find an open set U since the statement is true for germs at x. Thus we can find an open set U which works for all monomials of degree $\leq N$. By linearity U works for all polynomials of degree $\leq N$). Now consider a polynomial of the form $x^\alpha p$ where $\deg p = N$. $x^\alpha p$ can be written as a linear combination of the $x^\alpha p_i$'s and, hence, by induction as a linear combination of the p^i's themselves. Thus U works for all polynomials of degree $> N$ as well.

This proves that for all x' in U, $C_{x'}^\infty$ is formally generated by the p_i's as a module over $C_{y'}^\infty$. Therefore, by the Malgrange Preparation Theorem, it is actually generated by them. $\square$

In some sense, $\dim_{\mathbf{R}} \mathscr{R}_f(x)$ measures the *multiplicity of the point* x as a root of the equation $f(x) = a$. Over the complex numbers this vague statement can be made precise (see Remark 1 below), but over the reals we have to content ourselves with:

Proposition 2.4. *Let* $\dim_{\mathbf{R}} \mathscr{R}_f(x) = k$. *Then there exists a neighborhood,* U, *of* x *such that every* y *sufficiently close to* $f(x)$ *has at most* k *pre-image points in* U.

Proof. Choose U and $p_1, \ldots, p_k$ as in the proof of Proposition 2.3. Let $x_1, \ldots, x_r$ be pre-image points of y in U, and let $S = \{x_1, \ldots, x_r\}$. Let $C_S^\infty = \bigoplus_{i=1}^r C_{x_i}^\infty$ be germs of C^∞ functions on S, and let $\mathscr{R}_f(S)$ be the quotient ring $C_S^\infty / C_S^\infty f^* \mathscr{M}_y$. C_S^∞ is a finitely generated module over C_y^∞ with generators $p_1, \ldots, p_k$ (by the same reasoning as in the proof of Proposition 2.3), so $\mathscr{R}_f(S)$ is a finite dimensional vector space over **R** with the images of $p_1, \ldots, p_k$ as spanning vectors. On the other hand, the restriction map $\mathscr{R}_f(S) \mapsto \sum_{i=1}^r \mathscr{R}_f(x_i)$ is bijective; so $r \leq \dim_{\mathbf{R}} \mathscr{R}_f(S)$. $\square$

Remark 1. If X and Y are complex manifolds of the same dimension, and $f: X \to Y$ a holomorphic mapping satisfying (2.1) then a much sharper form of Proposition 2.4 is true. Namely, there exists an open nbhd U of x such that for every y close to $f(x)$ the sum

$$(2.5) \qquad \sum \dim_{\mathbf{C}} \mathscr{R}_f(z_i)$$

over the pre-image points of y contained in U is constant. (See [38].)

Remark 2. In the complex analytic category the converse of Proposition 2.4 is true; namely, if $f: X \to Y$ is locally "finite-to-one" then $\dim_{\mathbf{C}} \mathscr{R}_f(x) < \infty$. (See [14].)

We conclude this section by mentioning a result of Tougeron:

Theorem 2.6. *If* $\dim X \leq \dim Y$ *the finite maps are a residual subset of* $C^\infty(X, Y)$.

In fact, Tougeron proves they are much larger than just a residual set. The complement is, in a sense which we won't try to make precise, of "infinite codimension" in $C^\infty(X, Y)$. (See [51].)

Exercises

(1) Consider the Whitney map given by $f: (x_1, x_2) \to (x_1, x_1x_2 + \frac{1}{3}x_2^3)$ of $\mathbf{R}^2$ into $\mathbf{R}^2$. Given a in $\mathbf{R}^2$ what values can the sum

$$(2.7) \qquad \sum_{p \in f^{-1}(a)} \dim_{\mathbf{R}} \mathscr{R}_f(p),$$

take on?

(2) Consider the analogous problem for the complex Whitney map $f: (x_1, x_2) \to (x_1, x_1x_2 + \frac{1}{3}x_2^3)$ of $\mathbf{C}^2$ into $\mathbf{C}^2$.

(3) Call $\dim_{\mathbf{R}} \mathscr{R}_f(x)$ the *multiplicity* of the point x (with respect to the mapping $f: X \to Y$). Prove Proposition 2.4, counting pre-image points *with multiplicity*. (Hint: Confirm that we did prove this stronger result in the text.)

(4) Prove Tougeron's Theorem 2.6 for maps between two-manifolds using the Whitney theory.

(5) Let X and Y be manifolds of dimension n and let $f: (X, x) \to (Y, y)$ be a map exhibiting an S_i singularity. Show that $\dim_{\mathbf{R}} \mathscr{R}_f > i(i + 1)/2$. (Hint: Choose coordinates $x_1, \ldots, x_n$ centered at x and $y_1, \ldots, y_n$ centered at y such that f has the form

$$\begin{aligned} f^*y_j &= f_j \qquad j \leq i \\ f^*y_j &= x_j \qquad j > i \end{aligned}$$

where the leading term of f_j is quadratic. Show that $\mathscr{m}_f^2/\mathscr{m}_f^3$ is the space of all quadratic polynomials in $x_1, \ldots, x_i$ with the quadratic terms of $f_1, \ldots, f_i$ identified to zero.)

§3. Contact Classes and Morin Singularities

We shall now attempt to give some insight into the geometric content of the local ring $\mathscr{R}_f(p)$ defined in §2. This will be accomplished through the notion of contact equivalence.

Let Z be a manifold with p in Z. There is an obvious way to define a *submanifold germ of Z near p*; namely, two submanifolds are equivalent if they are identical in a nbhd of p and a submanifold germ is one of these equivalence classes. (Clearly we may think of a submanifold germ as a small piece of a submanifold.)

Definition 3.1. *Let A, B_1, and B_2 be equidimensional submanifold germs of Z near p. Then B_1 and B_2* have the same contact with *A if there exists a germ of a diffeomorphism $\phi : (Z, p) \to (Z, p)$ such that $\phi|A = id_A$ and $\phi(B_1) = B_2$.*

Next we define the local ring associated with the contact of two manifold germs A and B of Z near p. Define

$$\mathscr{I}_p(B) = \{g \in C_p^\infty(Z) \mid g(B) = 0\}.$$

Let $i_A : A \to Z$ be the canonical inclusion map. Define $\mathscr{I}_p(A, B)$ to be the ideal $i_A^*(\mathscr{I}_p(B))$ where $i_A^* : C_p^\infty(Z) \to C_p^\infty(A)$ is the ring homomorphism induced by i_A. Finally, let

$$\mathscr{R}_{A,B} = C_p^\infty(A)/\mathscr{I}_p(A, B).$$

It is clear that $\mathscr{R}_{A,B}$ is an invariant of the contact of B with A.

Definition 3.2 *$\mathscr{R}_{A,B}$ is called the* local ring of the contact of B with A.

Theorem 3.3 *Let A, B_1, and B_2 be equidimensional submanifold germs of Z near p. Then B_1 and B_2 have the same contact with A iff $\mathscr{R}_{A,B_1} = \mathscr{R}_{A,B_2}$.*

First we present three lemmas.

Lemma 3.4. *There exists a trivial tubular nbhd U of A in Z such that both B_1 and B_2 intersect the fiber of U at p transversely.*

Proof. We can choose a tubular nbhd U of A in Z which is a trivial vector bundle since A is a submanifold germ. Thus by choosing coordinates we may assume that $U = \mathbf{R}^k \times \mathbf{R}^l$ where $k = \dim A$, $l = \operatorname{codim} A$, A is identified with $\mathbf{R}^k \times \{0\}$, and p is identified with 0. In this local situation it is clear that we can rotate the $\mathbf{R}^l$ factor of U so that both B_1 and B_2 intersect $\{0\} \times \mathbf{R}^l$ transversely. □

Lemma 3.5. *Let G and H be linear maps of $\mathbf{R}^l \to \mathbf{R}^l$. Then there exists a linear map $F: \mathbf{R}^l \to \mathbf{R}^l$ such that $K = F(I_l - GH) + H$ is invertible.*

Proof. Choose subspaces V and W of $\mathbf{R}^l$ so that $V \oplus \operatorname{Ker} H = \mathbf{R}^l$ and $H(V) \oplus W = \mathbf{R}^l$. Choose F so that $F(V) = 0$ and $F: \operatorname{Ker} H \to W$ is an isomorphism. It is now easy to check that with these choices, $\operatorname{Ker} K = 0$. □

Lemma 3.6. *Let* $b: (\mathbf{R}^k, 0) \to (\mathbf{R}^l, 0)$ *be a map germ where* $b(x) = (b_1(x), \ldots, b_l(x))$. (*We assume that the coordinates on* $\mathbf{R}^k$ *are* x *and the coordinates on* $\mathbf{R}^l$ *are* y.) *Then* $\mathscr{I}_{(0,0)}(\mathbf{R}^k \times \{0\}, \textit{graph } b) = (b_1, \ldots, b_l)$ *where* $\mathbf{R}^k \times \{0\}$ *and graph* b *are submanifold germs of* $\mathbf{R}^k \times \mathbf{R}^l$ *near* $(0, 0)$.

Proof. The functions $b_i(x) - y_i$ clearly vanish on graph b_i in $\mathbf{R}^k \times \mathbf{R}^l$. Also $i^*_{R^k \times \{0\}}(b_i(x) - y_i) = b_i(x)$. So $\mathscr{I}_{(0,0)}(\mathbf{R}^k \times \{0\}, \text{graph } b) \supset (b_1, \ldots, b_l)$. Conversely, suppose that $f: \mathbf{R}^k \times \mathbf{R}^l \to \mathbf{R}$ vanishes on graph b. Then we may write $f(x, y) = (b_1(x) - y_1)f_1(x, y) + \cdots + (b_l(x) - y_l)f_l(x, y)$ where each f_i is a smooth function. (To see this, let $g(t) = f(x, (1 - t)y + tb(x))$ for t in $\mathbf{R}$. Then $f(x, y) = g(0) - g(1) = \int_1^0 dg/dt(t)\, dt$. Expanding dg/dt by the chain rule gives the desired result.) Thus $(i^*_{R^k \times \{0\}}f)(x) = b_1(x)f_1(x, 0) + \cdots + b_l(x)f_l(x, 0)$ and is in the ideal $(b_1, \ldots, b_l)$. □

Proof of Theorem 3.3. As noted above, it is easy to see that contact equivalent submanifold germs give rise to identical local rings. So we assume that $\mathscr{R}_{A,B_1} = \mathscr{R}_{A,B_2}$. Choose a tubular nbhd $U = \mathbf{R}^k \times \mathbf{R}^l$ of $A = \mathbf{R}^k \times \{0\}$ as in Lemma 3.4. The transversality assumptions imply that we can find smooth maps b^1 and $b^2 : \mathbf{R}^k \to \mathbf{R}^l$ so that $B_i = \text{graph } (b^i)$ (near p, of course). Let $b_1{}^i, \ldots, b_l{}^i$ be the coordinate functions of b^i. Then $\mathscr{I}_p(A, B_i) = (b_1{}^i, \ldots, b_l{}^i)$ by Lemma 3.6.

Now the assumption that $\mathscr{R}_{A,B_1} = \mathscr{R}_{A,B_2}$ implies that $\mathscr{I}_p(A, B_1) = \mathscr{I}_p(A, B_2)$. Thus the calculation that $\mathscr{I}_p(A, B_i) = (b_1{}^i, \ldots, b_l{}^i)$ implies that there exist smooth functions $g_{\alpha\beta}$ and $h_{\beta\gamma}$ where $1 \le \alpha, \beta, \gamma \le l$ so that

$$(*) \qquad b_\alpha{}^1 = \sum_{\beta=1}^{l} g_{\alpha\beta} b_\beta{}^2 \quad \text{and} \quad b_\beta{}^2 = \sum_{\gamma=1}^{l} h_{\beta\gamma} b_\gamma{}^1.$$

Let G and H denote the matrices $(g_{\alpha\beta})$ and $(h_{\beta\gamma})$. We claim that we may choose the $h_{\beta\gamma}$'s so that $H(x)$ is invertible for all x near 0. Using Lemma 3.5, choose a linear map F such that $K = F(I - GH) + H$ is invertible at zero and hence for all x near 0. A simple computation shows that $b_\beta{}^2 = \sum_{\gamma=1}^{l} k_{\beta\gamma} b_\gamma{}^1$ so we may replace H by K. Now define $\phi: U = \mathbf{R}^k \times \mathbf{R}^l \to U$ by $\phi(x, y) = (x, H(x)y)$. By $(*)$, $\phi: B_1 \to B_2$. Since ϕ is linear on fibers of U, $\phi | A = \phi | \mathbf{R}^k \times \{0\} = id_{\mathbf{R}^k \times \{0\}}$. Finally, ϕ is a diffeomorphism on a nbhd of p since $H(0)$ is invertible. The mapping ϕ shows that B_1 has the same contact with A as B_2. □

We now specialize this construction of contact equivalence to obtain results on $\mathscr{R}_f(p)$ where $f: (X, p) \to (Y, q)$.

Lemma 3.7. $\mathscr{R}_f(p) = (i)^*\mathscr{R}_{X \times \{q\}, \textit{graph } f}$ *where* $i: X \to X \times \{q\}$ *is the canonical map and the ambient manifold* Z *is taken to be* $X \times Y$.

Proof. Choose coordinates $x_1, \ldots, x_n$ near p and $y_1, \ldots, y_m$ near q. Then Lemma 3.6 states that

$$\mathscr{I}_{(p,q)}(X \times \{q\}, \text{graph} f) = (f_1, \ldots, f_m),$$

where $f_1, \ldots, f_m$ are the coordinate functions of f (having identified X with $X \times \{q\}$). Since $i^*(f_1, \ldots, f_m) = (f_1, \ldots, f_m)$ and $\mathscr{R}_f(p) = C_p^\infty(\mathbf{R}^n)/(f_1, \ldots, f_m)$ the lemma is proved. $\square$

We now make the obvious definition.

Definition 3.8. *Two map germs $f, g : (X, p) \to (Y, q)$ are* contact equivalent *if graph f and graph g have the same contact with $X \times \{q\}$ as submanifold germs of $X \times Y$ near (p, q).*

Notes. (1) Theorem 3.3 says that f and g are contact equivalent as germs near p iff $\mathscr{R}_f(p) = \mathscr{R}_g(p)$. We thus have a geometric interpretation of what it means for two map germs at a point to have the same local ring.

(2) We would like to generalize the definition of contact equivalence so that we can interpret what it means for two map germs to have isomorphic local rings. For example, we could weaken the definition of contact equivalence to allow germs of diffeomorphisms ϕ which leave A invariant and not demand that $\phi|A = id_A$. It is then easy to see that if graph f is "contact equivalent" to graph g with respect to $X \times \{q\}$ with this new definition, then $\mathscr{R}_f(p)$ is isomorphic to $\mathscr{R}_g(p)$. The problem is in the converse statement. Suppose that $\mathscr{R}_f(p) \cong \mathscr{R}_g(p)$, then what we would like to know is whether there exists a germ of a smooth diffeomorphism $\phi : (Z, p) \to (Z, p)$ such that ϕ^* induces the isomorphism between $\mathscr{R}_f(p)$ and $\mathscr{R}_g(p)$. The thrust of the proof of Theorem 3.3 is, of course, the construction of such a ϕ. Let us spend a moment to reflect on the problem. Suppose that there is no obstacle to lifting the isomorphism of $\mathscr{R}_f(p) \to \mathscr{R}_g(p)$ to an isomorphism $\psi : C_p^\infty(X) \to C_p^\infty(X)$. With a choice of coordinates we may assume that $\psi : C_0^\infty(\mathbf{R}^n) \to C_0^\infty(\mathbf{R}^n)$. Question: Is $\psi = \phi^*$? Since we know what $\psi(x_1), \ldots, \psi(x_n)$ are, there is only one possibility for ϕ, namely, $\phi(x) = (\psi(x_1), \ldots, \psi(x_n))$. It is easy to show that $\phi^* = \psi$ on any analytic function. The problem is that ψ is not uniquely defined on flat functions, so ϕ^* does not have to equal ψ. But we are saved by the Malgrange Preparation Theorem and Mather's Theorem (V, Theorem 1.2) that a stable map germ is determined by only a finite portion of its Taylor series. Thus to circumvent our problem we need only "jetify" our result and work with the local ring $\mathscr{R}_f(p)/\mathscr{M}_p^k(X)$ (for some appropriate k). It should be clear that any isomorphism between these finite dimensional local rings is induced by a smooth mapping.

We introduce some terminology. A *k-jet of a submanifold at p* of a manifold Z is an equivalence class of submanifold germs at p where two germs A and B are equivalent if every smooth function of $Z \to \mathbf{R}$ which when restricted to A vanishes to kth order at p also, when restricted to B, vanishes to kth order at p. To analyze submanifold k-jets at p, we can assume by choosing coordinates that $Z = \mathbf{R}^n$ and $p = 0$. Let $H_{k,l}^n$ be the set of k-jets of l-dimensional submanifolds of $\mathbf{R}^n$ at 0. First we claim that two equidimensional submanifold germs at 0, A and B, are in the same 1-jet equivalence class iff $T_0 A = T_0 B \subset \mathbf{R}^n$. This is clear since a function $\psi : Z \to R$ vanishes to first order at p in the A (respectively, B) directions iff $(d\psi)_0(T_0 A) = 0$ (respectively,

$(d\psi)_0(T_0B) = 0$). Thus we may identify $H^n_{1,l}$ with the Grassmann manifold of l-planes in n-space by assigning, to each A in $H^n_{1,l}$, T_0A in $G_{l,n}$. In this way $H^n_{1,l}$ is a manifold. In fact, we have the following:

Lemma 3.9. *$H^n_{k,l}$ is a smooth manifold. Moreover, $H^n_{k,l} \xrightarrow{\rho} G_{l,n}$ given by $\rho(A) = T_0A$ makes $H^n_{k,l}$ into a fiber bundle over $G_{l,n}$ whose typical fiber is the fiber F of the bundle $J^k(\mathbf{R}^l, \mathbf{R}^{n-l})_{0,0} \to J^1(\mathbf{R}^l, \mathbf{R}^{n-l})_{0,0}$.*

Proof. Recall how the manifold structure of $G_{l,n}$ is obtained. We assume that $\mathbf{R}^n$ comes equipped with an inner product. Then a nbhd of V in $G_{l,n}$ is given by $\mathcal{O}_V = \{W \in G_{l,n} \mid \pi_V : W \to V \text{ is a bijection}\}$ where $\pi_V : \mathbf{R}^n \to V$ is just orthogonal projection. We then identify Hom $(V, V^\perp)$ with $\mathcal{O}_V$ by the mapping $A \mapsto$ graph A. We claim that we can identify $\rho^{-1}(\mathcal{O}_V)$ with $\mathcal{O}_V \times F$ (where $V = \mathbf{R}^l$ in the definition of F). Given a submanifold germ A, there exists a unique germ of a smooth map $f: T_0A \to (T_0A)^\perp$ such that $A =$ graph f. Now just note that two submanifold germs A and B which are tangent at 0 yield the same submanifold k-jet iff the corresponding f and g satisfy $j^kf(0) = j^kg(0)$. (Note that since $A =$ graph f and $B =$ graph g, $j^1f(0) = j^1g(0) = 0$.) The "iff" is clear since $\psi: \mathbf{R}^n \to \mathbf{R}$ vanishes to kth order at 0 when restricted to A iff $j^k(\psi \cdot f)(0) = 0$, but $j^k(\psi \cdot f)(0) = j^k(\psi \cdot g)(0)$ iff $j^kf(0) = j^kg(0)$. Now let A be in $\rho^{-1}(\mathcal{O}_V)$. Then the maps $\pi_V : T_0A \to V$ and $\pi_{V^\perp} : (T_0A)^\perp \to V^\perp$ are bijections. The map $\sigma: \rho^{-1}(\mathcal{O}_V) \to \mathcal{O}_V \times F$ given by $A \mapsto (T_0A$, projection of $j^k(\pi_{V^\perp} \cdot f \cdot \pi_V^{-1})(0)$ into $F)$ where $A =$ graph f is a bijection. We topologize $H^n_{k,l}$ by demanding that all such σ be homeomorphisms and give $H^n_{k,l}$ a manifold structure by demanding that all such σ are diffeomorphisms. We leave it to the reader to check that everything works right on overlaps. □

Definition 3.2′. *Let A, B_1, and B_2 be equidimensional k-jet submanifold germs of Z near p. Then B_1 and B_2 are* contact equivalent with respect to A *if there exists a germ of a diffeomorphism $\phi : (Z, p) \to (Z, p)$ such that $\phi(A) = A$ and $\phi(B_1) = B_2$.*

Let $\mathscr{R}^k_{A,B} = C^\infty_p(A)/(\mathscr{I}_p(A, B) + \mathscr{M}^{k+1}_p(A))$. Then we have

Theorem 3.3′. *The k-jets of submanifolds B_1 and B_2 are contact equivalent with respect to A iff $\mathscr{R}^k_{A,B_1}$ is isomorphic with $\mathscr{R}^k_{A,B_2}$.*

Definition 3.7′. *Two k-jets of maps $j^kf(p)$ and $j^kg(p)$, both with target q, are* contact equivalent *iff graph f and graph g are contact equivalent with respect to $X \times \{q\}$ as k-jet submanifolds of $X \times Y$ at (p, q).*

Let $\mathscr{R}_f{}^k(p) = \mathscr{R}_f(p)/\mathscr{M}^{k+1}_p(X)$. If σ is a k-jet with source at p, then define $\mathscr{R}_\sigma = \mathscr{R}_f{}^k(p)$ where $j^kf(p) = \sigma$. Then the following is immediate.

Proposition 3.10. *Let σ and τ be in $J^k(X, Y)_{p,q}$. Then σ and τ are contact equivalent iff $\mathscr{R}_\sigma \cong \mathscr{R}_\tau$.*

Definition 3.11. *Let $\mathscr{R}$ be a local ring. The* contact class $S_{\mathscr{R}} \subset J^k(X, Y)$ *is given by*

$$S_{\mathscr{R}} = \{\sigma \in J^k(X, Y) \mid \mathscr{R}_\sigma \cong \mathscr{R}\}.$$

Remarks. (1) Let σ and τ be equivalent k-jets, then clearly $\mathscr{R}_\sigma \cong \mathscr{R}_\tau$ so σ and τ are contact equivalent. Thus the contact class of a singularity is an invariant of that singularity type.

(2) John Mather has proved that two stable k-jets are equivalent (with $k = \dim Y$) iff they are contact equivalent. This result clearly demonstrates the importance of contact classes. (For the details see [29].)

In the Thom-Boardman Theory one of the crucial technical tools is given by showing that the sets $S_{i,j,\ldots,k}$ are in fact submanifolds of the appropriate jet bundle. The same is true for contact classes.

Theorem 3.12. *Let $\mathscr{R}$ be a local ring. The contact class $S_{\mathscr{R}}$ is an immersed submanifold of $J^k(X, Y)$.*

Note. In fact, $S_{\mathscr{R}}$ is a submanifold but, as before, we shall not need this fact.

Proof. As in previous theorems of this type, the crucial part of the proof is in showing that $S_{\mathscr{R}} \cap J^k(X, Y)_{p,q}$ is a submanifold since $S_{\mathscr{R}}$ is clearly a subfiber bundle of $J^k(X, Y)$ over $X \times Y$. For this purpose, we may choose coordinates so that $X = \mathbf{R}^n$, $Y = \mathbf{R}^m$, and $p = q = 0$.

Now there is a natural action of the Lie group $G^k(\mathbf{R}^n)_0$ (=invertible k-jets on $\mathbf{R}^n$ at 0) on $H^n_{k,l}$. Let $(j^k\phi)_0$ be in $G^k(\mathbf{R}^n)_0$ and let A be in $H^n_{k,l}$. Then $\phi(A)$ is a submanifold germ of $\mathbf{R}^n$ at 0 in $H^n_{k,l}$. This gives a well-defined smooth action. Similarly, the action of $G^k(\mathbf{R}^n \times \mathbf{R}^m)_{(0,0)}$ on $H^{n+m}_{k,n}$ is a smooth action. Let G be the isotropy subgroup of this action whose fixed point is the submanifold germ $\mathbf{R}^n \times \{0\}$. By Theorem A.7 in the Appendix, G is also a Lie group.

It is now a tautology to see that the orbits of G acting on $H^n_{k,l}$ consist precisely of those k-jets which are contact equivalent. See Definitions 3.2′ and 3.7′. □

The other crucial data needed in our analysis of the Thom-Boardman singularities was the codimensions of the submanifolds $S_{i,j,\ldots,k}$. Here, too, that is the case. In general, it is a difficult combinatorial problem to compute the codimensions of the contact classes. To complete this section, we analyze one important class of examples.

The simplest local rings of finite dimension over $\mathbf{R}$ are the truncated polynomial rings $\mathbf{R}[t]/(t^{k+1})$. A smooth map $f: X \to Y$ has a *Morin singularity at p* if $\mathscr{R}_f(p) \cong \mathbf{R}[t]/(t^{k+1})$ for some k. We denote by S_{1_k} the contact class in $J^k(X, Y)$ determined by the ring $\mathbf{R}[t]/(t^{k+1})$. The main results about these singularities are due to B. Morin [39] and H. Levine [20, 21]. We will not discuss their results in complete detail, but will confine ourselves to the equidimensional case: $\dim X = \dim Y$.

We denote by $S_{1_k}(f)$ the points where f takes on a Morin singularity of type k; i.e., $S_{1_k}(f) = (j^kf)^{-1}(S_{1_k})$.

Proposition 3.13. *S_{1_k} is a submanifold of $J^k(X, Y)$ of codimension k. (Assuming that* $\dim X = \dim Y$.*)*

Corollary 3.14. *For a residual set of maps f, $S_{1_k}(f)$ is a submanifold of X of codimension k.*

Proof. Just apply the Thom Transversality Theorem. □

Proof of Proposition 3.13. It is enough to prove that $S_{1_k} \cap J^k(X, Y)_{p,q}$ is a codimension k submanifold of $J^k(X, Y)_{p,q}$. Moreover, we can assume that $X = Y = R^n$ and $p = q = 0$. Given σ_0 in S_{1_k} with source and target at 0 we will show that a nbhd of σ_0 in $S_{1_k} \cap J^k(R^n, R^n)_{0,0}$ is a codimension k submanifold. Clearly, we can assume σ_0 has the form $(x_1, \ldots, x_n) \mapsto (f(x), x_2, \ldots, x_n)$. Then σ_0 in S_{1_k} is equivalent to the condition

$$(*) \qquad \frac{\partial f}{\partial x_1}(0) = \frac{\partial^2 f}{\partial {x_1}^2}(0) = \cdots = \frac{\partial^k f}{\partial {x_1}^k}(0) = 0.$$

Now let $\pi: R^n \to R^{n-1}$ be the submersion $\pi(x_1, \ldots, x_n) = (x_2, \ldots, x_n)$ which induces a fiber mapping $\pi_*: J^k(R^n, R^n)_{0,0} \to J^k(R^n, R^{n-1})_{0,0}$. Let U be the open subset of $J^k(R^n, R^{n-1})_{0,0}$ consisting of all k-jets of maps of the form $(x_1, \ldots, x_n) \mapsto (f_2(x), \ldots, f_n(x))$ having the property that $dx_1, df_2, \ldots, df_n$ are linearly independent at 0. This is an open subset of $J^k(R^n, R^{n-1})_{0,0}$ containing the image of σ_0. To each σ in U we can associate an invertible k-jet $\tilde{\sigma}$ in $J^k(R^n, R^n)_{0,0}$—namely the k-jet of the map $(x_1, \ldots, x_n) \mapsto (x_1, f_2, \ldots, f_n)$.

Let Σ be the vector space of all polynomials of degree $\leq k$ in $x_1, \ldots, x_n$ with zero constant term. Σ is the "typical fiber" of the fibration π_*.

We will show that the fibration π_* is trivial over U by constructing an explicit trivialization

$$T: \Sigma \times U \cong (\pi_*)^{-1}(U).$$

Given p in Σ let k_p be the k-jet of the map $(x_1, \ldots, x_n) \mapsto (p, x_2, \ldots, x_n)$. Then we define $T(p, \sigma) \equiv k_p \cdot \tilde{\sigma}$ for p in Σ and σ in U. It is easy to see that this is a diffeomorphism between $\Sigma \times U$ and $(\pi_*)^{-1}(U)$.

Let Σ_{1_k} be the set of p in Σ satisfying $\dfrac{\partial p}{\partial x_1}(0) = \cdots = \dfrac{\partial^k p}{\partial {x_1}^k}(0) = 0$. These conditions are independent so Σ_{1_k} is a codimension k subspace of Σ. We let the reader check that T gives us a smooth identification

$$\Sigma_{1_k} \times U \underset{\to}{\cong} (\pi_*)^{-1}(U) \cap S_{1_k}. \qquad \text{Hint: use } (*) \quad \square$$

Finally we shall show that the Morin singularities are actually singularities of the type studied in Chapter VI. We recall that for a map $f: (X, p) \to (Y, q)$, p is in $S_1(f)$ iff f has corank 1 at p and that p is in $S_{1,1}(f)$ iff $f | S_1(f)$ has corank 1 at p (assuming—naturally—that $S_1(f)$ is a submanifold). We can continue this inductive construction of $S_{\overbrace{1,\ldots,1}^{k}}(f)$ as long as at each stage $S_{\overbrace{1,\ldots,1}^{k-1}}(f)$ is a submanifold of X. The Boardman Theorem states that this is the case for a residual set of f. We will prove this theorem in the special case of the Morin singularities by proving the following.

Proposition 3.15. *If* $j^k f \pitchfork S_{1_k}$ *for all* $k \leq \dim X + 1$, *then* $S_{1_k}(f) = S_{\overbrace{1,\ldots,1}^{k}}(f)$.

Proof. We need only prove this locally. Let p be in $S_{1_k}(f)$. By choosing coordinates on X based at p and on Y based at $f(p)$ we may assume that $p = f(p) = 0$ and that $f(x_1, \ldots, x_n) = (h(x), x_2, \ldots, x_n)$ where $\partial^s h/\partial x_1{}^s(0) = 0$ for $s \leq k$. In fact, $S_{1_k}(f)$ is given by the equations $\partial h/\partial x_1 = \cdots = \partial^k h/\partial x_1{}^k = 0$. (Recall the proof of Proposition 3.13.) We proceed by induction. In case $k = 1$, $S_1(f) = S_{1_1}(f)$ as both are given by the equation $\partial h/\partial x_1 = 0$. The transversality hypothesis guarantees that $S_1(f)$ is a submanifold. Now assume inductively that $S_{\overbrace{1,\ldots,1}^{k-1}}(f) = S_{1_{k-1}}(f)$ and that this set is a submanifold. Let q be in $S_{1_{k-1}}(f)$. We claim that q is in $S_{\overbrace{1,\ldots,1}^{k}}(f)$ iff $\partial/\partial x_1|_q$ is in $T_q S_{1_{k-1}}(f)$ since $\operatorname{Ker} d(f|S_{1_{k-1}}(f))_q = \operatorname{Ker}(df)_q \cap T_q S_{1_{k-1}}(f)$ and $\operatorname{Ker}(df)_q = (\partial/\partial x_1|_q)$. The proof is complete if we can show that $S_{\overbrace{1,\ldots,1}^{k}}(f) = \{q \in S_{1_{k-1}}(f) \mid \partial^k h/\partial x_1{}^k(q) = 0\}$ for then $S_{\overbrace{1,\ldots,1}^{k}}(f) = S_{1_k}(f)$ and the transversality hypothesis guarantees that these are submanifolds. Define $H\colon R^n \to R^{k-1}$ by $H(q) = (\partial h/\partial x_1(q), \ldots, \partial^{k-1}h/\partial x_1^{k-1}(q))$. Then clearly $H^{-1}(0) = S_{1_{k-1}}(f)$. Moreover H is a submersion at the points in $H^{-1}(0)$. This follows from the transversality hypothesis and the way $S_{1_{k-1}}$ is defined locally as a submanifold of $J^{k-1}(\mathbf{R}^n, \mathbf{R}^n)$. (Again see the proof of Proposition 3.13 and apply II, Lemma 4.3.) Since H is a submersion $T_q S_{1_{k-1}}(f) = \operatorname{Ker}(dH)_q$. It is a trivial calculation to see that $\partial/\partial x_1|_q$ is in $\operatorname{Ker}(dH)_q$ (for q in $H^{-1}(0)$) iff $\partial^k h/\partial x_1{}^k(q) = 0$. □

Exercises

(1) Prove that the mapping of $\mathbf{R}^3 \to \mathbf{R}^3$ is defined by the equations:

$$\begin{aligned} y_1 &= x_1x_2 + x_1{}^2x_3 + x_1{}^4 \\ y_2 &= x_2 \\ y_3 &= x_3. \end{aligned}$$

has a generic S_{1_3} singularity at the origin.

(2) In Exercise 1, what are the equations for the fold surface $S_1(f)$ and the locus of cusps, $S_{1,1}(f)$? Draw a sketch of them.

(3) For the map in Exercise 1 show that the image of $S_1(f)$ has the appearance of a "swallow's tail." (See Figure 5.)

(4) Show that the map $f\colon \mathbf{R}^n \to \mathbf{R}^n$ given by $f(x_1, \ldots, x_n) = (f_1(x), x_2, \ldots, x_n)$ takes on an S_{1_k} singularity transversely at 0 if

(a) $\dfrac{\partial f}{\partial x_1}(0) = \cdots = \dfrac{\partial^k f}{\partial x_1{}^k}(0) = 0$

and (b) $d\left(\dfrac{\partial f}{\partial x_1}\right)(0), \ldots, d\left(\dfrac{\partial^k f}{\partial x_1}\right)(0)$

are linearly independent. Hint: Look at the proof of Proposition 3.15.

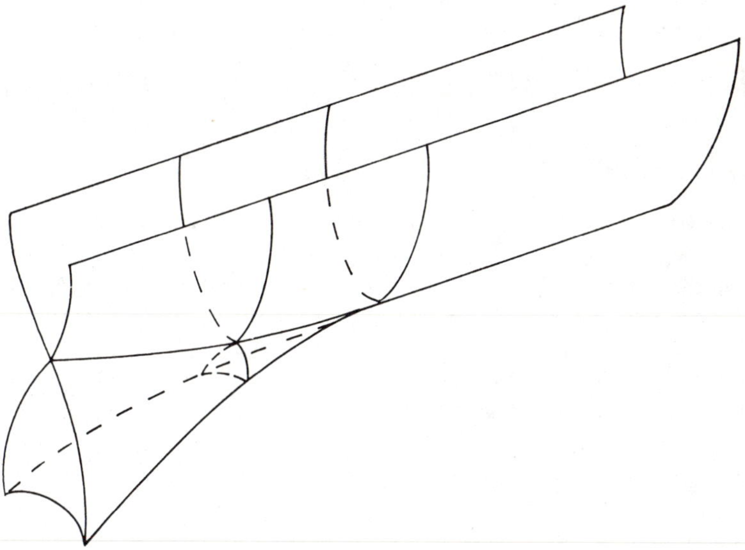

Figure 5: The Swallow's Tail

§4. Canonical Forms for the Morin Singularities

We will show that two Morin singularities are equivalent provided they are displayed transversally (i.e., provided the hypotheses of Proposition 3.5 are satisfied) and provided their local rings are isomorphic. We will prove this by showing that a Morin singularity which is displayed transversally has a simple canonical form. As in §3 we will just consider the equidimensional case: $\dim X = \dim Y$ (though we will discuss one non-equidimensional example at the end of this section.)

Theorem 4.1. *If $f: X \to Y$ satisfies the transversality condition: $j^k f \pitchfork S_{1_k}$, and $x_0 \in S_{1_k}(f)$, then there exist a coordinate system $x_1, \ldots, x_n$ centered at x_0 and a coordinate system $y_1, \ldots, y_n$ centered at $f(x_0)$ such that f has the form*

$$\begin{aligned} f^*y_1 &= x_2x_1 + \cdots + x_kx_1^{k-1} + x_1^{k+1} \\ f^*y_2 &= x_2 \\ &\vdots \\ f^*y_n &= x_n. \end{aligned} \tag{4.2}$$

The result is due to B. Morin (See [39]). Note that in dimension 2 it gives both Whitney canonical forms of Chapter VI, §2.

Proof. We can choose coordinates $x_1, \ldots, x_n$ centered at x_0 and $y_1, \ldots, y_n$ centered at $f(x_0)$, so that $f(x_1, \ldots, x_n) = (h(x), x_2, \ldots, x_n)$. Since x_0 is an S_{1_k} singularity the local ring $\mathscr{R}_f(x_0)$ is generated as a vector space over $\mathbf{R}$ by $1, x_1, \ldots, x_1^k$. By the Malgrange preparation theorem every germ of a function at x_0 can be written as a linear combination of $1, x_1, \ldots, x_1^k$ with smooth functions of the y's as coefficients; so in particular, we can write

$$x_1^{k+1} = f^*a_1 + (f^*a_2)x_1 + \cdots + (f^*a_{k+1})x_1^k \tag{4.3}$$

or

$$f^*a_1 = -(f^*a_2)x_1 - \cdots - (f^*a_{k+1})x_1^k + x_1^{k+1} \tag{4.3$'$}$$

the a_i's being smooth functions of y. Furthermore we can assume $a_{k+1} = 0$ (Proof: replace x_1 by $x_1 + (1/k)f^*a_{k+1}$ and leave $x_2, \ldots, x_k$ fixed.) By comparing the two sets of (4.3) we see that $a_1(0) = a_2(0) = \cdots = a_k(0) = 0$.

Let us now set $x_2 = \cdots = x_n = y_2 = \cdots = y_n = 0$ in (4.3) and expand both sides in powers of x_1. By assumption $f^*y_1 = h(x_1, 0, \ldots, 0) = cx_1^{k+1} + \cdots$, c being a nonzero constant and the dots indicating terms of degree $>k + 1$ in x_1. Therefore, if the x_1^{k+1} terms in (4.3) are to be equal, we must have $a_1(y_1, 0, \ldots, 0) = (1/c)y_1 + \cdots$. In particular $\partial a_1/\partial y_1 \neq 0$. This means that the map:

$$(y_1, \ldots, y_n) \mapsto (a_1(y), y_2, \ldots, y_n)$$

is a legitimate coordinate change in y space. In other words we can assume to begin with that our x_i's and y_i's satisfy:

$$\begin{aligned} &f^*y_1 = f^*a_2x_1 + \cdots + f^*a_kx_1^{k-1} + x_1^{k+1} \\ &f^*y_i = x_i \qquad i = 2, \ldots, n. \end{aligned} \tag{4.4}$$

(We have changed a_i to $-a_i$ to make the first line more visually appealing.)

Since the map f has the form of Exercise (4) of §3, the transversality condition says that

$$d\left(\frac{\partial f^*y}{\partial x_1}\right)(0), \ldots, d\left(\frac{\partial^k f^*y}{\partial x_1^k}\right)(0)$$

are linearly independent. This implies that

$$(dx_1)(0), d(f^*a_2)(0), \ldots, d(f^*a_k)(0)$$

are linearly independent. So this means that the differentials of the functions

$$a_2(0, y_2, \ldots, y_n), \ldots, a_k(0, y_2, \ldots, y_n)$$

are linearly independent at 0. Permuting, if necessary, the y_i's we can assume the matrix

$$\left(\frac{\partial a_i}{\partial y_j}(0)\right) \qquad 2 \leq i, j \leq k$$

is nonsingular. Therefore the mappings

$$(x_1, \ldots, x_n) \mapsto (x_1, f^*a_2, \ldots, f^*a_k, x_{k+1}, \ldots, x_n)$$

and

$$(y_1, \ldots, y_n) \mapsto (y_1, a_2, \ldots, a_k, y_{k+1}, \ldots, y_n)$$

are legitimate coordinate changes. With respect to the new coordinates f has the form (4.2). □

Morin obtained canonical forms for all the "Morin" singularities of §3, not just the equidimensional ones. Here we shall confine ourselves to discussing one nonequidimensional example, the "cross-cap" for maps of n manifolds into $2n - 1$ manifolds. This canonical form is due to Whitney [57], though the derivation we will give of it is due to Morin. First of all we need:

Definition 4.5. *Let X be an n manifold, Y a $2n - 1$ manifold, and $f: X \to Y$ a smooth map. A point x_0 in $S_1(f)$ is called a* cross-cap *if $j^1 f \pitchfork S_1$ at x_0.*

Note that for these dimensions, codim $S_1 = n$ (See VI, Proposition 1.1) so cross-caps occur as isolated points of X.

Whitney's result is:

Theorem 4.6. *If $f: X \to Y$ has a cross-cap at x_0, there exists a coordinate system $x_1, \ldots, x_n$ centered at x_0 and $y_1, \ldots, y_{2n-1}$ centered at $f(x_0)$ such that f has the form:*

$$\begin{aligned} f^*y_1 &= x_1^2 & & \\ f^*y_i &= x_i & i &= 2, \ldots, n \\ f^*y_{n+j} &= x_1 x_j & j &= 1, \ldots, n-1. \end{aligned} \tag{4.7}$$

Proof. We can choose coordinates $x_1, \ldots, x_n$ centered at x_0 and $y_1, \ldots, y_{2n-1}$ centered at $f(x_0)$ such that $f^*y_i = x_i$ for $i = 2, \ldots, n$. The set $S_1(f)$ is the locus of points for which

$$\frac{\partial f_1}{\partial x_1} = \frac{\partial f_{n+1}}{\partial x_1} = \cdots = \frac{\partial f_{2n-1}}{\partial x_1} = 0 \tag{4.8}$$

where $f_i = f^*y_i$. We let you check that the transversality assertion is equivalent to the assertion that

$$d\left(\frac{\partial f_1}{\partial x_1}\right), d\left(\frac{\partial f_{n+1}}{\partial x_1}\right), \ldots, d\left(\frac{\partial f_{2n-1}}{\partial x_1}\right) \tag{4.9}$$

are linearly independent at the points where (4.8) holds. (Hint: Use the "$D - BA^{-1}C$" lemma of Chapter II, Lemma 5.2. See also Exercise (7) of VI, §1.) This means in particular that one of the differentials (4.9) must be nonzero when evaluated on $\partial/\partial x_1$. By a linear transformation of the y coordinates we can arrange:

$$\frac{\partial^2 f_1}{\partial x_1^2} \neq 0, \qquad \frac{\partial^2 f_i}{\partial x_1^2} = 0 \quad \text{for} \quad i > n. \tag{4.10}$$

It is clear from (4.10) that the local ring $\mathcal{R}_f(x_0)$ is generated by 1 and x_1. By the Malgrange preparation theorem we can write

$$x_1^2 = f^*a_1 + (f^*a_2)x_1, \tag{4.11}$$

a_1 and a_2 being smooth functions of the y variables. If we make a change of x variables, substituting $x_1 + f^*a_2/2$ for x_1 and leaving the other x_i's fixed, we can make $a_2 = 0$ in (4.11); that is, we can assume

$$f^*a_1 = x_1^2. \tag{4.11$'$}$$

Now let us set $x_2 = \cdots = x_n = y_2 = \cdots = y_n = 0$ and expand the left hand side of (4.11)′ in powers of x_1. By (4.10), $f_1 = f^*y_1 = cx_1^2 + \cdots$ and $f_s(x_1, 0, \ldots, 0) = 0(x_1^3)$ for $s > n$, with $c \neq 0$, so we must have $a_1(y_1, 0, \ldots, 0) = (1/c)y_1$. In particular, $\partial a_1/\partial y_1 \neq 0$, so the map

$$(y_1, \ldots, y_{2n-1}) \mapsto (a_1(y), y_2, \ldots, y_{2n-1})$$

is a legitimate coordinate change. Replacing the old y coordinates by the new y coordinates, we have $f^*y_1 = x_1^2$ and we continue to have $f^*y_i = x_i$ for $2 \leq i \leq n$. The remaining coordinate functions $f_i = f^*y_i$, $i > n$, can be written in the form (using the Malgrange Theorem)

$$f_i = g_i(x_1^2, x_2, \ldots, x_n) + x_1 h_i(x_1^2, x_2, \ldots, x_n).$$

If we replace y_i by $y_i - g_i(y_1, \ldots, y_n)$ for $i = n + 1, \ldots, 2n - 1$ and leave the other y_i's as before, we obtain the system of equations

$$\begin{aligned} f^*y_1 &= x_1^2 \\ f^*y_i &= x_i \qquad i = 2, \ldots, n \\ f^*y_{n+j} &= x_1 h_{n+j}(x_1^2, x_2, \ldots, x_n) \qquad j = 1, \ldots, n-1. \end{aligned}$$

This is almost the form we want. In fact if we can show that the following are legitimate changes of coordinates:

$$(x_1, \ldots, x_n) \mapsto (x_1, h_{n+1}(x_1^2, \ldots, x_n), \ldots, h_{2n-1}(x_1^2, \ldots, x_n)),$$

and

$$(y_1, \ldots, y_{2n-1}) \mapsto (y_1, h_{n+1}(y), \ldots, h_{2n-1}(y), y_{n+1}, \ldots, y_{2n-1})$$

then we will have precisely the set of equations (4.7). To show this we must go back to the transversality condition (4.9). At $x_1 = 0$ this reduces to the condition that

$$dh_{n+1}(x_1^2, \ldots, x_n), \ldots, dh_{2n-1}(x_1^2, \ldots, x_n)$$

be linearly independent at 0, or, in other words, that the matrix

$$\left(\frac{\partial h_i}{\partial x_j}\right)_{2 \leq j \leq n}^{n+1 \leq i \leq 2n-1}$$

be nonsingular. This however is precisely what is needed to make the changes of coordinates above legitimate. □

Remark. We showed in Chapter II that for every n manifold X we can find an immersion $f: X \to \mathbf{R}^{2n}$. Cross caps arise as obstructions to lowering the dimension of this immersion by 1. In [60] Whitney proved that every n manifold can be immersed in $\mathbf{R}^{2n-1}$, the idea of the proof being to delete cross caps, two at a time, from a generic mapping. Pictures of cross-caps in

$\mathbf{R}^3$ can be found in classical books on topology in connection with the problem of immersing $\mathbf{P}^2$ topologically in $\mathbf{R}^3$ (See exercise 3.)

Exercises

(1) Let $f: X \to \mathbf{R}^{2n}$ be an immersion of the n manifold X. Let v be a regular value of the induced map

$$(df): TX \to \mathbf{R}^{2n}$$

and let $\pi: \mathbf{R}^{2n} \to \mathbf{R}^{2n-1}$ be a surjective linear map with v in its kernel. Show that $\pi \cdot f$ has no singularities except cross-caps (Compare with II, §1, Exercise 1.)

(2) Describe the image of the map $(x_1, x_2) \mapsto (x_1^2, x_2, x_1x_2)$. What are the images of the curves $x_1 = \text{const}$ and $x_2 = \text{const}$. Show that this map is 1-1 except along the "double line" $x_2 = 0$ (See Figure 6).

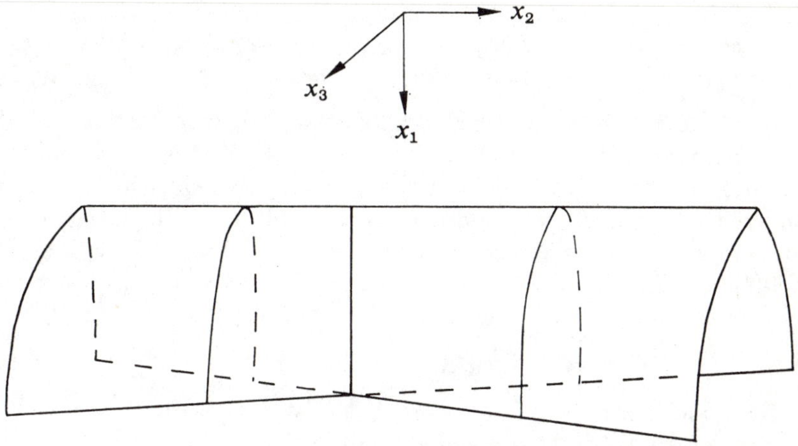

Figure 6: The Cross Cap

(3) Construct a topological immersion of $\mathbf{P}^2$ into $\mathbf{R}^3$ whose image is a "cross cap". (See Figure 7.)

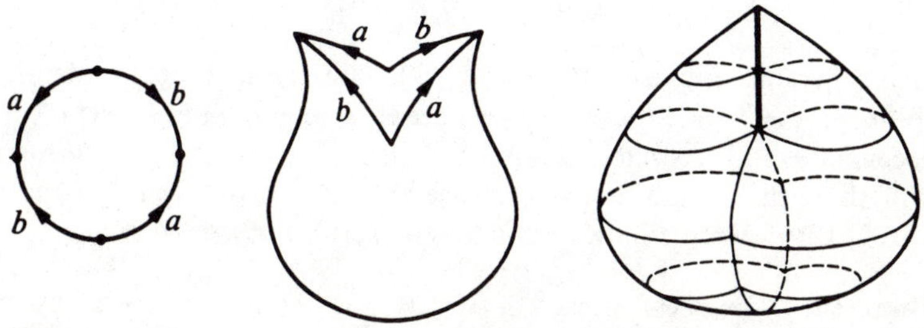

Figure 7: The Topological Cross Cap

§5. Umbilics

From Exercise 5 of §2 we know that S_2 singularities cannot occur with multiplicity 1, 2, or 3. An S_2 singularity which occurs with multiplicity 4 is called an *umbilic*†. Umbilics are the simplest S_2 singularities, and are the only ones that can occur stably in dimensions <6. We will begin our study of them by showing that for singularities of multiplicity 4, the only local rings that are allowable are those on the following list:

(5.1) $\mathbf{R}[t]/(t^4)$

(5.2) $\mathbf{R}[x, y]/(x^2 - y^2, xy)$

(5.3) $\mathbf{R}[x, y]/(x^2, y^2)$

(5.4) $\mathbf{R}[x, y]/(x^2, xy, y^3)$

(5.5) $\mathbf{R}[x, y, z]/(x^2, y^2, z^2, xy, xz, yz).$

In fact we will prove

Proposition 5.6. *Let $\mathscr{R}$ be a local ring over* $\mathbf{R}$ *with* $\dim_{\mathbf{R}} \mathscr{R} = 4$. *Then $\mathscr{R}$ is isomorphic to one of the rings on the above list.*

Proof. Let $\mathscr{M}$ be the maximal ideal of $\mathscr{R}$. Suppose first that $\dim_{\mathbf{R}} \mathscr{M}/\mathscr{M}^2 = 1$. Let t be an element of $\mathscr{M} - \mathscr{M}^2$. $\mathscr{M}^i = \{ct^i\} + \mathscr{M}^{i+1}$, so the mapping of $\mathbf{R}[t]$ into $\mathscr{R}$ is onto. The kernel is generated by a polynomial $p(t)$, which we can write as $t^k q(t)$, $q(t)$ having a nonzero constant term. The image of $q(t)$ in $\mathscr{R}$ is invertible, so the kernel is also generated by t^k. Hence $\mathscr{R}$ is isomorphic to $\mathbf{R}[t]/(t^k)$. Since $\dim_{\mathbf{R}} \mathscr{R} = 4$, $k = 4$. Next suppose $\dim_{\mathbf{R}} \mathscr{M}/\mathscr{M}^2 = 2$. We note that

(5.7) $$\dim_{\mathbf{R}} \mathscr{R} = \dim_{\mathbf{R}} \mathscr{R}/\mathscr{M} + \dim_{\mathbf{R}} \mathscr{M}/\mathscr{M}^2 + \cdots$$

by Nakayama's lemma. (If $\mathscr{M}^i = \mathscr{M}^{i+1}$ then $\mathscr{M}^i = 0$.) Therefore if $\mathscr{R}$ is 4 dimensional, $\dim \mathscr{M}^2/\mathscr{M}^3 = 1$ and $\mathscr{M}^i = 0$ for $i > 2$.

Consider now the bilinear map

$$\mathscr{M}/\mathscr{M}^2 \otimes \mathscr{M}/\mathscr{M}^2 \to \mathscr{M}^2$$

induced by the product operation on the local ring. We will, for the moment, fix a basis vector in $\mathscr{M}^2$ and regard this as a map:

(5.8) $$\mathscr{M}/\mathscr{M}^2 \otimes \mathscr{M}/\mathscr{M}^2 \to \mathbf{R}$$

i.e., as a symmetric bilinear form on $\mathscr{M}/\mathscr{M}^2$. This form cannot be identically zero otherwise $\mathscr{M}^2 = 0$; therefore there are three possibilities for it: it can be nondegenerate and definite, nondegenerate and indefinite, or degenerate. We will show that if the first is the case then $\mathscr{R}$ is isomorphic to the ring (5.2). In fact if (5.8) is definite we can find a basis for $\mathscr{M}/\mathscr{M}^2$ such that $x \cdot x =$

† Because the "umbilical points" of a surface in $\mathbf{R}^3$ are the points where its normal bundle map exhibits this kind of singularity.

$y \cdot y$ and $x \cdot y = 0$. Therefore, $\mathscr{R}$ is isomorphic to the polynomial ring in two variables divided by the ideal of relations $(x^2 - y^2, xy)$. A similar argument shows that if (5.8) is nondegenerate and indefinite then $\mathscr{R}$ is the ring (5.3) and if (5.8) is degenerate, $\mathscr{R}$ is the ring (5.4). Finally if $\dim \mathscr{M}/\mathscr{M}^2 = 3$, $\mathscr{M}^2 = 0$ by (5.7) and $\mathscr{R}$ is the ring (5.5). □

(5.1) is the local ring of an S_{1_3} singularity (a "swallow's tail"). In the equidimensional case, which is the only case that we shall consider here, the local ring (5.5) cannot occur (by Exercise 5 of §2.) (5.2) (5.3) and (5.4) are all possible candidates for umbilics.

Definition 5.9. *An umbilic is called* hyperbolic *if its local ring is* (5.3) *and* elliptic *if its local ring is* (5.2).

Our main theorem about umbilics will be that two generic umbilics are equivalent (as map germs) if and only if they are of the same type. For simplicity we shall just prove this for the elliptic and hyperbolic umbilics. (The parabolic case will be treated in the exercises.) Specifically, we will prove:

Theorem 5.10. *Let X and Y be n dimensional manifolds, $n \geq 4$. Let $f: X \to Y$ be a smooth map exhibiting either an elliptic or hyperbolic umbilic at x_0 in X. Suppose $j^1 f \pitchfork S_2$ at x_0. Then we can find coordinates, $x_1, \ldots, x_n$ centered at x_0 and $y_1, \ldots, y_n$ centered at $f(x_0)$ such that f has one of the following two canonical forms:*

Hyperbolic case

$$\begin{aligned} f^*y_1 &= x_1{}^2 + x_3x_2 \\ f^*y_2 &= x_2{}^2 + x_4x_1 \\ f^*y_3 &= x_3 \\ &\vdots \\ f^*y_n &= x_n \end{aligned} \tag{5.11}$$

Elliptic case

$$\begin{aligned} f^*y_1 &= x_1{}^2 - x_2{}^2 + x_3x_1 + x_4x_2 \\ f^*y_2 &= x_1x_2 + x_4x_1 - x_3x_2 \\ f^*y_3 &= x_3 \\ &\vdots \\ f^*y_n &= x_n \end{aligned} \tag{5.12}$$

Note that the assumption $n \geq 4$ is essential. The transversality condition cannot be satisfied in dimensions <4.

In the proof of (5.11) we will need the following

Lemma 5.13. *Let*

$$ax_1{}^2 + 2bx_1x_2 + cx_2{}^2 \tag{5.14}$$

be a quadratic form whose coefficients are smooth functions of a set of parameters: $z = (z_1, \ldots, z_m)$. *Suppose that when* $z = 0$ (5.14) *is equal to* $x_1^2 - x_2^2$. *Then one can find a rotation,*

$$S = \begin{pmatrix} s_{11} & s_{12} \\ s_{21} & s_{22} \end{pmatrix},$$

whose coefficients are smooth functions of z, *such that* $S(0)$ *is the identity, and such that in the rotated coordinate system* $(\bar{x}_1, \bar{x}_2)$ (*i.e.*, $\bar{x}_1 = s_{11}x_1 + s_{12}x_2$, $\bar{x}_2 = s_{21}x_1 + x_{22}x_2$) (5.14) *has the form* $\bar{a}\bar{x}_1^2 + \bar{b}\bar{x}_2^2$ *where* $\bar{a}(0) = 1$ *and* $\bar{b}(0) = -1$.

Proof. Solve for the eigenvalues of the matrix $A = \begin{bmatrix} a & b \\ b & a \end{bmatrix}$. Since A is close to $\begin{bmatrix} 1 & 0 \\ 0 & -1 \end{bmatrix}$ for z small the eigenvalues are distinct and are close to 1 and -1. Moreover they depend smoothly on z. For the eigenvalue close to 1 we can find an eigenvector $(1, \tau)$ with τ a smooth function of z and $\tau(0) = 0$. Since A is symmetric the other eigenvector will be $(-\tau, 1)$. Let S be the rotation

$$\frac{1}{\sqrt{1 + \tau^2}} \begin{pmatrix} 1 & \tau \\ -\tau & 1 \end{pmatrix}.$$

Note that $\bar{a}$ and $\bar{b}$ are the eigenvalues of A. □

We will now derive the normal form (5.11). Since x_0 is in S_2 we can choose a coordinate system $x_1, \ldots, x_n$ centered at x_0 and $y_1, \ldots, y_n$ centered at $f(x_0)$ such that f has the form

$$(x_1, \ldots, x_n) \mapsto (f_1(x), f_2(x), x_3, \ldots, x_n) \tag{5.15}$$

where the linear terms in f_1 and f_2 vanish and the quadratic terms are of the form

$$\begin{aligned} f_1 &= x_1^2 + \cdots \\ f_2 &= x_2^2 + \cdots \end{aligned} \tag{5.16}$$

the dots indicating quadratic terms like x_3x_1, x_3^2, etc. and higher order terms in all the x's. Therefore the local ring will be generated over **R** by 1, x_1, x_2, and x_1x_2. By the Malgrange preparation theorem every germ of a function at 0 can be written as a linear combination of 1, x_1, x_2 and x_1x_2 with smooth functions of y as coefficients. In particular we can write:

$$\begin{aligned} x_1^2 - x_2^2 &= f^*a_1 + f^*b_1x_1 + f^*c_1x_2 + f^*d_1x_1x_2, \quad \text{and} \\ x_1^2 + x_2^2 &= f^*a_2 + f^*b_2x_1 + f^*c_2x_2 + f^*d_2x_1x_2 \end{aligned} \tag{5.17}$$

where the a's, b's etc. are smooth functions of y vanishing at $y = 0$. Replacing x_1 by $x_1 + x_2f^*d_2/2$ and leaving the other coordinate fixed we can arrange that $d_2 \equiv 0$. Applying the lemma to $x_1^2 - x_2^2 - f^*d_1x_1x_2$ we can also arrange that $d_1 \equiv 0$. Note that since the change of coordinates $(x_1, x_2) \to (\bar{x}_1, \bar{x}_2)$ is given by a rotation we have that $x_1^2 + x_2^2 = \bar{x}_1^2 + \bar{x}_2^2$. Dropping the

$^-$'s on $\bar{x}_1$ and $\bar{x}_2$ and using the fact that $\bar{a}(0) = 1$ and $\bar{b}(0) = -1$ we may solve for x_1^2 and x_2^2 to get

$$x_1^2 = f^*\alpha_1 + f^*\beta_1 x_1 + f^*\gamma_1 x_2$$

and

$$x_2^2 = f^*\alpha_2 + f^*\beta_2 x_1 + f^*\gamma_2 x_2$$

the α's, β's and γ's being functions of y. One last simplification is possible. Replacing x_1 by $x_1 + f^*\beta_1/2$ and x_2 by $x_2 + f^*\gamma_2/2$ we can assume β_1 and γ_2 are zero. Therefore, with $\gamma = -\gamma_1$ and $\beta = -\beta_2$, we have

$$\begin{aligned} f^*\alpha_1 &= x_1^2 + f^*\gamma x_2 \\ f^*\alpha_2 &= x_2^2 + f^*\beta x_1 \end{aligned} \tag{5.18}$$

We continue, of course, to have $f^*y_i = x_i$ for $i > 2$ since we have not made any changes in these coordinates. Now we will set $x_3 = y_3 = \cdots = x_n = y_n = 0$ in (5.18). Comparing the quadratic terms on both sides and using (5.16) we see that:

$$\begin{aligned} \alpha_1(y_1, y_2, 0, \ldots, 0) &= y_1 + \cdots \\ \alpha_2(y_1, y_2, 0, \ldots, 0) &= y_2 + \cdots \end{aligned}$$

the dots indicating terms of degree >1 in y_1 and y_2. (Now in the coordinate changes above $\bar{x}_1 = x_1 + \cdots$ and $\bar{x}_2 = x_2 + \cdots$ where $\cdots$ stands for higher order terms so that (5.16) is still applicable.) This implies that the map

$$(y_1, \ldots, y_n) \mapsto (\alpha_1(y), \alpha_2(y), y_3, \ldots, y_n)$$

is a legitimate coordinate change. In the new coordinates we have

$$\begin{aligned} f^*y_1 &= x_1^2 + f^*\gamma x_2 \\ f^*y_2 &= x_2^2 + f^*\beta x_1 \\ f^*y_3 &= x_3 \\ &\vdots \\ f^*y_n &= x_n. \end{aligned} \tag{5.19}$$

This is nearly the canonical form we want. In fact if we can show that

$$(x_1, \ldots, x_n) \mapsto (x_1, x_2, f^*\gamma, f^*\beta, x_5, \ldots, x_n)$$

and

$$(y_1, \ldots, y_n) \to (y_1, y_2, \gamma, \beta, y_5, \ldots, y_n)$$

are legitimate coordinate changes, then in the new coordinates (5.19) will have the form (5.11). We will show that these coordinate changes are allowable precisely because of the transversality hypotheses. In fact letting $h_1(x)$ and $h_2(x)$ denote the right hand terms on the first two lines of (5.19), the set $S_2(f)$ is defined by the set of 4 equations

$$\frac{\partial h_1}{\partial x_1} = \frac{\partial h_2}{\partial x_1} = \frac{\partial h_1}{\partial x_2} = \frac{\partial h_2}{\partial x_2} = 0. \tag{5.20}$$

Use exercise 7, VI, §1 to show that the transversality hypothesis reduces to the assertion that on the set defined by (5.20) the differentials

$$d\left(\frac{\partial h_1}{\partial x_1}\right), \quad d\left(\frac{\partial h_2}{\partial x_1}\right), \quad d\left(\frac{\partial h_1}{\partial x_2}\right), \quad d\left(\frac{\partial h_2}{\partial x_2}\right) \tag{5.21}$$

are linearly independent. When we replace h_1 and h_2 by $x_1^2 + f^*\gamma x_2$ and $x_2^2 + f^*\beta x_1$ this condition reduces to the condition that

$$dx_1, dx_2, d(f^*\beta), d(f^*\alpha)$$

be linearly independent at 0 which is precisely what is needed to legitimatize the above coordinate changes.

This concludes our proof for the hyperbolic case of Theorem 5.10. The derivation of the canonical form (5.12) for the elliptic case is similar. We will indicate what changes need to be made in the proof above:

(1) Lemma 5.13 has to be replaced by the following "elliptic" analogue:

Lemma 5.22. *Let*

$$a(x_1^2 - x_2^2) + bx_2^2 \tag{5.23}$$

be a quadratic form whose coefficients are smooth functions of a set of parameters, z. Suppose that for $z = 0$ (5.23) *is just the form* $x_1^2 - x_2^2$. *Then there exists a function* λ *depending smoothly on z such that* $\lambda(0) = 1$ *and such that with respect to the coordinates* $\bar{x}_1 = \lambda x_1$, $\bar{x}_2 = (1/\lambda)x_2$ (5.23) *has the form* $\bar{a}(\bar{x}_1^2 - \bar{x}_2^2)$.

Proof. This is much easier than Lemma 5.13. Just take λ to be $(a/(a - b))^{1/4}$

(2) Now choose the coordinates (5.15) such that

$$f_1 = x_1^2 - x_2^2 + \cdots$$
$$f_2 = x_1x_2 + \cdots$$

and show that the local ring is generated over $\mathbf{R}$ by x_1, x_2 and x_2^2.

(3) Using the Generalized Malgrange Preparation Theorem show that

$$x_1^2 - x_2^2 = f^*a_1 + f^*b_1x_1 + f^*c_1x_2 + f^*d_1x_2^2$$

and

$$x_1x_2 = f^*a_2 + f^*b_2x_1 + f^*c_2x_2 + f^*d_2x_2^2$$

where the a_i's, b_i's etc. are smooth functions of y vanishing at 0. Replace x_1 by $x_1 + f^*d_2x_2$ to make $d_2 = 0$ and apply the lemma to make $d_1 = 0$.

(4) By a linear change of coordinates makes $c_1 = b_2$ and $b_1 = c_2$.

The rest of the proof is as before. □

Exercises

(1) Show that over the complex numbers there are just 4 local rings with $\dim_{\mathbf{C}} \mathscr{R} = 4$.

(2) Let z and w be complex numbers. Show that the map of $\mathbf{C}^2$ into $\mathbf{C}^2$ defined by

$$(z, w) \mapsto (z^2 + \bar{z}w, w) \tag{5.24}$$

has an elliptic umbilic at 0 when viewed as a map of $\mathbf{R}^4 \to \mathbf{R}^4$. (Hint: Compare with the canonical form (5.12).) For w fixed and real sketch the curve: $u = z^2 + \bar{z}w$, $|z| = \text{const.}$, in the u plane. Show that it has the appearance indicated in Figure 8.

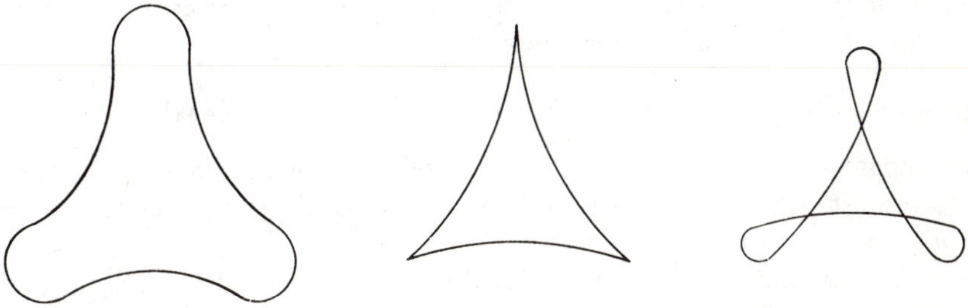

$|z| = \text{const} < \frac{|w|}{2}$ (a) $\qquad |z| = \frac{|w|}{2}$ (b) $\qquad |z| = \text{const} > \frac{|w|}{2}$ (c)

Figure 8

(3) Show that for the map (5.24) the image of S_1 in the 3 dimensional plane: $\operatorname{Im} w = 0$ has the appearance of the cusped surface depicted in

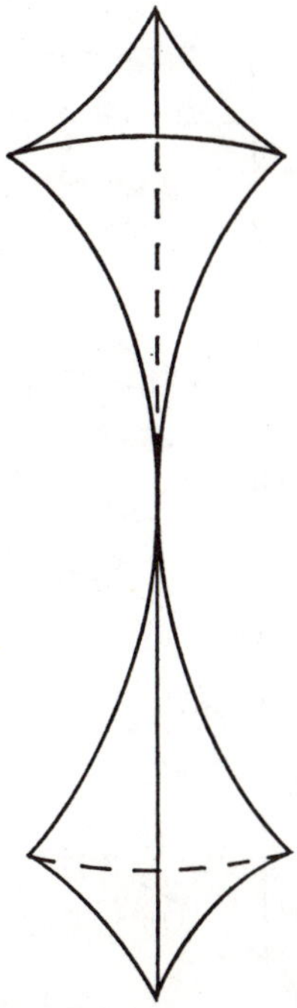

Figure 9: The elliptic Umbilic

Figure 9. (Hint: Show that S_1 is given by the equation $|z| = |w|/2$. Use the identity

$$\det \begin{pmatrix} \frac{\partial u_1}{\partial x_1} & \frac{\partial u_2}{\partial x_2} \\ \frac{\partial u_1}{\partial x_2} & \frac{\partial u_2}{\partial x_2} \end{pmatrix} = \left|\frac{\partial u}{\partial z}\right|^2 - \left|\frac{\partial u}{\partial \bar{z}}\right|^2$$

where $u = u_1 + iu_2$ and $z = x_1 + ix_2$. Now look at Figure 8(b).

(4) Let $f: \mathbf{R}^4 \to \mathbf{R}^4$ be the mapping

$$(x_1, x_2, x_3, x_4) \mapsto (x_1{}^2 + x_3x_2, x_2{}^2 + x_4x_1, x_3, x_4). \tag{5.25}$$

(Compare with 5.11.) Sketch the images of some of the lines $x_1 = \text{const.}$, $x_3 = \text{const.}$, $x_4 = \text{const.}$, and of some of the lines $x_2 = \text{const.}$, $x_3 = \text{const.}$, $x_4 = \text{const.}$

(5) Show that for the map (5.25) the image of S_1 in the 3 dimensional plane $y_4 = 0$ has the appearance of the surface depicted in Figure 10.

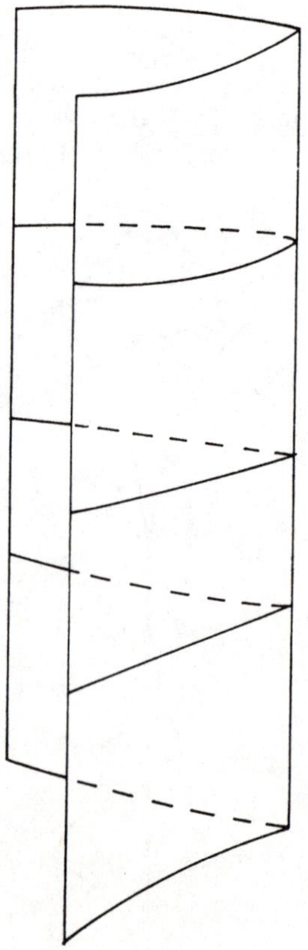

Figure 10: The Hyperbolic Umbilic

(6) Let $f: \mathbf{R}^n \to \mathbf{R}^n$ be the mapping

$$(x_1, \ldots, x_n) \mapsto (f_1(x), f_2(x), x_3, \ldots, x_n)$$

where

$$f_1 = \alpha_1(x_1, x_2) + \cdots$$

and

$$f_2 = \alpha_2(x_1, x_2) + \cdots$$

the dots indicating quadratic terms in x_1x_3, x_3^2, etc. plus higher order terms in $x_1, \ldots, x_n$. α_1 and α_2 are assumed to be homogeneous quadratic polynomials in x_1 and x_2 alone. Show

(a) 0 is a parabolic umbilic $\Leftrightarrow$ α_1 and α_2 have a common linear factor.
(b) $0 \in S_{2,1} \Rightarrow \alpha_i = c_i(s_1x_1 + s_2x_2)^2$, $i = 1, 2$.
(c) $0 \in S_{2,2} \Rightarrow \alpha_1 = \alpha_2 = 0$.

(7) Let Q be the vector space consisting of all pairs (α_1, α_2) where α_1 and α_2 are homogeneous quadratic polynomials in (x_1, x_2). Let P be the subset of Q consisting of all (α_1, α_2) for which α_1 and α_2 have a common linear factor. Let U be the subset of P consisting of all (α_1, α_2) for which α_1 and α_2 have a common quadratic factor (i.e., are constant multiples of each other.) Let W be the subset U consisting of all (α_1, α_2) for which

$$\alpha_i = c_i(s_1x_1 + s_2x_2)^2 \qquad i = 1, 2.$$

Prove:

(a) $P - U$ is a submanifold of Q of codimension 1.
(b) $U - W$ is a submanifold of Q of codimension 2.
(c) $W - \{0\}$ is a submanifold of Q of codimension 3.

(An elegant way to do this exercise is to define these sets using the resultant, $R(\alpha_1, \alpha_2)$, of the polynomials α_1 and α_2. See van der Waerden, [52], Vol. 1, Chapter IV, §27.)

(8) Let $\mathscr{R}$ be the local ring:

$$\mathbf{R}[x_1, x_2]/(x_1^2, x_2^3, x_1^2x_2, x_1x_2^2).$$

If X and Y are n dimensional manifolds show that the contact class, $S_{\mathscr{R}}$, in $J^2(X, Y)$ is a submanifold of codimension 7. If $f: X \to Y$ has the property: $j^1f \pitchfork S_2$, show that $x \in S_{2,1}(f) \Leftrightarrow j^2f(x) \in S_{\mathscr{R}}$. (Hint: Use Exercises 6 and 7, and the same kind of trick as in the proof of Proposition 3.13.)

(9) Let $\mathscr{R}$ be the local ring:

$$\mathbf{R}[x_1, x_2]/(x_1^3, x_1^2x_2, x_1x_2^2, x_2^3).$$

If X and Y are n dimensional manifolds show that the contact class $S_{\mathscr{R}}$ in $J^2(X, Y)$ is a submanifold of codimension 10. If $f: X \to Y$ has the property: $j^1f \pitchfork S_2$, show that $x \in S_{2,2}(f) \Leftrightarrow j^2f(x) \in S_{\mathscr{R}}$. Hint: Use exercises 6 and 7, and the same kind of trick as in the proof of Proposition 3.13.

(10) Show that the map $f: \mathbf{R}^4 \to \mathbf{R}^4$ by

$$(x_1, x_2, x_3, x_4) \to (x_1^2 + x_2^2, x_1x_3 + x_2x_4, x_3, x_4)$$

has an $S_{2,0}$ singularity at the origin and satisfies the transversality condition $j^1f \pitchfork S_2$. Show it is not an umbilic. Hint. Use Exercise 7, VI, §1.

Also show by an example that a small perturbation of f will yield an umbilic at 0.

(11) Let X and Y be n manifolds. Show that if $n < 6$, the set of maps, $f: X \to Y$, which exhibit only Morin singularities and umbilics, is residual. (Hint: Show that the phenomenon illustrated by exercise 10 occurs generically only in dimensions ≥ 6. Use part (b) of Exercise 7, and the same sort of trick as in the proof of Proposition 3.13.)

(12) Let S_E, S_H and S_P be the contact classes in $J^2(X, Y)$ associated with the rings (5.2), (5.3) and (5.4) respectively. Show that the codimension of the submanifolds S_H and S_E in $J^2(X, Y)$ is 4 and that S_P is a submanifold of codimension 5. (Hint: For S_P, use part (a) of Exercise 7.)

(13) Prove the following:

Theorem 5.26. *If $f: X \to Y$ has a parabolic umbilic at x_0 and $j^2f \pitchfork S_P$ at x_0, then there exist a system of coordinates $x_1, \ldots, x_n$ centered at x_0 and $y_1, \ldots, y_n$ centered at $f(x_0)$ such that f has the canonical form:*

$$(5.27)\qquad \begin{aligned} f^*y_1 &= x_1^2 + x_2x_3 + x_2^2x_4 \\ f^*y_2 &= x_1x_2 + x_2x_5 \\ f^*y_3 &= x_3 \\ &\vdots \\ f^*y_n &= x_n. \end{aligned}$$

(Note that for the transversality condition to hold the dimension of X must be ≥ 5.)

Hint: Assume f is in the form (5.15) with

$$\begin{aligned} f_1 &= x_1^2 + \cdots \\ f_2 &= x_1x_2 + \cdots \end{aligned}$$

the dots indicating terms in x_3x_1, x_4x_2, x_3^2 etc. and higher order terms. Show that the local ring is generated over **R** by 1, x_1, x_2, and x_2^2. Using the Malgrange Preparation Theorem show that

$$x_1^2 = f^*a_1 + f^*b_1x_1 + f^*c_1x_2 + f^*d_1x_2^2$$

and

$$x_1x_2 = f^*a_2 + f^*b_2x_1 + f^*c_2x_2 + f^*d_2x_2^2$$

the a's, b's etc. being smooth functions of y vanishing at 0. Use algebraic tricks to make $b_1 = b_2 = d_2 = 0$, and finally make a coordinate change in y_1, y_2 so that f has the form

$$\begin{aligned} f^*y_1 &= x_1^2 + f^*\beta x_2 + f^*\delta x_2^2 \\ f^*y_2 &= x_1x_2 + f^*\gamma x_2 \\ f^*y_3 &= x_3 \\ &\vdots \\ f^*y_n &= x_n. \end{aligned}$$

Finally use the transversality condition to show that β, δ and γ can be introduced as new coordinates in place of y_3, y_4 and y_5.

§6. Stable Mappings in Low Dimensions

Using the results of this chapter it is not hard to get a picture of what stable mappings look like in low dimensions. For simplicity we will restrict ourselves to the equidimensional case: dim X = dim Y. We will first make a list of the kinds of "nonremovable" singularities that can occur in dimensions ≤ 4. (By "nonremovable" we mean they can occur transversely, and therefore cannot be eliminated by small perturbations.)

$$\text{(6.1)}\qquad \begin{array}{ll} \textit{dim } 1 & S_1 \quad \text{(folds)} \\ \textit{dim } 2 & S_1, S_{1_2} \quad \text{(folds, cusps)} \\ \textit{dim } 3 & S_1, S_{1_2}, S_{1_3} \\ \textit{dim } 4 & \begin{cases} S_1, S_{1_2}, S_{1_3}, S_{1_4} \\ \text{elliptic and hyperbolic umbilics } (S_{2,0}) \end{cases} \end{array}$$

That the S_{1_k} singularities which we have listed are the only ones that can occur is clear from Corollary 3.14. In §5, exercises 11 and 12, we indicated how to prove analogous results for the S_2 singularities on the list above. S_3 singularities are, of course, removable as long as dim $X < 9$.

In particular, for dimensions ≤ 4, a stable map can only exhibit the above singularities; and, being stable, it must exhibit them transversely. Summarizing we have:

Proposition 6.2. *In dimensions* ≤ 4, *for a map germ to be stable it is necessary and sufficient that it exhibit only singularities on the list* (6.1), *and that it exhibit these singularities transversely.*

Remark. It is instructive to verify directly from Mather's criterion that the canonical forms described for the singularities in the list (6.1) which are given in Theorems 4.1 and 5.10 do indeed represent infinitesimally stable germs. For example, we verify this criterion for the Morin singularities where $f(x_1, \ldots, x_n) = (f_1(x), x_2, \ldots, x_n)$ and $f_1(x_1, \ldots, x_n) = x_2x_1 + \ldots + x_kx_1^{k-1} + x_1^{k+1}$. Locally the equation $\tau = (df)(\zeta) + \eta \cdot f$ translates to the system of functional equations

$$(*)\qquad \left[\begin{array}{l} \tau_1 = (x_2 + 2x_1x_3 + \cdots + (k-1)x_1^{k-2}x_k + (k+1)x_1^k)\zeta_1 \\ \qquad + x_1\zeta_2 + \cdots + x_1^{k-1}\zeta_k + \eta_1 \cdot f \\ \tau_2 = \zeta_2 + \eta_2 \cdot f \\ \vdots \\ \tau_n = \zeta_n + \eta_n \cdot f \end{array}\right.$$

where we must solve for the functions $\zeta_i(x)$ and $\eta_i(y)$ given the functions $\tau_i(x)$. By V, Theorem 1.2, we need only solve the equations (*) to order n and by Arnold's criterion (V, Proposition 1.13) we need only solve equations (*) when $\tau_l = x_s$ and $\tau_i = 0$ for $i \neq l$. When $l > 1$, let $\zeta_l = x_s$, $\zeta_i = 0$ $i \neq l$, and $\eta_i = 0$ for all i to solve (*). So we assume that $\tau_1 = x_s$, and $\tau_i = 0$ for $i > 1$. If $\tau_1 = x_1$, then let $\zeta_2 = 1$, $\eta_2 = -1$, $\zeta_1 = \zeta_3 = \cdots = \zeta_n = 0$, and $\eta_1 = \eta_3 = \cdots = \eta_n = 0$. If $\tau_1 = x_s$ for $s > 1$ then let $\eta_1(y) = y_s$ (so that $\eta_1 \cdot f(x) = x_s$),

$\zeta_1 = \cdots = \zeta_n = 0$, and $\eta_2 = \cdots = \eta_n = 0$. We leave it to the reader to check that umbilics are infinitesimally stability.

We will now try to find *global* criteria for stability. We made a tentative investigation of this problem in §5 of Chapter VI. We will apply our conclusions there to maps in low dimensions. There are at most six distinct types of singularities on the list (6.1); so for a map $f: X \to Y$, with $\dim X = \dim Y \leq 4$, we can partition the set of singularities of f into six disjoint subsets: $X_1, \ldots, X_6$. We showed in VI, §5 that for f to be stable it must satisfy the following "normal crossing" condition.

Condition NC. Given distinct points x_{i_r} in X_{i_r} $r = 1, \ldots, k$ such that $f(x_{i_1}) = \cdots = f(x_{i_k}) = y$ then the subspaces

$$(df)_{x_{i_r}}(T_{x_{i_r}}X_{i_r}) \quad \text{for} \quad r = 1, \ldots, k$$

of $T_y Y$ are in *general position.*

(See VI, Proposition 5.2.) Inter alia, this condition implies that f, restricted to each "stratum" X_i, is an immersion with normal crossings, and that the images of these strata intersect transversely. For example it implies that a point cannot simultaneously be the image of an umbilic and of an S_1 singularity.

Theorem 6.3. *Let* $\dim X = \dim Y \leq 4$ *and let* $f: X \to Y$ *be a map which exhibits only the singularities on the list* (6.1) *and exhibits these transversely. Then a necessary and sufficient condition for f to be stable is that it satisfy the condition NC described above.*

We will deduce this from a slightly more general result.

Theorem 6.4. *Let X and Y be n dimensional manifolds, and* $f: X \to Y$ *a map which is of rank* $\geq n - 1$ *everywhere. Then f is stable if and only if it satisfies the transversality conditions of Morin:*

$$j^k f \pitchfork S_{1_k} \qquad k = 1, \ldots, n + 1$$

and, in addition, satisfies the condition NC, for the stratification, $X_k = S_{1_k}(f)$, $k = 1, \ldots, n + 1$.

Proof. The necessity is obvious. To prove the sufficiency we only have to show that Mather's criterion for infinitesimal stability is true on the multijet level. (V, Theorem 1.6.)

The specific result we are going to prove is the following "canonical form" lemma.

Lemma 6.5. *Let* $f: X \to Y$ *be a map satisfying the hypotheses of Theorem 6.4. Let* $p_1, \ldots, p_s$ *be points of X such that* p_i *is in* $S_{1_{r_i}}(f)$ *for* $i = 1, \ldots, s$ *and such that* $f(p_1) = \cdots = f(p_s) = q$. *Then we can choose a coordinate system* $y_1, \ldots, y_n$ *centered at q and coordinate system* $x_1^{(i)}, \ldots, x_n^{(i)}$ *centered at each of the* p_i*'s such that f has the canonical form* (4.2) *simultaneously in each of these coordinate systems.*

Proof. For simplicity we will just consider the case $s = 2$ (to spare the reader rather than the authors; the general case is not any harder, but there are more indices to keep track of.) Let $p = p_1$ and $p' = p_2$; and for the moment let us just consider f in the vicinity of p. If p is an S_{1_k} singularity we can choose coordinate systems centered at p and at q such that f has the canonical form (4.2). Let x_1 be the first coordinate function in the coordinate system at p, and $y_1, \ldots, y_k$ the first k coordinate functions in the coordinate system about q. Then the tangent space to $S_{1_k}(f)$ at p is characterized by the equations

$$dx_1 = df^*y_2 = \cdots = df^*y_k = 0$$

(Compare with §3, Exercise 4) and the image space by the equations

$$dy_1 = dy_2 = \cdots = dy_k = 0. \tag{6.6}$$

If we make a similar choice of coordinates relative to p' and q, then the tangent space to $S_{k'}(f)$ at p' is characterized by the equations

$$dx'_1 = df^*y'_2 = \cdots = df^*y'_{k'} = 0$$

and the image space by the equations

$$dy'_1 = \cdots = dy'_{k'} = 0. \tag{6.6)'$$

By assumption the subspaces (6.6) and (6.6)′ are in general position. This means that the differentials $dy_1, \ldots, dy_k, dy'_1, \ldots, dy'_{k'}$ are linearly independent at q, and, therefore, that $y_1, \ldots, y_k, y'_1, \ldots, y'_{k'}$ can be introduced as the first $k + k'$ coordinate functions of some coordinate system. In this coordinate system we will simultaneously have the Morin canonical form for an S_{1_k} singularity at p and the Morin canonical form for an S_{1_k} singularity at p'. □

From this it is easy to prove Theorem 6.4. One merely checks that the multijet conditions of (†) in V, Theorem 1.6 are satisfied using the multijet canonical forms given in Lemma 6.5. The details are similar to the argument in the remark following proposition 6.2 and we leave them to the reader. To prove Theorem 6.3 we observe that since the condition NC is satisfied no multi-germ can have a source consisting of an umbilical point and another singular point since the umbilics occur as isolated points (in dimension 4) or not at all (in dimension <4). Therefore, to verify the multi-jet criterion for stability we only have to verify it for multi-jets involving singular points of type S_{1_k}; so we are back in the situation of Theorem 6.4. □

Appendix

§A. Lie Groups

The theorem that we need states that the orbits of a Lie group action are immersed submanifolds. First we define and sketch some facts about Lie groups.

Definition A.1. *Let G be both a group and a smooth manifold. Then G is a* Lie group *if the mapping of $G \times G \to G$ given by $(a, b) \mapsto ab^{-1}$ is smooth.*

Examples

(a) $\mathbf{R}^n$ where the group operation is addition.

(b) S^1 where the group operation is addition of angles.

More generally $T^n = S^1 \times \cdots \times S^1 = n$-torus is a Lie group where the group operation is coordinate-wise addition.

Note. The only abelian connected Lie groups are $\mathbf{R}^n \times T^m$ (no proof!).

(c) All matrix groups. For example $GL(n, \mathbf{R})$ = group of $n \times n$ invertible real matrices, $0(n)$ = group of $n \times n$ orthogonal matrices, and $SL(n, \mathbf{R})$ = group of $n \times n$ real matrices with determinant equal to 1. All of these groups are submanifolds of $\mathbf{R}^{n^2}$. (See Theorem A.7.) Also $GL(n, \mathbf{C})$ = group of $n \times n$ invertible matrices with complex entries. Here we view $GL(n, \mathbf{C})$ as a submanifold of $\mathbf{R}^{n^2} \oplus \mathbf{R}^{n^2}$.

(d) Let X be a manifold with p in X. Let σ be a k-jet ($k > 0$) in $J^k(X, X)_{p,p}$. Then σ is *invertible* if any representative of σ is a diffeomorphism on a nbhd of p. The invertible k-jets form a group under composition and a manifold since they are an open subset of $J^k(X, X)_{p,p}$. We shall denote the set of invertible k-jets at p by $G^k(X)_p$. To see that $G^k(X)_p$ is a Lie group, we choose coordinates near p and inspect $G^k(\mathbf{R}^n)_0$. Further we may identify $J^k(\mathbf{R}^n, \mathbf{R}^n)_{0,0}$ with polynomial functions from $\mathbf{R}^n \to \mathbf{R}^n$ of degree $\leq k$ mapping 0 to 0. Under this identification $G^k(\mathbf{R}^n)_0$ is the open subset of polynomial mappings f for which $(df)_0$ is nonsingular. Here we see that the group operation is given by composition of the polynomial mappings but throwing away all terms in the composition of degree $> k$. This is clearly a smooth operation. It is also not hard to see that the mapping $b \mapsto b^{-1}$ in $G^k(\mathbf{R}^n)_0$ is a smooth operation.

The tangent space to a point in a smooth manifold is always locally diffeomorphic to the manifold (using chart mappings). On a Lie group, G, there is a naturally defined identification $\exp: T_eG \to G$ which is a diffeomorphism on a nbhd of 0. We shall construct this mapping.

Let v be in T_eG. Then v along with the group action defines a vector field on G. For a in G let $L_a: G \to G$ be defined by $L_a(g) = a \cdot g$. Clearly L_a is a diffeomorphism. Define $\zeta_a{}^v = (dL_a)_e(v)$. The smoothness of the group action guarantees that ζ^v is a vector field on G. Also ζ^v satisfies $(dL_a)_b\zeta_b{}^v = \zeta_{a\cdot b}^v$ for all a, b in G.

Definition A.2. *A vector field ζ satisfying $(dL_a)\zeta = \zeta$ for all a in G is called* left invariant.

Lemma A.3. *Let ζ be a left-invariant vector field on G. Then ζ is the infinitesimal generator of a globally defined one parameter group.*

Proof. Let $\psi_t: U \to G$ ($|t| < \varepsilon$, U a nbhd of e in G) be a locally defined one parameter group near e in G as given by I, Lemma 6.2. As we saw in the case that G is compact (I, Theorem 6.5), the trick in showing that there is a globally defined one-parameter group is to show that ψ_t is globally defined on G for $|t| < \varepsilon$. Now since ζ is left-invariant $(dL_g)\zeta = \zeta$. Thus $(d/dt)g\psi_t(a) = (d/dt)\psi_t(ga)$ for all g, a, and ga in U. Hence (*) $g\psi_t(a) = \psi_t(ga)$. In particular, $\psi_t(g) = g\psi_t(e)$ for g in U. Thus we can clearly extend ψ_t to be globally defined and smooth on G. The left invariance of ζ guarantees that ψ_t ($|t| < \varepsilon$) is still a one-parameter group for ζ on all of G. □

Remark. Given a vector v in T_eG, let ζ^v be the left invariant vector field that it generates. Let ψ^v be the globally defined one parameter group whose existence is assured by the last Lemma. We can think of ψ as a mapping of $(T_eG) \times G \times \mathbf{R} \to G$ given by $(v, g, t) \mapsto \psi_t{}^v(g)$. Thus we have the following:

Proposition A.4. *The mapping ψ is smooth and satisfies*
(1) $\psi_t{}^v(ga) = g\psi_t{}^v(a)$, *and*
(2) $\psi_s{}^{tv} = \psi_{ts}{}^v$.

Proof. For fixed v, ψ^v is just a one-parameter group and is thus smooth. Varying v just varies the initial conditions to the first order system of ODE's which define ψ^v. Since solutions to such a system vary smoothly with the initial conditions, ψ is a smooth mapping. Note that (1) is just a restatement of (*) in the proof of Lemma A.3. For (2), note that $\psi_s{}^{tv}$ and $\psi_{st}{}^v$ are both one-parameter groups on G for fixed t and v. Now $\psi_s{}^{tv}$ has infinitesimal generator ζ^{tv} and

$$\frac{d}{ds}\psi_{st}{}^v(g)\bigg|_{s=0} = t\frac{d}{dr}\psi_r{}^v(g)\bigg|_{r=0} = t\zeta_g{}^v = \zeta_g{}^{tv}.$$

Thus the infinitesimal generator of $\psi_{st}{}^v$ is also ζ^{tv}. Applying the fact that one-parameter groups are unique we have $\psi_s{}^{tv} = \psi_{st}{}^v$. □

Define $\exp: T_eG \to G$ by $\exp(v) = \psi_1{}^v$

Theorem A.5. *$\exp: T_eG \to G$ is smooth and is a diffeomorphism on a nbhd of* 0. *In fact $(d\exp)_0 = identity$. (Note: we identify $T_0(T_eG)$ with T_eG.)*

Proof. Clearly exp is smooth. Using (2) in the last Proposition, we have

$$(d\exp)_0(v) = \frac{d}{dt}(\exp tv)\bigg|_{t=0} = \frac{d}{dt}\psi_1{}^{tv}(e)\bigg|_{t=0} = \frac{d}{dt}\psi_t{}^v(e)\bigg|_{t=0} = \zeta_e{}^v = v.$$

So $(d\exp)_0 = id_{T_eG}$. □

Exercise: We may identify $T_eGL(n, \mathbf{R})$ with $M(n, \mathbf{R})$ = vector space of $n \times n$ real matrices. Using this identification show that $\exp A = \operatorname{Exp} A$ where A is in $N(n, \mathbf{R})$ and

$$\operatorname{Exp} A = \sum_{i=0}^{\infty} \frac{A^i}{i!}.$$

Corollary A.6. *Let V and W be subspaces of T_eG such that $V \oplus W = T_eG$. Define $\gamma: T_eG \to G$ by $\gamma(v, w) = \exp v \cdot \exp w$ for v in V and w in W. Then ϕ is a diffeomorphism on a nbhd of* 0 *in T_eG with a nbhd of e in G.*

Proof. Certainly γ is smooth. Moreover $(d\gamma)_0|V = (d\exp)_0|V = id_V$ by Theorem A.5. Similarly for W. So $(d\gamma)_0$ is invertible. □

Definition A.5. *Let G be a Lie group. Then $H \subset G$ is a* Lie subgroup *if*
(i) *H is a subgroup of G;*
(ii) *H is an immersed submanifold of G; and*
(iii) *H is a Lie group with the group operation assumed in* (i) *and the manifold structure assumed in* (ii).

Note. Lie subgroups are *not*, in general, submanifolds. For example, let $G = T^2$ viewed as the decomposition space $\mathbf{R}^2/\mathbf{Z}^2$ where $\mathbf{Z}^2$ = the subgroup of integer lattice points in $\mathbf{R}^2$. Let H' be a line in $\mathbf{R}^2$ through the origin with irrational slope and let $H = \pi(H')$ where $\pi: \mathbf{R}^2 \to T^2$ is the obvious projection. Then $\pi|H'$ is a 1:1 immersion so that H is an immersed submanifold and a Lie subgroup. But H is not a submanifold of T^2 since H is dense.

One of the more interesting facts about Lie subgroups which indicates the strong connection between the geometry and algebra on a Lie group is the following.

Theorem A.7. *Let H be a subgroup of the Lie group G which is a topologically closed subset. Then H is a Lie subgroup.*

Remark. The content of this theorem is that any closed subgroup of a Lie group is an immersed submanifold and thus a submanifold.

First some lemmas.

Lemma A.8. *Let $|\ |$ be a norm on T_eG. Suppose that $v_1, v_2, \ldots$ is a sequence of nonzero vectors in T_eG such that* $\operatorname{Lim}_{i\to\infty} v_i = 0$, $\exp v_i \in H$ *for all i, and* $\operatorname{Lim}_{i\to\infty} (1/|v_i|)v_i = \bar{v}$. *Then* $\exp t\bar{v} \in H$ *for all t in* $\mathbf{R}$.

Proof. $\operatorname{Lim}_{i\to\infty} (t/|v_i|)v_i = t\bar{v}$. Choose integers k_i so that $k_i|v_i| \mapsto t$. Then $\exp(k_iv_i) \mapsto \exp(t\bar{v})$. But

$$\exp(k_iv_i) = \psi_1^{k_iv_i}(e) = \psi_{k^i}^{v_i}(e) = (\psi_1^{v_i}(e))^{k_i} = (\exp v_i)^{k_i}$$

is in H, using Proposition A.4. Since H is closed $\exp(t\bar{v})$ is in H. □

Lemma A.9. *Let* $V = \{v \in T_eG \mid \forall t \in \mathbf{R}, \exp(tv) \in H\}$. *Then V is a vector subspace of T_eG.*

Proof. Clearly V is closed under scalar multiplication so we need only show that V is closed under addition. Let v and w be in V and suppose $v + w \neq 0$. Consider $\exp(tv)\cdot\exp(tw)$ which is in H. Using Theorem A.5, we see that for all small t there exists a unique $f(t)$ such that $\exp(tv)\cdot\exp(tw) = \exp f(t)$. Moreover f is a smooth curve in G.

Using Proposition A.4,

$$\exp(tv)\cdot\exp(tw) = \psi_1^{tv}(e)\cdot\psi_1^{tw}(e) = \psi_t^{v}(e)\cdot\psi_t^{w}(e) = \psi_t^{v}(\psi_t^{w}(e)).$$

So

$$\frac{d}{dt}\exp(tv)\cdot\exp(tw)\bigg|_{t=0} = \frac{d}{dt}\psi_t^{v}(e)\bigg|_{t=0} + \frac{d}{dt}\psi_t^{w}(e)\bigg|_{t=0} = v + w.$$

Now

$$\frac{d}{dt}\exp t(v + w)\bigg|_{t=0} = v + w.$$

So in local coordinates near e, $\exp(f(t)) - \exp t(v + w) = 0(t)$. Since exp is a diffeomorphism near 0, $\operatorname{Lim}_{t\to 0} f(t)/t = v + w$. Apply Lemma A.8 with $v_i = f(1/i)$ and $\bar{v} = (v + w)/|v + w|$ to show that $v + w$ is in V. ▯

Proof of Theorem A.7. Let W be a vector space complement to the V of Lemma A.9 in T_eG. Consider the local diffeomorphism $\gamma: T_eG \to G$ as in Corollary A.6. We claim $\exp(V)$ is a nbhd of e in H. Suppose not. Then there exists a sequence $h_1, h_2, \ldots$ of points in H with $\operatorname{Lim}_{i\to\infty} h_i = e$ such that $h_i \notin \exp(V)$. Choose points v_i in V and w_i in W such that $\exp v_i\cdot\exp w_i = h_i$. Thus $\exp w_i$ is in H for all i and by restricting to a subsequence we may assume that $w_i/|w_i| \mapsto w$ in W. Apply Lemma A.8 to show that w is in V. But then $w \in V \cap W = \{0\}$ and $|w| = 1$ is a contradiction. Thus $\exp V$ is a nbhd of 0 in H. Hence there is an open nbhd of 0 in V mapped diffeomorphically onto an open nbhd of e in H by exp. $\exp^{-1}$ is then a chart for the manifold structure of H near e. Via the translations L_g we can obtain an atlas of charts for H and H is a manifold. The inclusion mapping of $H \to G$ is clearly an immersion. ▯

We need one more Theorem before getting to the result mentioned in the beginning of this appendix. Let H be a closed subgroup of G. Then the space of cosets G/H has a natural Hausdorff topology—namely, the weakest topology which makes the obvious projection $\pi: G \to G/H$ continuous. Since H is also a Lie subgroup we can say more.

Theorem A.10. *Let H be a closed subroup of a Lie group G. Then G/H is a smooth manifold with* $\dim G/H = \dim G - \dim H$. *Moreover $\pi: G \to G/H$ is smooth and for each q in G/H, there exists a nbhd Q of q and a smooth mapping $\tau: Q \to G$ such that $\pi\cdot\tau = id_Q$.*

Proof. Let W be a vector space complement to T_eH in T_eG. Let U be a nbhd of 0 in T_eG such that $\exp U$ is a diffeomorphism. Then $\pi\cdot\exp: U \cap W \to G/H$ is a local homeomorphism. Certainly $\pi\cdot\exp(U \cap W)$ is open in G/H. It is easy to compute $(\pi\cdot\exp \mid U \cap W)^{-1}$ and to check continuity. This is a chart for G/H near the identity coset. Use the translation L_g to move this chart to the coset gH. In this way G/H is a manifold and $\dim G/H = \dim W = \dim G - \dim H$. Moreover π is smooth. The smooth mapping $\exp(\pi \cdot \exp \mid U \cap W)^{-1}$ is the desired local "section". □

Definition A.11. *Any manifold of the form G/H with the differentiable structure as given in Theorem A.10 is called a* homogeneous space.

Example. Let $G = GL(n, \mathbf{R})$ and fix a k-plane V in $\mathbf{R}^n$. Let $H = \{A \in GL(n, \mathbf{R}) \mid A(V) = V\}$. Clearly H is a closed subgroup of G. So G/H is a differentiable manifold. In fact, $G/H = G_{k,n}$—the Grassmann manifold of k-planes in n-space.

Definition A.12. *Let X be a smooth manifold and let G be a Lie group.*

(a) An action of G on X *is a homomorphism:* $\tau: G \to \text{Diff}(X)$ *such that the mapping $G \times X \to X$ given by $(g, x) \mapsto \tau(g)(x)$ is smooth.*

(b) *Let p be in X. Denote by $\mathcal{O}_p \equiv \{\tau(g)(p) \mid g \in G\}$ the* orbit *(of the action of G on X)* through p.

(c) *Let $H_p \equiv \{g \in G \mid \tau(g)(p) = p\} =$ the* isotropy subgroup *at p.*

Note. H_p is a closed subgroup of G and thus a Lie subgroup of G.

Example. Let X be a compact manifold. Let ζ be a vector field on X and let ψ_t be the corresponding globally defined one parameter group on X. Then the mapping $t \mapsto \psi_t$ defines an action of $\mathbf{R}$ on X. The orbits of this action are just the integral curves of the vector field. Conversely, an action of $\mathbf{R}$ on X is just a one-parameter group.

Theorem A.13. *Let G be a Lie group acting on a manifold X. Then the orbits of the action of G on X are immersed submanifolds of X.*

Proof. Denote the action of G on X by ρ and let p be in X. Consider the mapping $\sigma: G \to X$ given by $\sigma(g) = \rho(g)(p)$. Clearly $\text{Im}\,\sigma = \mathcal{O}_p$. Moreover σ is constant on cosets of G/H_p and so induces a 1:1 onto mapping $\lambda: G/H_p \to \mathcal{O}_p$. Locally $\lambda = \sigma \cdot \tau$ where τ is the local "section" given in Theorem A.10. So λ is a 1:1 smooth mapping of the manifold G/H_p onto $\mathcal{O}_p$. We claim that λ is an immersion and thus that $\mathcal{O}_p$ is an immersed submanifold of X. It is enough to show that $(d\lambda)_{\bar{e}}$ is 1:1 where $\bar{e} = eH_p$ in G/H_p since the mapping $L_g: G \to G$ induces a smooth mapping $S_g: G/H_p \to G/H_p$ and $(d\lambda)_{\bar{g}} = (d\rho(g))_p \cdot (d\lambda)_{\bar{e}} \cdot (dS_g)_{\bar{g}}^{-1}$.

Let w be in $T_{\bar{e}}G/H_p$ such that $(d\lambda)_{\bar{e}}(w) = 0$. Then $(d\sigma)_e(v) = 0$ where $v = (d\tau)_{\bar{e}}(w) \in T_eG$, since $\lambda = \sigma \cdot \tau$ near $\bar{e}$. Let $\eta_t(q) = \rho(\psi_t{}^v(e))$. Then η_t is a one-parameter group on X since $\psi_{t+s}^v(e) = \psi_t{}^v(e) \cdot \psi_s{}^v(e)$ using Proposition A.4. Let ζ be the infinitesimal generator of η_t. Since $\eta_t(p) = \sigma(\psi_t{}^v(e))$, we see that

$$\zeta_p = \frac{d}{dt}\eta_t(p)\bigg|_{t=0} = \frac{d}{dt}\sigma \cdot \psi_t{}^v(e)\bigg|_{t=0} = (d\sigma)_p(v) = 0.$$

Applying Note (2) after I, Theorem 6.4, we see that $\eta_t(p) = p$ for all t. Thus $\phi(t)$ is in H_p for all t and $v = (d\tau)_{\bar{e}}(w)$ is in T_eH.

So $w = (d\pi)_e(d\tau)_{\bar{e}}(w) = 0$ and $(d\lambda)_{\bar{e}}$ is injective. □

Note. An orbit is *not*, in general, a submanifold of X. For example, it is easy to construct an action of $\mathbf{R}$ on T^2 whose orbit through (0, 0) is dense.

If G is a Lie group, let $\mathring{G}$ denote the connected component of e in G. Clearly $\mathring{G}$ is a Lie subgroup of G.

Lemma A.14. *Let G be a Lie group acting on a manifold X and let p be in X. The connected component of $\mathcal{O}_p$ containing p is $\mathring{G}\cdot p \equiv \{\tau(g)(p) \mid g \in \mathring{G}\}$ where τ denotes the action.*

Note. We speak of the topology on $\mathcal{O}_p$ induced from G/H_p by λ, not the topology induced on $\mathcal{O}_p$ by X.

Proof. Clearly $\mathring{G}\cdot p$ is connected. If G' is any other connected component of G, then $G'\cdot p \cap \mathring{G}\cdot p = \varnothing$ or $\mathring{G}\cdot p$. So $\mathring{G}\cdot p$ is both open and closed in $\mathcal{O}_p$ and is thus a component of $\mathcal{O}_p$. Clearly p is in $\mathring{G}\cdot p$. □

Bibliography

1. R. Abraham, J. Robbin. *Transversal Mappings and Flows*. W. A. Benjamin, Inc. New York 1967.

2. J. Frank Adams. *Lectures on Lie Groups*. W. A. Benjamin, Inc., New York 1969.

3. L. Ahlfors. *Complex Analysis*. 2nd Edition, McGraw Hill, Inc., New York 1966.

4. V. I. Arnold. Singularities of Smooth Mappings, *Russian Math. Surveys* (1968), pp. 1–43. Vol. 23, No. 1, Jan.–Feb. 1968. (Translated from *Uspehi Mat. Nauk*. **23** (1968), pp. 3–44.)

5. M. Atiyah. *K-Theory*. W. A. Benjamin, Inc., New York 1967.

6. J. M. Boardman. Singularities of differentiable maps. *Publ. Math. I.H.E.S.* **33** (1967), pp. 21–57.

7. A. Borel. *Linear Algebraic Groups*. W. A. Benjamin, Inc., New York 1969.

8. J. Dieudonné. *Foundations of Modern Analysis*. Academic Press, New York 1960.

9. G. Glaeser. Fonctions composées differentiables. *Ann. Math.* **77** (1963), pp. 193–209.

10. R. C. Gunning and H. Rossi. *Analytic Functions of Several Complex Variables*. Prentice Hall Englewood Cliffs, New Jersey 1965.

11. A. Haefliger. Les singularities des applications differentiables. *Seminaire H. Cartan E.N.S.* 1956/57 Exposé N° 7.

12. ——. Quelques Remarks sur les applications differentiables d'une surface dans le plan. *Ann. Inst. Fourier* **10** (1960), pp. 47–60.

13. —— and A. Kosinski. Un theoreme de Thom sur les singularities des applications differentiables. *Seminaire H. Cartan E.N.S.* 1956/57 Exposé N° 8.

14. M. Hervé. Several complex variables, local theory. Published for the Tata Institute, Bombay by the *Oxford University Press*, London, New York 1963.

15. K. Hoffman and R. Kunze. *Linear Algebra.* 2nd Edition, Prentice-Hall, Inc., Englewood Cliffs, New Jersey 1971.

16. W. Hurewicz. *Lectures on Ordinary Differential Equations.* The MIT Press, Cambridge, Mass. 1958.

17. S. Lang. *Introduction to Differentiable Manifolds.* Wiley (Interscience), New York 1962.

18. H. I. Levine. *Singularities of Differentiable Mappings.* Liverpool Singularities, pp. 1–89.

19. ——. Mappings of manifolds into the plane. *Amer. J. Math.* **88** (1966), pp. 357–365.

20. ——. The Singularities of S_1^q. *Ill. J. Math.* **8** (1964), pp. 152–168.

21. ——. Eliminations of cusps. *Topology* **3** (1965), p. 263.

22. B. Malgrange. *The Preparation Theorem for Differentiable Functions.* In "Differential Analysis" papers presented at Bombay Colloquium, 1964. Oxford 1964, pp. 203–208.

23. ——. Le théorème de préparation en géometrie différentiable. *Sém. H. Cartan* **15** (1962–63), exps. 11, 12, 13, 22.

24. ——. Ideals of Differentiable Functions. Oxford University Press 1966.

25. J. F. Marsden. *Hamiltonian Mechanics, Infinite Dimensional Lie Groups, etc.* Berkeley Notes, December 1969.

26. J. N. Mather. Stability of C^∞ Mappings. I: The division theorem. *Annals of Math.*, Vol. 87, No. 1, Jan. 1968, pp. 89–104.

27. ——. Stability of C^∞ Mappings. II: Infinitesimal stability implies stability. *Annals of Math.*, Vol. 89, No. 2, March 1969, pp. 254–291.

28. ——. Stability of C^∞ Mappings. III: Finitely determined map germs. *Publ. Math. I.H.E.S.* **35** (1968), pp. 127–156.

29. ——. Stability of C^∞ Mappings. IV: Classification of stable germs by *R*-algebras. *Publ. Math. I.H.E.S.* **37** (1969), pp. 223–248.

30. ——. Stability of C^∞ Mappings. V: Transversality. *Advances in Mathematics*, Vol. 4, No. 3, June 1970, pp. 301–336.

31. ——. Stability of C^∞ Mappings. VI: The nice dimensions. *Liverpool Singularities* I, pp. 207–253.

32. ——. On Nirenberg's Proof of the Malgrange Preparation Theorem. *Liverpool Singularities I*, pp. 116–120.

33. ——. Springer-Verlag 1974.

34. M. Menn. *Singular Manifolds*. Brandeis University (Doctoral Dissertation) 1968.

35. J. Milnor. *Lecture Notes on Differential Topology*. Mimeographed Notes, Princeton 1958.

36. ——. Morse Theory. *Annals of Math. Studies* **51**, Princeton University Press, Princeton, New Jersey 1963.

37. ——. *Topology from the Differential Viewpoint*. The University Press of Virginia, Charlottesville 1965.

38. ——. Singular points of complex hypersurfaces. *Annuals of Math. Studies* **61**, Princeton University Press, Princeton, New Jersey 1968.

39. B. Morin. Formes canoniques des singularities d'une application différentiable. *Comptes Rendus Acad. Sci.*, Paris 260, 5662–5665, 6503–6506 (1965).

40. J. Munkres. Obstructions to the Smoothing of Piecewise Differentiable Homeomorphisms. *Annals of Math.*, Vol. 72, No. 3 Nov. 1960.

41. L. Nirenberg. A Proof of the Malgrange Preparation Theorem. *Liverpool Singularities* **1**, pp. 97–105.

42. *Proceedings of Liverpool Singularities Symposium I.* Lecture Notes in Mathematics **192**, Springer-Verlag, New York 1971.

43. *Proceedings of Liverpool Singularities Symposium II.* Lecture Notes in Mathematics **209**, Springer-Verlag, New York 1971.

44. A. Sard. The Measure of Critical Values of Differentiable Maps. *Bull. A.M.S.*, Vol. 48, pp. 883–890 (1942).

45. S. Sternberg. *Lectures on Differential Geometry*. Prentice-Hall, Inc., Englewood Cliffs, New Jersey 1964.

46. R. Thom. Les singularités des applications différentiables, *Ann. Inst. Fourier t.* **6** (1955–56), pp. 43–87.

47. ——. Stabilité Structurelle et Morphogénèse. W. A. Benjamin, Inc., Reading, Mass. 1972.

48. ——. Local Properties of Differentiable Mappings. *Differential Analysis.* Oxford Press, London 1964, pp. 191–202.

49. ——. Un lemme sur les applications différentiable. *Bol. Soc. Math. Mexicana* (1956), pp. 59–71.

50. J. C. Tougeron. Idéaux de Fonctions Différentiables, Ergebnisse Band 71, Springer-Verlag, New York 1972.

51. ——, Idéaux de fonctions différentiables I. *Ann. Inst. Fourier*, Grenoble (1968), pp. 177–240.

52. B. L. van der Waerden. *Modern Algebra.* 2nd Edition, Vol. 1, Frederick Ungar Publishing Co., New York 1964.

53. C. T. C. Wall. Lectures on C^∞ Stability and Classification. *Proceedings of Liverpool Singularities I*, Springer Lecture Notes 192, pp. 178–206.

54. H. Whitney. Elementary Structure of Real Algebraic Varieties. *Ann. Math.* **66** (1957), pp. 545–556.

55. ——. Differentiable even functions. *Duke Journal of Mathematics*, Vol. 10, 1943, pp. 159–160.

56. ——. Singularities of mappings in Euclidean spaces. In *Symposium internacional de topologia algebraica*, pp. 285–301, Universidad Nacional Autonoma de Mexico and UNESCO, Mexico City 1958.

57. ——. The general type of singularity of a set of $2n - 1$ smooth functions of n variables. *Duke Journal of Mathematics*, Ser. 2, 45 (1944), pp. 220–293.

58. ——. On singularities of mappings of Euclidean spaces I, Mappings of the plane into the plane. *Ann. of Math.* **62** (1955), pp. 374–410.

59. ——. Analytic extensions of differentiable functions defined in closed sets. *Trans. Amer. Math. Soc.* **36** (1), 1934, pp. 63–89.

60. ——. The singularities of smooth n-manifolds into $(2n - 1)$-space. *Ann. of Math.* **45** (1944), pp. 247–293.

61. O. Zariski and P. Samuel. *Commutative Algebra*, Vol. 2. D. Van Nostrand Company, Inc. 1960, Princeton.

SYMBOL INDEX

Graduate Texts in Mathematics

continued from page ii

48 SACHS/WU. General Relativity for Mathematicians.
49 GRUENBERG/WEIR. Linear Geometry. 2nd ed.
50 EDWARDS. Fermat's Last Theorem.
51 KLINGENBERG. A Course in Differential Geometry.
52 HARTSHORNE. Algebraic Geometry.
53 MANIN. A Course in Mathematical Logic.
54 GRAVER/WATKINS. Combinatorics with Emphasis on the Theory of Graphs.
55 BROWN/PEARCY. Introduction to Operator Theory I: Elements of Functional Analysis.
56 MASSEY. Algebraic Topology: An Introduction.
57 CROWELL/FOX. Introduction to Knot Theory.
58 KOBLITZ. p-adic Numbers, p-adic Analysis, and Zeta-Functions. 2nd ed.
59 LANG. Cyclotomic Fields.
60 ARNOLD. Mathematical Methods in Classical Mechanics.
61 WHITEHEAD. Elements of Homotopy Theory.
62 KARGAPOLOV/MERZLJAKOV. Fundamentals of the Theory of Groups.
63 BOLLABÁS. Graph Theory.
64 EDWARDS. Fourier Series. Vol. I. 2nd ed.
65 WELLS. Differential Analysis on Complex Manifolds. 2nd ed.
66 WATERHOUSE. Introduction to Affine Group Schemes.
67 SERRE. Local Fields.
68 WEIDMANN. Linear Operators in Hilbert Spaces.
69 LANG. Cyclotomic Fields II.
70 MASSEY. Singular Homology Theory.
71 FARKAS/KRA. Riemann Surfaces.
72 STILLWELL. Classical Topology and Combinatorial Group Theory.
73 HUNGERFORD. Algebra.
74 DAVENPORT. Multiplicative Number Theory. 2nd ed.
75 HOCHSCHILD. Basic Theory of Algebraic Groups and Lie Algebras.
76 IITAKA. Algebraic Geometry.
77 HECKE. Lectures on the Theory of Algebraic Numbers.
78 BURRIS/SANKAPPANAVAR. A Course in Universal Algebra.
79 WALTERS. An Introduction to Ergodic Theory.
80 ROBINSON. A Course in the Theory of Groups.
81 FORSTER. Lectures on Riemann Surfaces.
82 BOTT/TU. Differential Forms in Algebraic Topology.
83 WASHINGTON. Introduction to Cyclotomic Fields.
84 IRELAND/ROSEN. A Classical Introduction to Modern Number Theory.
85 EDWARDS. Fourier Series: Vol. II. 2nd ed.
86 VAN LINT. Introduction to Coding Theory.
87 BROWN. Cohomology of Groups.
88 PIERCE. Associative Algebras.
89 LANG. Introduction to Algebraic and Abelian Functions. 2nd ed.
90 BRØNDSTED. An Introduction to Convex Polytopes.
91 BEARDON. On the Geometry of Discrete Groups.
92 DIESTEL. Sequences and Series in Banach Spaces.

93 DUBROVIN/FOMENKO/NOVIKOV. Modern Geometry — Methods and Applications Vol. I.
94 WARNER. Foundations of Differentiable Manifolds and Lie Groups.
95 SHIRYAYEV. Probability, Statistics, and Random Processes.
96 CONWAY. A Course in Functional Analysis.
97 KOBLITZ. Introduction in Elliptic Curves and Modular Forms.
98 BRÖCKER/tom DIECK. Representations of Compact Lie Groups.
99 GROVE/BENSON. Finite Reflection Groups. 2nd ed.
100 BERG/CHRISTENSEN/RESSEL. Harmonic Analysis on Semigroups: Theory of Positive Definite and Related Functions.
101 EDWARDS. Galois Theory.
102 VARADARAJAN. Lie Groups, Lie Algebras and Their Representations.
103 LANG. Complex Analysis. 2nd. ed.
104 DUBROVIN/FOMENKO/NOVIKOV. Modern Geometry — Methods and Applications Vol. II.
105 LANG. $SL_2(\mathbf{R})$.
106 SILVERMAN. The Arithmetic of Elliptic Curves.
107 OLVER. Applications of Lie Groups to Differential Equations.
108 RANGE. Holomorphic Functions and Integral Representations in Several Complex Variables.
109 LEHTO. Univalent Functions and Teichmüller Spaces.